MATHEMATICS IN INDUSTRY   **6**

Vincenzo Capasso
Jacques Périaux
Editors

Multidisciplinary Methods for Analysis Optimization and Control of Complex Systems

With 91 Figures

 Springer

Editors

Vincenzo Capasso

MIRIAM & Department of Mathematics
University of Milano
Via C. Saldini 50
20133 Milano, Italy
vincenzo.capasso@unimi.it

Jacques Périaux

Pole Scientifique – Dassault Aviation
UPMC
Dassault Aviation
78 Quai Marcel Dassault
99214 Saint-Cloud, France
jacques.periaux@dassault-aviation.fr

Library of Congress Control Number: 2004112289

Mathematics Subject Classification (2000):
60K35, 65MXX, 65YXX, 68TXX, 73KXX, 76-XX, 86-XX, 90BXX, 93-XX

ISBN 3-540-22310-X Springer Berlin Heidelberg New York

Springer is a part of Springer Science+Business Media

springeronline.com

© Springer-Verlag Berlin Heidelberg 2005
Printed in Germany

Typeset by the authors using a Springer TEX macro-package
Cover design: *design & production* GmbH, Heidelberg
Printed on acid-free paper   46/3142YL - 5 4 3 2 1 0

Preface

The Summer School has been dedicated to one of the proponents and first Chairman of the Strategy Board of MACSI-net, the late Jacques Louis Lions (see the dedication by Roland Glowinski).

MACSI-net is a European Network of Excellence, where both enterprises and university institutions co-operate to solve challenging problems to their mutual benefit. In particular the network focuses on strategies to enhance interactions between industry and academia. The aim is to help industry (in particular SMEs) alert academia about industrial needs in terms of advanced mathematical and computational methods and tools. The network is multidisciplinary oriented, combining the power of applied mathematics, scientific computing and engineering, for modeling and simulation. It was set up by a joint effort of ECCOMAS and ECMI European associations.

This particular event, occurred during March 17-22, 2003, was a joint effort of the Training Committee (chaired by VC) and Industrial Relations Committee (chaired by JP) to alert both Academia and Industry about the increasing role of Multidisciplinary Methods and Tools for the design of complex products in various areas of industrial interest. This increasing complexity is driven by societal constraints to be satisfied in a simultaneous and affordable way. The mastering of complexity implies the sharing of different tools by different actors which require much higher level of communication between culturally different people. The school offered to young researchers the opportunity to be exposed to the presentation of real industrial and societal problems and the relevant innovative methods used; the need of further contributions from mathematics to improve or provide better solutions had also been considered.

Important examples of such interdisciplinary needs came from Environmental problems (pollution of coastal waters due to oil spill, air quality of big cities, noise reduction, air/ocean interface, etc); Quality of Life (Bioengineering of tissues and prostheses, air traffic control for safety, Immunology, etc.); Material science and metallurgy (polymers, steel, etc); Aeronautics (multidisciplinary design optimisation, including aerodynamic efficiency, drag reduction, aero- and vibroacoustics, aeroelasticity - flutter, thermal flows, UAV trajectory control, etc)

The need of facing multidisciplinary optimisation methods for multiphysics problems raises hard problems with respect to the following aspects: Multiple scales (homogenisation, boundary layers, asymptotics, etc); Variable stochastic geometries (interfaces, domain decomposition, etc); Nonlinear reaction-advection-diffusion systems; Stochastic differential systems; Innovative inverse and optimisation algorithms because of multicriteria related to multisystem approach (evolutionary algorithms such as genetic algorithms, ant systems, game theory, hybrid methods, etc) ; Optimal control (level set –Hamilton-Jacobi– methods, nature inspired methods such as ant systems, hybrid methods, etc.); Strong coupling of several software (numerical analysis algorithms and CAD /CAM geometry tools); Mathematical methods for visualisation.

This large spectrum of methods and tools presented during the school had the scope to emphasise to young researchers, either mathematicians or engineers, the need to deepen specific areas of competence with the added value of communication in an interdisciplinary setting; i.e. the vertical expertise in a single discipline has to be revisited in an horizontal multidisciplinary application area .

This event has performed along the main scopes of MACSI net to stimulate and increase communication among the academic and the industrial communities with respect to the strategic goals of EU Framework Programmes, and in particular to anticipate preparation of young researchers in order to make them ready to collaborate on multidisciplinary oriented projects.

With respect to the scientific and technological content, the school offered an unexpected integration of scientific excellence of the speakers on multidisciplinary topics covered, and the class of participants who attended all the lectures with great level of interaction. This brought to a pleasant atmosphere for exchanging information and cross knowledge. This was also favoured by the choice of the venue and local hospitality.

Fruitful links have been established between lecturers and young researchers of different European universities for future collaboration.

The volume is divided in two parts; the first part including material of the lectures delivered by the main lecturers mainly focussing on methods; the second part including a series of case studies, illustrating the power of multimethods in facing challenging complex problems arising in real applications.

We are sure that the quality and diversity of the material presented here for further dissemination, will be of great use and stimulus , especially among young scientists.

We hope to have fulfilled the task of contributing to the implementation of the vision expressed by Professor J.L.Lions in his written address to the kick off meeting of the MACSInet pointing out the importance of interfaces in challenging multidisciplinary problems of industry and society.

Thanks are due to all participants, and in particular to the significant number of speakers that have so enthusiastically participated in the event. We are deeply sorry that, for independent bureaucratic reasons, a couple of relevant contributions could not be included.

We will not forget to thank the Coordinator of the Network, Professor Robert M. Mattheij, for his continued support (also in the nontrival management of the administrative aspects), and of course the European Union for the financial support of this successful event.

We also wish to thank the assistance of the staff of MIRIAM (the Milan Research Centre for Industrial and Applied Mathematics), and in particular of Dr Daniela Morale, without whom the school could impossibly have run so smoothly.

Vincenzo Capasso (Milano)
Jacques Périaux (Paris)

May, 2004

Dedicated to Jacques-Louis Lions

Ladies and Gentlemen, Dear Mrs Lions

Thank you for attending the *"Jacques-Louis Lions Summer School"*. Considering the temperature we have known these last few days calling this event a Winter School would have been more appropriate. Beside the participants our thanks go the organizers, namely Vincenzo Capasso and Jacques Periaux, and to their collaborators, for their choice of Montecatini and for the warm atmosphere they have been able to built in such a very short time. As many of you know, Professor Jacques-Louis Lions was very found of Italy. Indeed, some of his most important scientific work was done in collaboration with celebrated Italian mathematicians such as E. Magenes, G. Prodi, and G. Stampacchia. He was also a member of the Italian Academy of Sciences and of the Accademia Pontificia. Concerning J.L. Lions relations with MACSInet let me say that he was one of the founders of MACSInet, being a strong believer of the mutual benefits that Industry and Mathematics can bring to each other. I have the privilege to have been his collaborator for more than thirty years and I remember his pride when the group he was leading at IRIA (I was then his deputy) got its first contract from Industry (by the way, it was about solving the Maxwell equations in a car alternator by a finite element method). A most famous Italian global scientist, Leonardo da Vinci, said something like: *"True science has to take inspiration from the real world and its results have to be validated by experiments."* From that point of view, Jacques-Louis Lions was a true follower of Leonardo in the sense that for him Mathematics (at least the ones he was doing) had to be related to the real world and, if possible, applicable not only to one but to several important practical problems. Indeed, J.L. Lions was a master at going from the particular to the general and the other way back. At that stage, I cannot resist telling you what Professor J.L. Lions, V. Capasso and I have in common. What we have in common is CERFACS ! (CERFACS stands for European Center for Research and Advanced Training in Scientific Computing).

In the early 90's Professor J.L. Lions was Chairman of the board of CERFACS and was looking for a new C.E.O. (Directeur General in French). Since he liked very much V.Capasso for his warm personality and scientific talents, he offered him the position. Professor Capasso almost said yes, but ultimately had to say no for personal reasons. I was offered the position, accepted the offer, and never regretted it (I kept it during three years after which I decided to return to the U.S.; I was maybe missing the turbulence associated to the U.S. president of that time).

I am pretty sure that J.L. Lions would have loved the program of this Summer / Winter School since it includes topics on Optimization, Control, Atmospheric / Ocean coupling, Environmental Sciences, Aerospace, Fluid Dynamics, all topics to which he greatly contributed. Professor J.L. Lions' son, Pierre-Louis, was supposed to be with us this week, but could not do it for several reasons, among them:

1) he will have today his first meeting at College de France, a prestigious institution w here he was elected last year;

2) the second reason is that P.L. Lions has to meet this week with the scientific advisor of the French president Jacques Chirac to discuss some important issues related to Science and Technology.

Professor P.L. Lions deeply apologizes for his absence and wishes us good luck. We have, however, the pleasure and honor to have Mrs. J.L. Lions with us this week.

A last word: I personally miss Professor Jacques-Louis Lions as do his former collaborators, students, colleagues, friends, and indeed all of us, since he was a kind of light-house, in fact a sun (or a star) illuminating the whole field of Computational and Applied Mathematics.

Montecatini, March 20, 2003

Roland Glowinski

Contents

List of Contributors

Didier Auroux
Laboratoire J.A. Dieudonné,
Université de Nice
Sophia-Antipolis, Parc Valrose,
F-06108 Nice Cedex 2, France
auroux@math.unice.fr

Eiichi Baba
Hiroshima University
1-3-2 Kagamiyama
Higashi-Hiroshima, Japan 739-8511
ebaba@hiroshima-u.ac.jp

Alexandre M. Bayen
Hybrid Systems Laboratory
Department of Aeronautics
and Astronautics
Stanford University, Stanford, CA
bayen@stanford.edu

Jacques Blum
Laboratoire J.A. Dieudonné,
Université de Nice
Sophia-Antipolis, Parc Valrose,
F-06108 Nice Cedex 2, France
jblum@math.unice.fr

Marco A. Boschetti
Department of Computer Science
University of Bologna, Italy
marco.boschetti@unibo.it

Vincenzo Capasso
MIRIAM &
Department of Mathematics
University of Milano
Via C. Saldini, 50, Milano, Italy
capasso@mat.unimi.it

Edward Dean
Department of Mathematics
University of Houston
Houston, Texas
77204-3008, USA
dean@math.uh.edu

Heinz W. Engl
Johann Radon Institute for Computational and Applied Mathematics
Austrian Academy of Sciences
A-4040 Linz, Austria
heinz.engl@oeaw.ac.at

Roland Glowinski
Department of Mathematics
University of Houston
Houston, Texas
77204-3008, USA
roland@math.uh.edu

Luis F. Gonzalez
School of Aerospace, Mechanical
and Mechatronic Engineering
The University of Sydney
Sydney, NSW 2006, Australia
gonzales@aeromech.usyd.edu.au

Lorenzo Hector Juarez
Departamento de Matemáticas
Universidad Autónoma
Metropolitana-Iztapalapa
Iztaplapa, D. F. , 09340, Mexico
hector@math.uh.edu

Hideo Kawarada
Faculty of Distribution
and Logistics Systems
Ryutsu Keizai University
Hirahata 120, Ryuugasaki
Ibaraki, Japan
kawarada@rku.ac.jp

Philipp Kügler
Institut für Industriemathematik
Johannes Kepler Universität
A-4040 Linz, Austria
kuegler@indmath.uni-linz.ac.at

Vittorio Maniezzo
Department of Mathematics
University of Bologna
Bologna, Italy
maniezzo@csr.unibo.it

Ian M. Mitchell
Computer Science Department
University of British Columbia
Vancouver, BC, CANADA, V6T 1Z4
mitchell@cs.ubc.ca

Daniela Morale
MIRIAM &
Department of Mathematics
University of Milano,
Via C. Saldini, 50, Milano, Italy
morale@mat.unimi.it

Meeko K. M. Oishi
Hybrid Systems Laboratory
Department of Aeronautics
and Astronautics
Stanford University
Stanford, CA 94305-4035
moishi@stanford.edu

Mitsumasa Okada
Department of Material Science
and Chemical System
Hiroshima University
Kagamiyama1-4-1
Higashi-Hiroshima, 739-8527 Japan
okada@environ.jp

L. Padovan
Dipartimento di Energetica
Università degli Studi di Trieste,
Trieste, Italy

Tsorng-Whay Pan
Department of Mathematics
University of Houston
Houston, Texas, 77204-3008, USA
pan@math.uh.edu

Valentino Pediroda
Dipartimento di Energetica
Università degli Studi di Trieste
Trieste, Italy
pediroda@units.it

Jacques Périaux
Pole Scientifique – Dassault Aviation
UPMC
Dassault Aviation
78 Quai Marcel Dassault
99214 Saint-Cloud, France
jacques.periaux@dassault-
aviation.fr

Carlo Poloni
Dipartimento di Energetica
Università degli Studi di Trieste
Trieste, Italy
poloni@univ.trieste.it

Claire J. Tomlin
Hybrid Systems Laboratory
Department of Aeronautics
and Astronautics

Stanford University
Stanford, CA, 94305-4035
tomlin@stanford.edu

Eric J. Whitney
School of Aerospace, Mechanical
and Mechatronic Engineering
The University of Sydney
Sydney, NSW 2006, Australia
eric@aeromech.usyd.edu.au

Part I

Lectures

Nonlinear Inverse Problems: Theoretical Aspects and Some Industrial Applications

Heinz W. Engl[1,2] and Philipp Kügler[2]

[1] Johann Radon Institute for Computational and Applied Mathematics, Austrian
    Academy of Sciences, A–4040 Linz, Austria `heinz.engl@oeaw.ac.at`
[2] Institut für Industriemathematik, Johannes Kepler Universität, A–4040 Linz,
    Austria `kuegler@indmath.uni-linz.ac.at`

1 Introduction

Driven by the needs from applications both in industry and other sciences, the field of inverse problems has undergone a tremendous growth within the last two decades, where recent emphasis has been laid more than before on nonlinear problems. This is documented by the wide current literature on regularization methods for the solution of nonlinear ill-posed problems. Advances in this theory and the development of sophisticated numerical techniques for treating the direct problems allow to address and solve industrial inverse problems on a level of high complexity.

Inverse problems arise whenever one searches for causes of observed or desired effects. Two problems are called inverse to each other if the formulation of one problem involves the solution of the other one. These two problems then are separated into a direct and an inverse problem. At first sight, it might seem arbitrary which of these problems is called the direct and which one the inverse problem. Usually, the direct problem is the more classical one. E.g., when dealing with partial differential equations, the direct problem could be to predict the evolution of the described system from knowledge of its present state and the governing physical laws including information on all physically relevant parameters while a possible inverse problem is to estimate (some of) these parameters from observations of the evolution of the system; this is called "parameter identification". Sometimes, the distinction is not so obvious: e.g., differentiation and integration are inverse to each other, it would seem arbitrary which of these problems is considered the direct and the inverse problem, respectively. But since integration is stable and differentiation is unstable, a property common to most inverse problems, one usually considers integration the direct and differentiation the inverse problem. Note also that integration is a smoothing process, which is inherently connected with the instability of differentiation.

Other important classes of inverse problems are

- *(Computerized) tomography* (cf. [Nat86]), which involves the reconstruction of a function, usually a density distribution, from values of its line integrals and is important both in medical applications and in nondestructive testing [ELR96b]. Mathematically, this is connected with the inversion of the Radon transform.
- *Inverse scattering* (cf. [CK92], [Ram86]), where one wants to reconstruct an obstacle or an inhomogeneity from waves scattered by those. This is a special case of *shape reconstruction* and closely connected to shape optimization [HN88]: while in the latter, one wants to construct a shape such that some outcome is optimized, i.e., one wants to reach a *desired* effect, in the former, one wants to determine a shape from measurements, i.e., one is looking for the cause for an *observed* effect. Here, uniqueness is a basic question, since one wants to know if the shape (or anything else in some other kind of inverse problem) can be determined uniquely from the data ("identifiability"), while in a (shape) optimization problem, it might even be advantageous if one has several possibilities to reach the desired aim, so that one does not care about uniqueness there.
- *Inverse heat conduction problems* like solving a heat equation backwards in time or "sideways" (i.e., with Cauchy data on a part of the boundary) (cf. [ER95], [BBC85]).
- *Geophysical inverse problems* like determining a spatially varying density distribution in the earth from gravity measurements (cf. [ELR96a]).
- Inverse problems in *imaging* like deblurring and denoising (cf. [BB98])
- *Identification of parameters* in (partial) differential equations from interior or boundary measurements of the solution (cf. [BK89], [Isa98]), the latter case appearing e.g. in *impedance tomography*. If the parameter is piecewise constant and one is mainly interested in the location where it jumps, this can also be interpreted as a shape reconstruction problem (cf. [IN99]).

Detailed references for these and many more classes of inverse problems can be found e.g. in [EHN96], [Eng93], [EG87], [Gro93], [Lou899], [Kir96], [Hof99], [CER90].

The mathematical formulation of inverse problems leads to models that typically are *ill-posed*: According to Hadamard, a mathematical problem is called well-posed if

- for all admissible data, a solution exists,
- for all admissible data, the solution is unique and
- the solution depends continuously on the data.

If one of these properties is violated, the problem is called ill-posed. Neither existence nor uniqueness of a solution to an inverse problem are guaranteed. As mentioned, non-uniqueness is sometimes of advantage, then allowing to choose among several strategies for obtaining a desired effect. In practical applications, one never has exact data, but only data perturbed by noise are available due to errors in the measurements or also due to inaccuracies the model itself. Even if their deviation from the exact data is small, algorithms developed for well-posed problems then fail in case of a violation of the third Hadamard condition if they do not address the instability, since data as well as round-off errors may then be amplified by an arbitrarily large factor. In order to overcome these instabilities one has to use *regularization* methods, which in general terms replace an ill-posed problem by a family of neighboring well-posed problems.

In this survey paper, we concentrate on regularization techniques for solving inverse and ill-posed problems that are nonlinear. We formulate these problems in functional analytic terms as nonlinear operator equations. Nevertheless, we start with the theory for linear problems in order to familiarize the reader with basic properties and definitions, then also relevant for the discussion of nonlinear problems. The latter will address both theoretical and computational aspects of two popular classes of regularization methods, namely Tikhonov regularization and iterative techniques. Finally, we present in some detail examples for nonlinear inverse problems appearing in iron and steel production as well as in quantitative finance and show how regularization methods were used for solving them in a numerically stable way.

2 Regularization Methods

A prototype for linear inverse problems are linear integral equations of the first kind such as

$$\int_G k(s,t)x(t)\,dt = y(s) \quad (s \in G) \tag{1}$$

with $k \in L^2(G \times G), y \in L^2(G)$. A case of special importance is that k actually depends on $s - t$, i.e., y is a convolution of x and k; solving (1) is then called *deconvolution*. For this and a collection of other linear inverse problems from various application fields we refer to [Eng93]. A simple parameter identification problem serves as our prototype example of a nonlinear inverse problem: In physical or technical applications, the physical laws governing the process may be known in principle, while actual values of some of the physical parameters in these laws are often unknown. For instance, in describing the heat conduction in a material occupying a three dimensional domain Ω whose temperature is kept zero at the boundary, the temperature distribution u after a sufficiently long time is modeled by

$$-\nabla \cdot (q(x)\nabla u) = f(x) \quad x \text{ in } \Omega \tag{2}$$
$$u = 0 \quad \text{on } \partial\Omega,$$

where f denotes internal heat sources and q is the spatially varying heat conductivity. If one cannot measure q directly, one can try to determine q from internal measurements of the temperature u or from boundary measurements of the heat flux $q\frac{\partial u}{\partial n}$. One refers to this inverse problem also as an *indirect measurement* problem. Parameter identification problems like (2) appear, e.g., in geophysical applications and in non-destructive material testing. Note that (2) with unknown q is nonlinear since the relation between this parameter and the solution u, which serves as the data in the inverse problem, is nonlinear even if the direct problem of computing u with given q is linear.

Both these inverse problems turn out to be ill-posed, for obtaining a solution in a (numerically) stable way, one has to develop regularization methods. We will use (1) and (2) for illustrating regularization for linear and nonlinear inverse problems, respectively. Regularization methods replace an ill-posed problem by a family of well-posed problems, their solution, called *regularized solutions*, are used as approximations to the desired solution of the inverse problem. These methods always involve some parameter measuring the closeness of the regularized and the original (unregularized) inverse problem, rules (and algorithms) for the choice of these *regularization parameters* as well as convergence properties of the regularized solutions are central points in the theory of these methods, since only they allow to finally find the right balance between stability and accuracy (see below).

While the theory of regularization methods for linear ill-posed problems is by now rather comprehensive, it is still evolving and far from complete in the nonlinear case. Though we mainly focus on nonlinear inverse problems, we begin our survey with the theory for linear problems in order to give a first introduction into basic perceptions and terminologies.

2.1 Linear Inverse Problems

Starting point for our discussion is the operator equation

$$Tx = y, \tag{3}$$

where T denotes a bounded linear operator acting between Hilbert spaces X and Y. As concept of solution we use that of a best-approximate solution, which is the minimizer of the residual $\|Tx - y\|$, i.e., a least squares solution, that minimizes $\|x\|$ among all minimizers of the residual. Our goal is to approximate the best-approximate solution of (3) in the situation that the exact data y are possibly not known precisely and only perturbed data y^δ with

$$\|y - y^\delta\| \le \delta \tag{4}$$

are available. Here, δ is called the noise level. Note that we use a deterministic error concept by assuming a bound in the Hilbert space norm and also considering convergence there. A different approach which is followed recently in connection with "uncertainty" is to use a stochastic error concept and to consider convergence with respect e.g. to the Prokhorov metric on a space of probability measures, see e.g. [EW85], [BB01].

The operator $T^\dagger$ which maps the exact data $y \in D(T^\dagger)$ to the best-approximate solution of (3) is known as the Moore-Penrose (generalized) inverse of T (see, e.g., [Gro77], [Nas76]) and its domain is given by

$$D(T^\dagger) = R(T) \dotplus R(T)^\perp.$$

The Moore-Penrose inverse is bounded, i.e., the problem of determining the best-approximate solution of (3) is stable if and only if $R(T)$ is closed. Otherwise, solving (3) is ill-posed, the first and the third of Hadamard's conditions being violated even if we consider the best-approximate solution since $D(T^\dagger)$ is only dense in Y. The range $R(T)$ is non-closed especially if T is compact with $\dim R(T) = \infty$, for which the injectivity of a compact operator T is a sufficient condition if X is infinite dimensional. Since an integral operator like the one in (1) is compact under the conditions on k mentioned there, integral equations of the first kind with non-degenerate kernels are a prototype of ill-posed problems.

In the ill-posed case, $T^\dagger y^\delta$ cannot serve as a reliable approximation of $T^\dagger y$ due to the unboundedness. Instead, we are looking for a regularized solution which depends continuously on the noisy data (such that it can be computed in a stable way) and converges to $T^\dagger y$ as the noise level tends to zero with the regularization parameter properly chosen. We explain the construction of a regularization method for the important special case of a compact operator T and refer to [EHN96] for the non-compact situation.

If T is a compact operator, there exists a singular system $(\sigma_i; u_i, v_i)_{i \in N}$, which is defined as follows: With $T^* : Y \to X$ denoting the adjoint operator of T (introduced via the requirement that for all $x \in X$ and $y \in Y$, $\langle Tx, y \rangle = \langle x, T^*y \rangle$ holds), the $(\sigma_i^2)_{i \in N}$ are the non–zero eigenvalues of the self–adjoint operator T^*T (and also of TT^*), written down in decreasing order with multiplicity, $\sigma_i > 0$. Furthermore, the $(u_i)_{i \in N}$ are a corresponding complete orthonormal system of eigenvectors of T^*T (which spans $\overline{R(T^*)} = \overline{R(T^*T)}$), and the $(v_i)_{i \in N}$ are defined via

$$v_i := \frac{Tu_i}{\|Tu_i\|}.$$

As in the finite–dimensional situation (recall the singular value decomposition of a matrix), the $(v_i)_{i \in N}$ are a complete orthonormal system of eigenvalues of TT^* and span $\overline{R(T)} = \overline{R(TT^*)}$. This translates into the formulas

$$Tu_i = \sigma_i v_i \tag{5}$$

$$T^* v_i = \sigma_i u_i \tag{6}$$

$$Tx = \sum_{i=1}^{\infty} \sigma_i \langle x, u_i \rangle v_i \quad (x \in X) \tag{7}$$

$$T^* y = \sum_{i=1}^{\infty} \sigma_i \langle y, v_i \rangle u_i \quad (y \in Y), \tag{8}$$

where these infinite series converge in the Hilbert space norms of X and Y, respectively; (7) and (8) are called "singular value expansion" and are the infinite–dimensional analogues of the singular value decomposition.

If (and only if) T has a finite–dimensional range, T has only finitely many singular values, so that all infinite series involving singular values degenerate to finite sums. However, if there are infinitely many singular values (the generic case), they accumulate (only) at 0, i.e.,

$$\lim_{i \to \infty} \sigma_i = 0.$$

Since for $y \in D(T^\dagger)$, which holds if and only if the Picard Criterion

$$\sum_{n=1}^{\infty} \frac{|\langle y, v_n \rangle|^2}{\sigma_n^2} < \infty$$

is satisfied, the best-approximate solution of (3) has the series representation

$$T^\dagger y = \sum_{n=1}^{\infty} \frac{\langle y, v_n \rangle}{\sigma_n} u_n,$$

we see why (2.1) turns (3) into an ill-posed problem: errors in the Fourier components of y with respect to v_n, i.e., in $\langle y, v_n \rangle$, are multiplied by $\frac{1}{\sigma_n}$, a factor growing to infinity for $n \to \infty$ due to (2.1) (if $\dim R(T) = \infty$). Thus, especially errors in Fourier components of the data for large n, usually termed as "high frequency errors", are strongly amplified. Also, the faster the decay of the σ_n, the stronger the error amplification, which also allows to quantify ill-posedness: one usually distinguishes between mildly, i.e., $\sigma_n = \mathcal{O}(n^{-\alpha})$ (with $\alpha > 1$), and severely, i.e., $\sigma_n = \mathcal{O}(e^{-n})$, ill-posed problems. Although, as we will see, the connection between the ill-posedness of a nonlinear problem and of its linearization is not as close as one might expect, this is usually also used as a (rough) quantification of ill-posedness in the nonlinear situation via the decay rate of the singular values of the linearization.

Regularization methods now are techniques that can handle these problems. In the linear compact case, they can be constructed and analyzed based on the singular value expansion: From (5)–(8) and (2.1) we see that

$$T^\dagger y = \sum_{n=1}^{\infty} \frac{\langle T^* y, u_n \rangle}{\sigma_n^2} u_n$$

holds. The basic idea for deriving a regularization method is to replace the amplification factors $\frac{1}{\sigma_n^2}$ by a filtered version $U(\alpha, \sigma_n^2)$, where the filter function $U(\alpha, \cdot)$ is piecewise continuous on $[0, +\infty[$ for a regularization parameter $\alpha > 0$ and converges to $\frac{1}{\lambda}$ as $\alpha \to 0$. This allows to introduce the regularized solution

$$x_\alpha := \sum_{n=1}^{\infty} U(\alpha, \sigma_n^2) \cdot \langle T^* y, u_n \rangle u_n$$

or

$$x_\alpha^\delta := \sum_{n=1}^{\infty} U(\alpha, \sigma_n^2) \cdot \langle T^* y^\delta, u_n \rangle u_n$$

in case of perturbed data y^δ fulfilling (4).

The conditions on the family $\{(U(\alpha, \lambda), \alpha > 0\}$ under which x_α in fact converges to $T^\dagger y$ are stated in

Theorem 1. Let, for an $\varepsilon > 0$, $U : R^+ \times [0, \sigma_1^2 + \varepsilon] \to R$ fulfill the following assumptions

$$\text{for all } \alpha > 0, U(\alpha, .) \text{ is piecewise continuous;} \tag{9}$$

$$\text{there is a } C > 0 \text{ such that for all} \tag{10}$$

$$(\alpha, \lambda), |\lambda \cdot U(\alpha, \lambda)| \le C \text{ holds}$$

$$\text{for all } \lambda \ne 0, \lim_{\alpha \to 0} U(\alpha, \lambda) = \frac{1}{\lambda}. \tag{11}$$

Then, for all $y \in D(K^\dagger)$,

$$\lim_{\alpha \to 0} U(\alpha, T^* T) T^* y = T^\dagger y.$$

holds.

Furthermore, we have the following stability estimate for the regularized solutions:

Theorem 2. Let U be as in Theorem 1, x_α and x_α^δ be defined by (2.1)–(2.1). For $\alpha > 0$, let

$$g_U(\alpha) := \sup\{|U(\alpha, \lambda)|/\lambda \in [0, \sigma_1^2]\}.$$

Then

$$\|x_\alpha - x_\alpha^\delta\| \le \delta \cdot \sqrt{C g_U(\alpha)}$$

holds.

Different choices of the filter function $U(\alpha, \lambda)$ now lead to different regularization methods. The probably simplest choice satisfying the assumptions of Theorem 1 is

$$U_\alpha(\lambda) := \frac{1}{\alpha + \lambda},$$

then leading to *Tikhonov regularization*

$$x_\alpha^\delta = \sum_{n=1}^{\infty} \frac{\sigma_n}{\alpha + \sigma_n^2} \langle y^\delta, v_n \rangle u_n = (\alpha I + T^*T)^{-1} T^* y^\delta.$$

Here, the regularized solution x_α^δ can also be characterized in variational form as minimizer of the functional

$$x \to \|Tx - y^\delta\|^2 + \alpha \|x\|^2, \tag{12}$$

which allows to carry this method over to nonlinear problems (although the proofs of its properties are then different since the spectral theoretic foundation is lacking there). Without the additional penalty term $\alpha \|x\|^2$, this would be called "output least squares minimization", but would be unstable. Instead of $\alpha \|x\|^2$, one uses also penalty terms of the more general form $\alpha \|Lx\|^2$ with a suitable (usually differential) operator L; the method will then approximate a least squares solution minimizing $\|Lx\|$.

Another widely used method is the truncated singular value expansion with

$$U(\alpha, \lambda) = \begin{cases} \frac{1}{\lambda} & \lambda \geq \alpha \\ 0 & \lambda < \alpha. \end{cases}$$

This yields

$$x_\alpha^\delta = \sum_{n=1 \, \sigma_n^2 \geq \alpha} \frac{\langle y_\delta, v_n \rangle}{\sigma_n} u_n,$$

where the small singular values ("high frequencies") are filtered out by a "low–pass filter". This method is in some sense optimal (cf. [EHN96]), but can only be used if the singular value expansion is readily available. Finally, we consider the method of asymptotic regularization which is based on the initial value problem

$$u_\delta'(t) + T^*T u_\delta(t) = T^* y^\delta, \quad t \in R_0^+, \tag{13}$$

$$u_\delta(0) = 0, \tag{14}$$

where $u_\delta : R_0^+ \to X$. A motivation for looking at this initial value problem is that when the solution becomes stationary, it solves the *(Gaussian) normal equation*

$$T^*Tx = T^*y. \tag{15}$$

Thus, in the noise free case, one expects the solution u of this initial value problem to tend, as $t \to \infty$, to the best-approximate solution $T^\dagger y$, which is characterized by (15). The regularized solution is now defined as

$$x_\alpha^\delta = u_\delta(\frac{1}{\alpha})$$

and can be put into the general framework via

$$x_\alpha^\delta = U(\alpha, T^*T)T^*y^\delta$$

with

$$U(\alpha, \lambda) = \int_0^{\frac{1}{\alpha}} e^{-\lambda s}\, ds.$$

The regularization effect in this method is obtained by integrating the initial value problem not up to infinity but only up to an abscissa $1/\alpha$. The method can also be understood as a continuous version of the iterative Landweber method to be discussed in Section 2.2 since the latter can be derived from solving (13) by the forward Euler method with step size α, i.e.,

$$u_\delta(t + \alpha) \sim u_\delta(t) + \alpha T^*(y^\delta - Tu_\delta(t)). \tag{16}$$

We will see that there, the regularization is achieved by stopping the iteration at a specific iteration index, which is a discrete analogue to stopping the integration of (13) early.

In any regularization method, the *regularization parameter* α plays a crucial role. As can be seen for instance from (2.1), its choice always represents a compromise between accuracy and stability: if α is too large, the series (2.1) is truncated too early, leading to a poor approximation of $T^\dagger y$. On the other hand, if α is chosen too small, possible data errors may already be amplified too strongly. For choosing the parameter, there are two general classes of options: *A-priori rules* define the regularization parameter as a function of the noise level only, i.e., $\alpha = \alpha(\delta)$, while in *a-posteriori rules*, α depends both on the noise level and the actual data, i.e., $\alpha = \alpha(\delta, y^\delta)$. An example for the latter is the so-called *discrepancy principle*, where α is chosen such that

$$\|Tx_\alpha^\delta - y^\delta\| = C\delta \tag{17}$$

holds (with some $C > 1$). Note that in (17) the determination of the parameter amounts to solving a nonlinear equation.

One can show (see [EHN96]) that *error-free* strategies, where $\alpha = \alpha(y^\delta)$ does not depend on δ, cannot lead to convergence as $\delta \to 0$ in the sense that

$\lim_{\delta \to 0} x_\alpha^\delta = T^\dagger y$ for all y^δ satisfying (4) and all $y \in D(T^\dagger)$. Since this is only an asymptotic statement, these techniques may still occasionally work well for a fixed noise level $\delta > 0$, see [HH93]. However, the knowledge and use of a bound for the data error as in (4) is necessary for the construction of regularization methods based on a sound theoretical foundation. The error-free strategies include the popular methods of generalized cross-validation ([Wah90]) and the *L-curve method* ([HO93]); for its non–convergence, see [EG94] and [Vog96].

Crucial questions in applying regularization methods are convergence *rates* and how to choose regularization parameters to obtain optimal convergence rates. By convergence rates we mean rates for the worst-case error

$$\sup\{\|x_\alpha^\delta - T^\dagger y\| \mid \|y - y^\delta\| \le \delta\}$$

for $\delta \to 0$ and $\alpha = \alpha(\delta)$ or $\alpha = \alpha(\delta, y^\delta)$ chosen appropriately. For an ill–posed problem, no uniform rate valid for all $y \in Y$ can be given, convergence of any method for an ill-posed problem can be arbitrarily slow ([Sch85]), rates can only be obtained on compact subsets of X (cf. [LY98]), i.e., under additional assumptions on the solution $T^\dagger y$. For instance, under a *source condition* (with $\nu > 0$)

$$T^\dagger y \in R((T^*T)^\nu), \tag{18}$$

which can be (due to the fact that usually T is smoothing) thought of as an (abstract) a–priori smoothness condition, Tikhonov regularization converges with the rate

$$\|x_\alpha^\delta - T^\dagger y\| = O(\delta^{\frac{2\nu}{1+2\nu}}) \tag{19}$$

for the a–priori choice

$$\alpha \sim \delta^{\frac{2}{1+2\nu}}$$

and $\nu \le 1$. This (as it turns out, optimal under (18)) rate is also achieved with the a–posteriori parameter choice (17), but only for $\nu \le \frac{1}{2}$. For a–posteriori parameter choice rules that always lead to optimal convergence rates see [EG88] and [Rau84].

The typical total error behavior of a regularization method is shown in Figure 1: the regularization error $\|x_\alpha - T^\dagger y\|$ goes to 0 as $\alpha \to 0$, while the propagated data error $\|x_\alpha - x_\alpha^\delta\|$ grows without bound as $\alpha \to 0$. The difficulty in optimally choosing the regularization parameter is that the curves in Figure 1 are not computable.

For numerically solving an inverse problem, any regularization method has to be realized in finite-dimensional spaces. In fact, a regularization effect can

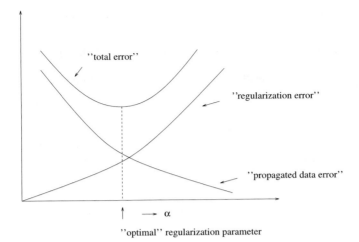

Fig. 1. Typical error behaviour

already be obtained by a finite-dimensional approximation of the problem, where the approximation level plays the role of the regularization parameter. Projection methods based on this regularizing property contain the least squares projection, where the minimum norm solution of (3) is sought in a finite-dimensional subspace X_n of X, and the dual least-squares method, where (3) is projected onto a finite-dimensional subspace Y_n of Y before the minimum norm solution of the resulting equation is computed, see [Nat77], [GN88], [Eng82]. However, error estimates for the case of noisy data and numerical experience show that at least for severely ill-posed problems the dimension of the chosen subspace has to be low in order to keep the total error small. Hence, for obtaining a reasonable accuracy, projection methods should be combined with an additional regularization method, e.g., with one of those discussed before, see [EN88], [PV90].

Methods which are closely related to regularization methods are *mollifier methods*, where one looks for a smoothed version of the solution, cf. [LM90]. Based on these ideas, one can develop methods where solutions of linear inverse problems can be computed fast by applying an "approximate inverse", which is usually an integral operator whose kernel, the "reconstruction kernel", can be precomputed (see EHN184). This is in turn closely related to the "linear functional strategy" from e.g. [And86]. While these are still linear methods, there are also *nonlinear methods* for solving linear ill–posed problems, e.g., the *Backus–Gilbert method* ([BG67], [KSB88]) and the *conjugate gradient method* ([Bra87], [Lou87], [Han95]). For details about these methods and other aspects we do not touch here (like the non–compact case and numerical aspects in the framework of combining regularization in Hilbert space with projection into finite-dimensional spaces) see [EHN96]. We mention in closing that there is a close connection between regularization and approximation by neural networks (cf. [BE00] and the references quoted there).

2.2 Nonlinear Inverse Problems

After this first introduction to the field of inverse and ill-posed problems we now turn to the nonlinear situation. There we will again meet many of the topics mentioned in the previous section, i.e., we will address rules for the choice of the regularization parameter, observe again the typical error behavior from Figure 1 and discuss convergence rates. However, since the tools of spectral theory are no longer available, the construction and especially the analysis of regularization methods for nonlinear ill-posed problems becomes much harder.

Nonlinear inverse problems can be cast into the abstract framework of nonlinear operator equations

$$F(x) = y, \tag{20}$$

where F acts between two Hilbert spaces X and Y. The basic assumptions for a reasonable theory are that F is continuous and is weakly sequentially closed, i.e., for any sequence $x_n \subset \mathcal{D}(F)$, $x_n \rightharpoonup x$ in X and $F(x_n) \rightharpoonup y$ in Y imply that $x \in \mathcal{D}$ and $F(x) = y$. (cf. [EKN89]). As opposed to the linear case, F is usually not explicitly given, but represents the operator describing the direct (also sometimes called "forward") problem. Considering for instance the parameter identification problem (2), the *parameter-to-output map* F maps the parameter q onto the solution u_q of the state equation or the heat flux $q\frac{\partial u_q}{\partial n}$. In inverse scattering, the operator F maps the shape of a scattering body onto the scattered far field pattern.

Neither existence nor uniqueness of a solution to (20) are guaranteed. Assuming for simplicity that the exact data y are attainable ,i.e., that (20) in fact admits a solution and that the underlying model is thus correct, we again introduce a generalized solution concept (see [BEGNS94] for the non-attainable case): For $x^* \in X$, we call a solution $x^\dagger$ of (20) which minimizes $\|x - x^*\|$ among all solutions an x^*-*minimum–norm solution* $(x^* - MNS)$ of (20). The element x^* should include available a–priori information like positions of singularities in x if they happen to be available and will also be part of solution algorithms, see below.

In the following, we slur over the issue of uniqueness and consider problem (20) to be ill-posed if its solution does not depend continuously on the data y. Although, as mentioned above, the degree of ill-posedness of a nonlinear problem is frequently characterized via the decay of the singular values of its linearization, this is not always appropriate: It is shown in [EKN89] that a nonlinear ill-posed problem may have a well-posed linearization and that well-posed nonlinear problems may have ill-posed linearizations. If one accepts a quantification of ill-posedness via the linearization, then, e.g., inverse scattering is severely ill-posed with $\sigma_n = \mathcal{O}(\frac{1}{n!}\left(\frac{k}{n}\right))$, where k denotes the wave number (cf. [CK92]). Since the linearization is used in most numerical algorithms, this certainly makes sense.

As in the linear case, compactness of F, together with, e.g., the (local) injectivity of the operator again serves as a sufficient condition for the ill-posedness of (20), see [EKN89]:

Proposition 1. *Let F be a (nonlinear) compact and continuous operator, and let $\mathcal{D}(F)$ be weakly closed. Furthermore, assume that $F(x^\dagger) = y$ and that there exists an $\varepsilon > 0$ such that $F(x) = \hat{y}$ has a unique solution for all $\hat{y} \in \mathcal{R}(F) \cap U_\varepsilon(y)$. If there exists a sequence $\{x_n\} \subset \mathcal{D}(F)$ satisfying*

$$x_n \rightharpoonup x^\dagger \quad \text{but} \quad x_n \nrightarrow x^\dagger$$

then F^{-1} - defined on $\mathcal{R}(F) \cap U_\varepsilon(y)$ - is not continuous in y.

Note that if $\mathcal{D}(F)$ happens to be compact then F^{-1} is continuous as soon as it exists due to the Arzela-Ascoli Theorem. This property allows to regularize a nonlinear ill-posed problem by simply restricting the domain of F to a compact set; however, this usually does not yield qualitative stability estimates. In the following, we survey two widely used approaches for solving nonlinear inverse problems in a stable way, namely Tikhonov regularization and iterative regularization methods.

Tikhonov Regularization

In Tikhonov regularization, problem (20) with data satisfying (4) is replaced by the minimization problem

$$\|F(x) - y^\delta\|^2 + \alpha\|x - x^*\|^2 \to \min, \quad x \in \mathcal{D}(F), \tag{21}$$

where $x^* \in X$ is an initial guess for a solution of (20), motivated from the linear case, see (12). For a positive regularization parameter α, minimizers always exist under the above-mentioned assumptions on F but need not be unique, whence we call *any* global minimizer of (21) a regularized solution x_α^δ. One can show that x_α^δ depends continuously on the data for α fixed and that x_α^δ converges towards a solution of (20) in a set-valued sense with $\alpha(\delta) \to 0$ and $\delta^2/\alpha(\delta) \to$ as δ tends to zero, see [EKN89].

The basic result on convergence rates for Tikhonov regularization is

Theorem 3. *Let $x^\dagger$ be an element in the interior of the convex domain $\mathcal{D}(F)$. Furthermore, let F be Fréchet differentiable with*

$$\|F'(x^\dagger) - F'(x)\| \leq C\|x^\dagger - x\|$$

in a neighborhood of $x^\dagger$. If there exists an element $w \in Y$ satisfying the source condition

$$x^\dagger - x^* = (F'(x^\dagger)^* F'(x^\dagger))^\nu w \tag{22}$$

for some $\nu \in [1/2, 1]$ with $C\|w\| < 1$, then the (a priori) parameter choice

$$\alpha \sim \delta^{\frac{2}{2\nu+1}} \tag{23}$$

yields

$$\|x_\alpha^\delta - x^\dagger\| = \mathcal{O}(\delta^{\frac{2\nu}{2\nu+1}}).$$

In (22), $F'(x^\dagger)^*$ denotes the Hilbert space adjoint of the Fréchet-derivative. Formally, the source condition as well as the obtained rate correspond to (18) and (19). Again, (22) is an abstract smoothness condition on the difference between the true solution $x^\dagger$ and the a-priori guess x^* used in (21), once more supporting the importance of the choice of the latter. Since (22) also plays a crucial role in the convergence analysis of iterative regularization methods, we postpone a further discussion. For the proof and variants of Theorem 3, see [EHN96].

The disadvantage of rule (23) is that the parameter depends on the smoothness index ν of the exact solution $x^\dagger$ which is not known in practice. A slight variant of Tikhonov regularization which allows to prove the (then no longer always optimal) rate $\mathcal{O}(\sqrt{\delta})$ as long as (22) holds with $\nu \geq 1/2$ for the choice $\alpha(\delta) = \mathcal{O}(\delta^2)$ (now independent of the unknown ν) can also be found in [EHN96]. Turning to a-posteriori rules, the use of the discrepancy principle, where $\alpha(\delta, y^\delta)$ is defined as the solution of

$$\|F(x_\alpha^\delta) - y^\delta\| = C\delta, \tag{24}$$

is rather problematic since - in the nonlinear situation - problem (24) only admits a solution under severe additional assumptions, see [KS85]. For a (quite complicated) a posteriori strategy that always leads to optimal rates see [SEK93].

In (21), it is not obligatory to use the norm induced by the inner product in X as penalty term. Other possibilities include *maximum entropy* regularization

$$\|F(x) - y^\delta\|^2 + \alpha \int_\Omega x(t) \log \frac{x(t)}{x^*(t)} \, dt \quad \to \min,$$

see [EL93], [Egg93], [EL96], [LA96], or *bounded variation* regularization

$$\|F(x) - y^\delta\|^2 + \alpha \int_\Omega |\nabla x(t)|\, dt \;\rightarrow\; \min, \tag{25}$$

which enhances sharp features in x as needed in, e.g., image reconstruction, see [Rud94], [NS98], [Sch02].

With respect to the numerical implementation of Tikhonov regularization one can relax the task of exactly solving problem (21) to looking for an element $x^\delta_{\alpha,\eta}$ satisfying

$$\|F(x^\delta_{\alpha,\eta}) - y^\delta\|^2 + \alpha\|x^\delta_{\alpha,\eta} - x^*\|^2 \leq \|F(x) - y^\delta\|^2 + \alpha\|x - x^*\|^2 + \eta$$

for all $x \in \mathcal{D}(F)$ with η a small positive parameter, see [EKN89]. Tikhonov regularization combined with finite dimensional approximation of X (and of F, see also Section 2.2) is discussed e.g. in [Neu89], [NS90].

However, finding a global minimizer (even only approximately) to a nonlinear optimization problem is in general not an easy task. Numerical experience shows that the functional in (21), which is in general not convex (unlike (12) in the linear case) has usually many local minima in which a descent method tends to get stuck if the underlying problem is ill-posed. Since furthermore the determination of an appropriate regularization parameter α can require high computational efforts, iterative regularization methods are an attractive alternative.

Iterative Methods

A first candidate for solving (20) in an iterative way could be Newton's method

$$x_{k+1} = x_k + F'(x_k)^{-1}(y - F(x_k)), \tag{26}$$

starting from an initial guess x_0. Even if the iteration is well-defined and $F'(\cdot)$ is invertible for every $x \in \mathcal{D}(F)$, the inverse is usually unbounded for ill-posed problems (e.g. if F is continuous and compact). Hence, (26) is inappropriate since each iteration means to solve a linear ill-posed problem, and some regularization technique has to be used instead. For instance, Tikhonov regularization applied to the linearization of (20) yields the Levenberg Marquardt method (see [Han97])

$$x_{k+1} = x_k + (F'(x_k)^*F'(x_k) + \alpha_k I)^{-1}F'(x_k)^*(y - F(x_k)), \tag{27}$$

where α_k is a sequence of positive numbers. Augmenting (27) by the term

$$-(\alpha_k I + F'(x_k)^*F'(x_k))^{-1}\alpha_k(x_k - x^*)$$

for additional stabilization gives the *iteratively regularized Gauss-Newton method* (see [Bak92], [BNS97])

$$x_{k+1} = x_k + (F'(x_k)^* F'(x_k) + \alpha_k I)^{-1} [F'(x_k)^*(y - F(x_k)) - \alpha_k(x_k - x^*)].$$
$$(28)$$

Usually, x^* is taken as x_0, but this is not necessary. As (27) and (28), most iterative methods for solving the nonlinear ill-posed problem (20) are based on solving the normal equation

$$F'(x)^*(F(x) - y) = 0 \qquad (29)$$

via successive iteration starting from x_0. Equation (29) is the first-order optimality condition for the nonlinear output least-squares problem

$$\frac{1}{2}\|y - F(x)\|^2 \to \min, \quad x \in \mathcal{D}(F). \qquad (30)$$

Alternatively to Newton type methods like (27) and (28), methods of steepest descent like the *Landweber iteration*

$$x_{k+1} = x_k + F'(x_k)^*(y - F(x_k)), \qquad (31)$$

see [HNS95], are used, where the negative gradient of the functional in (30) determines the update direction for the current iterate. From now on, we shall use x_k^δ in our notation of the iterates in order to take possibly perturbed data y^δ with (4) into account.

In the ill-posed case, due to the instability inherent in (20), it is common to all iterative methods that the iteration must not be arbitrarily continued. Instead, an iterative method only then can become a regularization method, if it is stopped "at the right time", i.e., only for a suitable stopping index k_*, the iterate $x_{k_*}^\delta$ yields a stable approximation to the solution $x^\dagger$ of (20). Due to the ill-posedness, a mere minimization of (30), i.e., an ongoing iteration, leads to unstable results and to a typical error behavior as shown in Figures 2 and 3, compare also to Figure 1. While the error in the output decreases as the iteration number increases, the error in the parameter starts to increase after an initial decay.

Again, there are two classes of methods for choosing the regularization parameter, i.e., for the determination of k_*, namely a-priori stopping rules with $k_* = k_*(\delta)$ and a-posteriori rules with $k_* = k_*(\delta, y^\delta)$. Once more, the discrepancy principle, where k_* now is determined by

$$\|y^\delta - F(x_{k_*}^\delta)\| \le \tau\delta < \|y^\delta - F(x_k^\delta)\|, \quad 0 \le k < k_*, \qquad (32)$$

for some sufficiently large $\tau > 0$ is a widely used representative for the latter. As opposed to (24) for Tikhonov regularization, (32) now is a rule easy to implement, provided that an estimate for the data error as in (4) is available. The discrepancy principle for determining the index k_* is based on stopping as soon as the residual $\|y^\delta - F(x_k^\delta)\|$ is in the order of the data error, which is

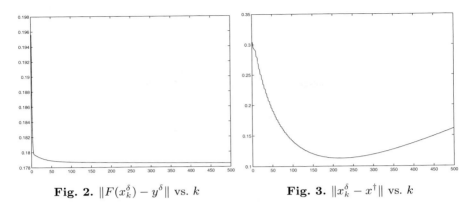

Fig. 2. $\|F(x_k^\delta) - y^\delta\|$ vs. k **Fig. 3.** $\|x_k^\delta - x^\dagger\|$ vs. k

somehow the best one should expect. For solving (20) when only noisy data y^δ with (4) are given, it would make no sense to ask for an approximate solution $\tilde{x}$ with $\|y^\delta - F(\tilde{x})\| < \delta$, the price to pay would be instability.

Iterative regularization methods are also used for linear inverse problems such as (3). Some of them then can even be analyzed by means of the general theory provided by Theorems 1 and 2. For instance, the "linear version" of Landweber iteration (31), i.e.,

$$x_{k+1}^\delta = x_k^\delta + T^*(y^\delta - Tx_k^\delta),$$

see also (16), is represented by the filter function

$$U(k, \lambda) = \sum_{j=0}^{k-1}(1 - \lambda)^j$$

with k^{-1} playing the role of α. However, for nonlinear problems (20) the investigation of an iterative method is much more complicated and has mostly to be done for each class of methods individually. Some kind of general framework is provided in [ES00].

Theoretical studies of iterative methods for nonlinear ill-posed problems based on a fixed point formulation of the nonlinear problem (20) under contractivity and nonexpansivity assumptions on the fixed point operator can be found in [Vas87], [Vas92], [VA95], [Vas98]. The Mann iteration and its variants, see [Man53], [OG69], [Gro72] are further popular methods for solving fixed point equations. There, the basic principle is to feed the fixed point operator with an weighted average of the previous iterates. For ill-posed problems, it has been analyzed in [EL01], [KL03].

Though it is also possible to formulate the methods (31), (27) and (28) as fixed point operations, the necessary assumptions on the corresponding fixed point operators may become too restrictive for a reasonable theory (see [Sch95], [ES00]) such that also alternative approaches for their investigation are considered. In [HNS95], the convergence analysis of the Landweber iteration (31) is carried out under the following assumptions: for a ball $\mathcal{B}_\rho(x_0)$ of radius ρ around x_0 with

$$\mathcal{B}_\rho(x_0) \subset \mathcal{D}(F), \tag{33}$$

the Fréchet-differentiable forward operator F is required to satisfy

$$\|F(\tilde{x}) - F(x) - F'(x)(\tilde{x} - x)\| \le \eta \|F(\tilde{x}) - F(x)\|, \quad x, \tilde{x} \in \mathcal{B}_\rho(x_0) \tag{34}$$

with $\eta < 1/2$. If furthermore, the Fréchet derivative is locally bounded by one, i.e.,

$$\|F'(x)\| \le 1, \quad x \in \mathcal{B}_\rho(x_0), \tag{35}$$

at least local convergence of the iterates x_k to a solution of (20) in $\mathcal{B}_{\rho/2}(q_0)$ can be guaranteed. Together with (33) these assumptions also guarantee that all iterates x_k remain in $\mathcal{D}(F)$, which makes the iteration well-defined. In case of noisy data y^δ not belonging to the range of F, the iterates x_k^δ cannot converge. Still, condition (34) again forces the iterates x_k^δ to remain in $\mathcal{D}(F)$ and allows a stable approximation $x_{k_*}^\delta$ of a solution to (20), provided that the iteration is terminated after $k_* = k_*(\delta, y^\delta)$ steps according to the discrepancy principle (32) with τ satisfying

$$\tau > 2\frac{1+\eta}{1-2\eta} > 2. \tag{36}$$

In order to fulfill (35) for a (locally) bounded Fréchet derivative, one eventually has to rescale (20), i.e., instead to consider

$$\lambda F(x) = \lambda y. \tag{37}$$

If λ is chosen appropriately, then (35) holds, while condition (34) is scaling invariant, meaning that the requirement $\eta < 1/2$ cannot be weakened. At the first glance, condition (34) looks like a standard closeness assumption. In [HNS95] it is compared to the weaker Fréchet estimate

$$\|F(\tilde{x}) - F(x) - F'(x)(\tilde{x} - x)\| \le C\|\tilde{x} - x\|^2$$

for a Lipschitz continuous F', in [Sch95] even a geometric interpretation is given. Summarizing, (34) can be seen as a nonlinearity condition on F, actually requiring that the nonlinearity of F must not be too strong. Note that a linear operator F clearly would satisfy this condition. However, the conditions (35) and (34) only yield stability, i.e., the regularized solution $x_{k_*}^\delta$ depends continuously on y^δ, and convergence, i.e., $x_{k_*}^\delta \to x^\dagger$ for $\delta \to 0$, of the Landweber method, but not a convergence rate.

As already mentioned, for any iterative regularization method, the rate of convergence of $x_k \to x^\dagger$ for $k \to \infty$ (in case of exact data) or $x_{k_*}^\delta \to x^\dagger$ for $\delta \to 0$ may be arbitrarily slow. As in Tikhonov regularization, convergence rate estimates can only be obtained under a source condition accompanied by additional assumptions. For Landweber iteration, a typical rate result reads as (see [HNS95])

Theorem 4. *Assume that $x^\dagger$ is a solution of (20) in $\mathcal{B}_{\rho/2}(x_0)$ and that F satisfies (35), (34) and*

$$F'(x) = R_x F'(x^\dagger), \quad x \in \mathcal{B}_\rho(x_0), \tag{38}$$

where $\{R_x \mid x \in \mathcal{B}(x_0)\}$ is a family of bounded linear operators $R_x : Y \to Y$ with

$$\|R_{\tilde{x}} - I\| \leq C\|\tilde{x} - q^\dagger\|, \quad \tilde{x} \in \mathcal{B}_\rho(x_0) \tag{39}$$

for a positive constant C. If $x^\dagger - x_0$ fulfills the source condition (22) with some $\nu \in (0, 1/2]$ and $\|w\|$ sufficiently small, then

$$\|x^\dagger - x_{k_*}^\delta\| = \mathcal{O}(\delta^{\frac{2\nu}{2\nu+1}}), \tag{40}$$

where $x_{k_*}^\delta$ is defined according to the discrepancy principle (32).

Conditions (38) and (39) mean that the derivative of F at any point $x \in \mathcal{B}_\rho(x_0)$ can be decomposed into $F'(x^\dagger)$ and an operator which is bounded and boundedly invertible on the range of $F'(x^\dagger)$, such that for the linearized problem, the part which changes with the linearization point is well-posed. For a linear operator F we would have $R_x = I$, therefore (38) can be considered as a further restriction on the nonlinearity of F.

The convergence (rate) results for (31) given in [HNS95] are reproven in [DES98], where the assumptions (38) and (39) are replaced by a Newton-Mysovskii condition on F, i.e.,

$$\|(F'(x) - F'(x^\dagger))F'(x^\dagger)^\sharp\| \leq C_{NM}\|x - x^\dagger\|, \quad x \in \mathcal{D}(F). \tag{41}$$

Here $F'(x^\dagger)^\sharp$ denotes a left inverse of $F'(x^\dagger)$. Furthermore, a logarithmic type source condition

$$\exists w \in Y : \quad x^\dagger - x_0 = g_p(F'(x^\dagger)^* F'(x^\dagger))w \tag{42}$$

with

$$g_p(\lambda) := \begin{cases} \left(\ln \frac{\exp(1)}{\lambda}\right)^{-p} & \text{for } 0 < \lambda \leq 1 \\ 0 & \text{else} \end{cases}$$

is used in [DES98] in order to derive the rate

$$\|x^\dagger - x_{k_*}^\delta\| = \mathcal{O}(-\ln \delta)^{-p}).$$

The motivation for this is that source conditions of the type (22) are too restrictive for problems where the operator $F'(x)$ is strongly smoothing, i.e., for severely ill-posed problems. There, (42) is more appropriate since it gives rise to interpretable conditions and still allows to give a rate estimate, although a slower one. Discussions of (42) and (2.2) are especially led in [Hoh97] in the context of severely ill-posed inverse scattering problems. Further variants of Landweber iteration are discussed in [Sch98] and [Sch95].

In the field of Newton type iteration methods, the Levenberg Marquardt iteration (27) has been analyzed in [Han97]. There, convergence and stability of the scheme in combination with the discrepancy principle was proven essentially under the assumption that

$$\|F(x) - F(\tilde{x}) - F'(\tilde{x})(x - \tilde{x})\| \le \tilde{C}\|F(x) - F(\tilde{x})\|\|x - \tilde{x}\| \quad x, \tilde{x} \in \mathcal{B}(x^\dagger), \tag{43}$$

if the parameter α_k is chosen such that

$$\|y^\delta - F(x_k^\delta) - F'(x_k^\delta)(x_{k+1}^\delta - x_k^\delta)\| \le \rho\|y^\delta - F(x_k^\delta)\|$$

is satisfied with some $\rho < 1$. A convergence rate result for (27) is still missing. This is different to the iteratively regularized Gauss-Newton method (28) discussed [BNS97]. There, the source condition (22) already is needed in order to obtain stability and convergence of the iterates with the sequence of regularization parameters chosen as

$$\alpha_k > 0, \quad 1 \le \frac{\alpha_k}{\alpha_{k+1}} \le r, \quad \lim_{k \to \infty} = 0$$

for some $r > 1$. If the iteration is terminated a priori after k_* steps with

$$\delta \sim \alpha_{k_*}^{\nu+1/2},$$

the Lipschitz continuity of the Fréchet derivative F' suffices to prove convergence (rates) with $\nu \in [1/2, 1]$ in (22) (and in (40)). Using the discrepancy principle (32) as stopping rule, convergence (rates) for $\nu \in [0, 1/2]$ are obtained if F satisfies

$$F'(\tilde{x}) = R(\tilde{x}, x)F'(x) + Q(\tilde{x}, x)$$
$$\|I - R(\tilde{x}, x)\| \le C_R \qquad \tilde{x}, x \in \mathcal{B}_{2\rho}(x_0) \tag{44}$$
$$\|Q(\tilde{x}, x)\| \le C_Q\|F'(x^\dagger)(\tilde{x} - x)\|$$

with ρ, C_R and C_Q sufficiently small. Similar to (38) and (39), these conditions guarantee that the linearization is not too far away from the nonlinear operator. At first sight, Newton type methods would be considered to converge much faster than Landweber iteration; this is of course true in the sense that an approximation to a solution of (20) with a given accuracy can be obtained by fewer iteration steps. However, since a single iteration step in (27) or (28) is more expensive than in (31) and also since the instability shows its effect earlier in Newton type methods so that the iteration has to be stopped earlier, it cannot be said that Newton type methods are in general preferable for ill-posed problems to the much simpler Landweber method.

Nearly all assumptions in the theory sketched above as well as the iteration schemes themselves are formulated in terms of the Fréchet derivative and its adjoint operator. We show how Landweber iteration is realized for our prototype parameter identification problem (2). The ideas presented then also apply in a similar way to other methods as (27) and (28).

In a first step, we translate the problem into a Hilbert space framework and therefore consider the underlying partial differential equation in its weak operator formulation

$$A(q)u = \hat{f} \tag{45}$$

with $A(q) : H_0^1(\Omega) \to H^{-1}(\Omega)$ and $\hat{f} \in H^{-1}(\Omega)$ defined by

$$(A(q)u, v) = \int_\Omega q(x)\nabla u \nabla v \, dx \quad \text{and}(\hat{f}, v) = \int_\Omega f v \, dx.$$

For a set $\mathcal{D}(F) \subset X = H^s(\Omega)$ (with $s > d/2$ where d is the dimension of Ω) of admissible parameters q, the direct problem (45) admits a unique solution $A(q)^{-1}\hat{f} \in H_0^1(\Omega)$ which will be denoted by by u_q in order to emphasize its dependence on q. If we regard for simplicity the case of distributed L^2-temperature measurements, the parameter identification problem can put into the form (20) with

$$F : \mathcal{D}(F) \subset X \to Y = L^2(\Omega), q \to Eu_q \tag{46}$$

and $y = Eu_{q^\dagger}$, where $E : H_0^1(\Omega) \to L^2(\Omega)$ is the embedding operator.

For given $q \in \mathcal{D}(F)$, a formal linearization of the direct problem (45) in direction $p \in X$ yields

$$A(q)u_q'p = -A(p)u_q, \tag{47}$$

where the right-hand side is due to the linearity of $A(\cdot)$ with respect to q. Therefore, the Fréchet derivative of (46) is given by

$$F'(q) : X \to Y, \ p \to Eu_q'p,$$

where $u_q'p \in H_0^1(\Omega)$ denotes the solution of (47), i.e.,

$$u_q'p = -A(q)^{-1}A(p)u_q. \tag{48}$$

Hence, if we build the inner product in (31) with an arbitrary test function $p \in X$, the k-th iteration step becomes (where we omit E for reasons of readability)

$$(q_{k+1}^\delta, p) = (q_k^\delta, p) + (F'(q_k^\delta)^*(y^\delta - u_{q_k^\delta}), p) \tag{49}$$

$$(q_{k+1}^\delta, p) = (q_k^\delta, p) + (y^\delta - u_{q_k^\delta}, F'(q_k^\delta)p)$$

$$(q_{k+1}^\delta, p) = (q_k^\delta, p) - (y^\delta - u_{q_k^\delta}, A(q_k^\delta)^{-1}A(p)u_{q_k^\delta}). \tag{50}$$

Because of

$$(y^\delta - u_{q_k^\delta}, A(q_k^\delta)^{-1}A(p)u_{q_k^\delta}) = (A(q_k^\delta)^{-1^*}(y^\delta - u_{q_k^\delta}), A(p)u_{q_k^\delta}),$$

the iteration can also be written as

$$(q_{k+1}^\delta, p) = (q_k^\delta, p) - (w_k, A(p)u_{q_k^\delta})$$

$$= (q_k^\delta, p) - \int_\Omega p(x)\nabla w_k \nabla u_{q_k^\delta} \, dx, \tag{51}$$

where w_k denotes the solution of the linear *adjoint* problem

$$A(q_k^\delta)^* w_k = y^\delta - u_{q_k^\delta}. \tag{52}$$

Hence, each iteration step in (31) requires to solve the direct problem (45) in order to obtain $u_{q_k^\delta}$ and the adjoint problem (52) with the residual $y^\delta - u_{q_k^\delta}$ as right-hand side. Eventually, the update according to (51) can be numerically realized as follows: If $\{p_1, p_2, ..., p_n\}$ is an n-dimensional basis of the parameter space $X_n \subset X$ with $\mathbf{q}_k^\delta$ denoting the vectorial representation of q_k^δ, then (51) means to solve the linear system

$$M\mathbf{s}_k^\delta = \mathbf{r}_k^\delta,$$

where M is the Gramian Matrix

$$M(i, j) = (p_i, p_j)$$

and the vector $\mathbf{r}_k$ is defined via

$$r_k^\delta(j) = \int_\Omega p_j(x)\nabla w_k \nabla u_{q_k^\delta} \, dx,$$

and to update the parameter via

$$\mathbf{q}_{k+1}^\delta = \mathbf{q}_k^\delta + \mathbf{s}_k^\delta.$$

Note that the approach (50) would require to solve n linear problems (47), clearly showing the advantages of (51) which gets by with solving a single problem (52).

Returning to our general discussion, we have already indicated with (13) that some iterative regularization methods can also be derived from certain initial value problems, which are then in turn called *continuous iteration methods*. For nonlinear problems, some of these methods are analyzed and related to their discrete analogues (especially (28)) in [ARS00] and [KNR02]. The asymptotic regularization method

$$u'_\delta(t) = F'(u_\delta(t))^*(y^\delta - F(u_\delta(t))), \tag{53}$$

see (13), is studied in the nonlinear setting in [Tau94]; it is also called inverse scale-space method in the context of imaging problems, see [SG01], [Sch03], [RSW00]. In [LS03], it is shown that (53) applied to (20), where F is the concatenation of a forward operator and a certain projection operator, can in fact be considered as a level set method. Level set methods, see [OF02], [Set99] have been successfully used for shape reconstruction problems e.g. in [San96], [OS01], [Bur03], their role as regularization methods for inverse problems has been analyzed in [Bur01].

As in Tikhonov regularization, a practical realization of an iterative method requires to take into account that in general the forward operator F cannot be exactly evaluated due to the nonlinearity and that only a sequence of approximations F_n with

$$\|F_n(x) - F(x)\| \le \varepsilon_n, \quad x \in \mathcal{B}_\rho(x^\dagger),$$

is available, where ϵ_n denotes a (given) locally uniform approximation quality. For instance, F_n could represent an iterative solver for the direct problem with n denoting its iteration index. Then, an obvious question is how to link the inner iteration, i.e., the approximation level ε_n, to the index k of the inverse iteration in an efficient way. Such issues are explicitly addressed by multilevel techniques, see [Sch98b], [DES98] and [Ram99]. The basic principle is to start the iteration scheme on a rather rough approximation level in order to initially keep the computational efforts low. Of course - for actually approximating a solution of (20) - the quality of F_n has to be gradually increased, and in order to guarantee the desired regularization properties, the choice of ε_n has to be closely coupled to the outer iteration index k and the noise level of the data. A methodically different approach that is also based on a hierarchy of different approximation levels is to solve (20) (directly) via multigrid methods, see [Kin92], [Kal01].

Not only the evaluation of F but also that of F' and of its adjoint operator may cause computational problems and hence call for modifications of standard iteration methods. Especially in case of an already nonlinear direct problem, the basic assumptions of the convergence theory, which all involve F', may become too restrictive, or the adjoint problem gets too complicated though it is linear. Hence, one might think of replacing F' in the iteration process by another linear operator which is easier to handle. Based on this idea, the convergence analysis of an iterative method related to the Landweber method (31) has been performed in [Sch95] with a modification of the nonlinearity condition (34). In the context of distributed parameter identification, an iteration operator is constructed in [Küg03] such that the desired regularization properties already follow from the unique solvability of the direct problem and the differentiability assumptions on F get redundant. This also might serve as a basis for the development of iterative routines for the identification of parameters that appear in variational inequalities, which is a prime example for a nonlinear and non-differentiable inverse problem.

3 Some Industrial Applications

In this section, we report about some inverse problems posed by industrial partners; the companies involved in the first two projects, which are connected with iron and steel making, were VOEST Alpine Stahl and VAI, both global players based in Linz. We close with a current example from quantitative finance. All problems are nonlinear inverse problems and have been attacked with regularization methods as described in this paper. Although usually, practical problems are too complicated for an application of a mathematical theory in a way where all the assumptions can be checked, a sound theory is indispensable in order to know what effects are to be expected and how to tackle them. This is especially true for the instabilities associated with inverse problems: only a sound theory can act as a guideline on how to solve inverse problems in a stable and efficient way.

3.1 Determining the Inside Lining of an Iron Making Furnace

Our first example is concerned with the determination of the inside lining of a furnace from temperature measurements in its wall (cf. [RW98]). The blast furnace process is still the most important technology for producing iron from ore, although more recent alternative like the COREX process are available (cf. [Sch00] for a discussion of a mathematical model for this process, which is an adaptation of the kinetic model of the blast furnace process discussed in See [DZSKFES98], see also PATENT PCT/EP00/01463).

Due to variations in the working conditions of the furnace, slag may deposit on its sides and on its bottom, increasing the thickness of the lining. At the same time, the wall of the blast furnace is subject to physical and chemical wear, that the resulting erosion of the bricks causes the thickness of the lining to decrease. Therefore, the inner contour of the furnace has to be observed in order to avoid a breakthrough which would cause extensive damage and would also be highly dangerous. On the other hand, one should not stop the process before it is actually necessary for obvious economic reasons.

A direct observation of the wall from the inside of the furnace is obviously impossible, the temperatures inside being up to $1500°C$. Hence, one has to reliably calculate the thickness of the wall in order to stop the process at the right time. For that purpose, temperature and heat flux measurement devices are placed inside the wall when the lining is bricked. Based on these measurements, one wants to calculate the shape of the wall on the inner, inaccessible side of the furnace.

The problem is an inverse heat conduction problem and is severely ill-posed. The first step is the development of a mathematical model that allows the numerical simulation of the process, i.e., a model for the direct problem. Since the furnace is (essentially) rotationally symmetric, the wall of the furnace in the region considered can be modeled in cylindrical coordinates. The (practically justified) assumptions of stationarity (due to the long time scales of changes in the shape of the inner lining)and rotational symmetry lead to the following nonlinear elliptic equation in the radial and height variables (the temperature u does not depend on the angular variable):

$$\frac{\partial}{\partial r}(\lambda \frac{\partial u}{\partial r}) + \frac{\lambda}{r} \cdot \frac{\partial u}{\partial r} + \frac{\partial}{\partial z}(\lambda \frac{\partial u}{\partial z}) = 0 \quad \text{in } \Omega, \tag{54}$$

where Ω denotes a radial cross-section of the wall (see Figure 4) with the boundary $\Gamma = \Gamma_1 \cup \Gamma_2 \cup \Gamma_3 \cup \Gamma_4$.

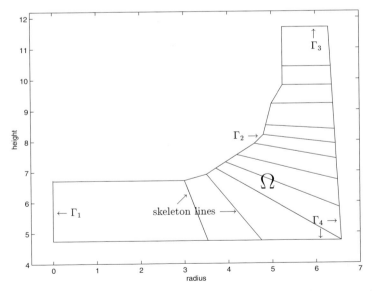

Fig. 4. Sketch of the furnace: two-dimensional, rotationally symmetric

The heat conductivity depends on the material, which changes over Ω (since Ω is covered by different types of brick and possibly slag), and hence on position, and also on the temperature:

$$\lambda = \lambda(r, z, u).$$

This temperature-dependence cannot be neglected and makes even the direct problem nonlinear. Due to rotational symmetry and the fact that the lining continues past Γ_3, we have the boundary conditions

$$\frac{\partial u}{\partial n} = 0 \quad \text{on} \quad \Gamma_1 \quad \text{and on} \quad \Gamma_3.$$

The outer surface is cooled by water with temperature T_w, which leads to the boundary condition

$$-\lambda \frac{\partial u}{\partial n} = \alpha_0 \cdot (u - T_w) \quad \text{on} \quad \Gamma_4$$

with a (measurable) heat transfer coefficient α_0. At the inner surface of the wall, an analogous condition

$$-\lambda \frac{\partial u}{\partial n} = \alpha \cdot (u - T_i) \quad \text{on} \quad \Gamma_2 \tag{55}$$

holds, where the heat transfer coefficient α depends on the actual material present at Γ_2 and T_i is the interior temperature of the furnace (i.e., of molten iron).

Now, if Γ_2, i.e., the shape of the inside lining, and hence Ω were known, we could solve the direct problem (54)-(55) in order to compute the temperature field u in Ω. Especially, we could predict the temperatures

$$\tilde{u}_j := u(x_j), j \in \{1, \cdots, m\} \tag{56}$$

measured by thermo couples located at finitely many points $x_j \in \Omega$. In other words, given Γ_2, we could evaluate the nonlinear forward operator

$$F : \Gamma_2 \to \tilde{u}. \tag{57}$$

The inverse problem now is to determine the inner contour Γ_2 and hence Ω from measurements of $\tilde{u}_j$.

For this problem, which could also be viewed as a shape reconstruction problem, the issue of uniqueness is of relevance since one wants to find a unique inner lining which gives information about the current status of the furnace, and not several (or even infinitely many) possible inner contours. In the setup based on (57) uniqueness cannot be expected, since only finitely many data

$\tilde{u}_1, \cdots, \tilde{u}_m$ are available for the determination of the curve Γ_2. One could consider the corresponding infinite-dimensional problem of determining Γ_2 from temperature values at a whole curve inside Ω or, which is more practical, describe Γ_2 by finitely many parameters. In [RW98], Γ_2 has been described by the distances p_i from the outer contour Γ_4 along the skeleton lines (see Figure 4). If we denote, for $p = (p_1, \cdots, p_n)$, by u_p the temperature field according to (54)-(55) with Γ_2 (and hence Ω) determined by p, then our inverse problem can be reformulated as the least squares problem

$$\Phi(p) := \sum_{j=1}^{m} (u_p(x_j) - \tilde{u}_j)^2 \to \min., \quad p \in C, \qquad (58)$$

where C symbolizes constraints on p, see [RW98]. But even in this formulation the lack of uniqueness remains: Figure 5 shows two domains Ω which

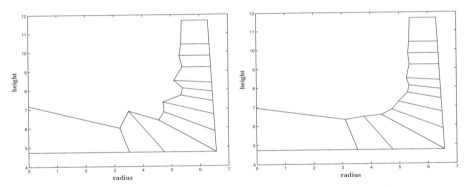

Fig. 5. Two possible solutions, the first one being not very realistic, while the second one is physically reasonable

are markedly different, but nearly yield the same value for Φ. This example indicates not only non-uniqueness, but also the inherent instability of the problem. In order to deal with this situation, the optimization problem (58) has to be regularized and - based on physical considerations - a-priori smoothness requirements for Γ_2 have to be incorporated. In [RW98], Tikhonov regularization has been applied for stably solving the inverse problem and the least squares minimization problem (58) has been replaced by

$$\sum_{j=1}^{m} (u_p(x_j) - \tilde{u}_j)^2 + \alpha \sum_{j=1}^{n+1} (\psi_j(p) - \psi_{j-1}(p))^2 \to \min., p \in C.$$

Thereby, $\psi_0(p) = \psi_{n+1}(p) = \frac{\pi}{2}$ and $\psi_j(p)(j = 1, \cdots, n)$ denotes the angle between the j-tn skeleton line and Γ_2 (see Figure 4). This regularization term penalizes oscillations in Γ_2 as those shown in the left picture of Figure 5 and enforces uniqueness. Figure 6 illustrates the results obtained by this regularization method for simulated data.

In each picture, the boxes show the measurement points, the dashed line is the starting contour, the continuous line denotes the regularized solution and the third line (---)is the "true solution ". The latter was used to generate the data before they were corrupted by 10 % noise. These calculations, which were done for tuning the regularization parameter by experiment, show that without regularization, no useful solution can be obtained, while regularization yields quite good results (in consideration of the high noise level). The value $\alpha = 1225$ has then been used for solving the inverse problem with the real world data from the industrial company supposed to be of the same noise level. Of course, this shows the discrepancy between theory and practice: In the preceding sections, we wrote about strategies for determining the regularization parameter, and now we say that we used a value of 1225.

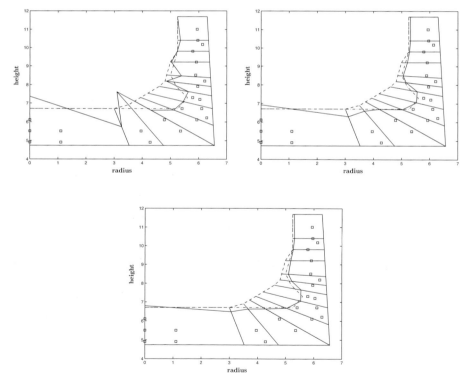

Fig. 6. Solutions with simulated data and different regularization parameters, 10% noise

The fact is that in a real practical situation like here, one frequently has too little information (especially about the noise level) to actually use the nice methods provided by the theory. As said above in connection with "error free strategies", one should not try to solve an inverse problem without any idea about the noise level. In the problem just described, one might have and will then use a rough upper bound for the noise level; but with such a rough bound, a regularization parameter based on an a-posteriori strategy will probably not be better than one determined by numerical experiments with simulated data, which is what is frequently used in practice. In our opinion this does not make the mathematical theory superfluous, since the theory indicates what is important and in which way e.g. a tuning of the regularization parameter by numerical experiment should be done.

We finally remark that the same practical problem was also treated in [TMO98] for the simple case of a constant heat conductivity; instead of Tikhonov regularization, Kalman filtering was used there.

3.2 An Inverse Solidification Problem in Continuous Casting of Steel

Our example deals with the continuous casting process for the production of steel, where liquid steel is solidified in a continuous way before the hot rolling process. This process poses a lot of mathematical modelling and numerical simulation problems, also inverse problems, about two of which we shortly report here (cf. [GBEM95]). At the top end of a continuous caster, liquid steel is cooled in a water–cooled mould to form a solidified shell which can support the liquid pool at the mould exit. Typical temperatures at the end of the mould are 1100 °C at the strand surface and 1550 °C in the center of the strand. Since steel does not solidify at a fixed temperature, but over a temperature interval, there is a mushy region where the steel is neither completely solid nor completely liquid. After the mould, the cooling process is continued in secondary cooling zones by spraying water onto the strand. Near the point of final solidification, called the crater end, segregation and thermal stresses may increasingly occur because of strongly different temperature gradients in the interior and on the boundary of the strand. These effects would then drastically decrease the quality of the final product. In order to counteract the segregation, the strand thickness at the end of the crater is slightly reduced by compression. For a successful application of this technique named soft-reduction one has to ensure by an appropriate secondary cooling strategy that the final solidification takes place within the soft-reduction zone.

Hence, in this inverse problem we have to find the cause, namely the secondary cooling of the strand by cooling water, for a desired effect, namely the crater end to remain within the soft-reduction zone all the time or as long as possible. Here, a unique solution is not necessary since it may even be of advantage to choose among several cooling strategies. However, the main difficulty lies in the fact that in practice the casting speed is not always constant; e.g., when the width is changed, the process is slowed down and then accelerated again. If the speed changes, no cross-section of the steel strand can be considered on its own, which significantly increases the complexity of the problem even though only the location of the crater end has to be controlled. For a constant casting speed, the more general inverse problem of controlling the whole solidification front by secondary cooling was considered already much earlier by several industries and their research partners, see e.g. [EL88] for our contribution. As mentioned, the complexity of the direct (and hence also the inverse) problem is much lower, since for a constant casting speed, only one cross section needs to be considered as it runs through the caster.

Again, the first step is to derive a mathematical model for the direct problem. In [GBEM95], it is described in Lagrangian coordinates: The strand of thickness d moves in the casting direction z with the casting speed $v(t)$, y is the direction of width and x the direction of thickness of the strand in which the spray cooling takes place. The amount of cooling water with temperature U_w sprayed in the secondary cooling region enters into the boundary condition via a heat transfer function $g(z,t)$, furthermore the cooling due to radiation is considered with U_α denoting the temperature of the surrounding air. The material parameters of steel primarily relevant for the problem are the thermal conductivity $k(u)$, the density $\rho(u)$ and the specific heat $c(u)$, which all depend on the temperature $u = u(x, z, t)$. σ is the Stefan–Boltzmann constant and ϵ a further material parameter that depends on the steel grade. Under symmetry assumptions on the cooling and the initial temperature f and under the (physically justified) neglection of the heat conduction in casting direction, the temperature field is described by the following spatially two-dimensional problem

$$[k(u)u_x]_x = \rho(u)c(u)u_t \tag{59}$$

$$u(x, z = -\textstyle\int_0^t v(\tau)\,d\tau, t) = f(x,t) \tag{60}$$

$$u_x(d/2, z, t) = 0 \tag{61}$$

$$k(u(0, z, t))u_x(0, z, t) = g(z + \textstyle\int_0^t v(\tau)\,d\tau, t)(u(0, z, t) - U_w) \tag{62}$$
$$+ \sigma\epsilon(u^4(0, z, t) - U_a^4).$$

Here, the Eulerian coordinate 0 describes the lower end of the mould which is passed at time t by the cross–section with Lagrangian coordinate $z = -\int_0^t v(\tau)\,d\tau$. The temperature $f(x,t)$ at the end of the mould in (60) for this cross–section can be computed by calculating the temperature in the mould. It is assumed that at the beginning of the mould there is a constant initial temperature. The function f depends on time since the casting speed is not constant, so that different cross–sections remain in the mould for different amounts of time and thus cool down in a different way. $z + \int_0^t v(\tau)\,d\tau$ is the Eulerian coordinate of the cross–section with Lagrangian coordinate z at time t. This is used to compute the cooling for the cross–section z in (62).

If the material parameters k, ρ and c, the casting speed v and the initial temperature f are known, the direct problem is to calculate the temperature field u satisfying (59)–(62) for a given heat transfer function $g(z,t)$. In fact, the temperature and - as a consequence - the point of complete solidification can be determined for each cross section z separately. For mathematical questions like existence and uniqueness of a solution of (59)–(62) we refer to [Gre98].

Given a prescribed soft-reduction zone, the inverse problem now is to find a heat transfer function g such that with this cooling, the resulting solution of the direct problem is such that the crater end of each cross–section of the strand remains in the soft–reduction zone. The amount of water sprayed onto the strand in the secondary cooling zones has to be constant in each cooling zone (there are typically six cooling zones) for technical reasons but can vary in time. Therefore, one has to admit piecewise constant, i.e., non–smooth heat transfer functions g, which makes the mathematical theory even for the direct problem a bit complicated (see [Gre98]). On the other hand, as a function of time, g should not jump too often, since each jump reflects a change in the cooling setup. Furthermore, upper and lower bounds for the total amount of water sprayed onto the strand have to be taken into account.

Though in the direct problem each cross-section z can be considered separately, they have to be treated simultaneously in the inverse problem due to the varying casting speed. The basic idea now is to penalize cross-sections that are not in the soft-reduction zone when they become solid, to integrate the penalties (which are a function of z) over all relevant cross-sections and then to minimize the resulting functional (which is a function of the heat transfer function g) over all admissible heat transfer functions g. To make things more tractable by approximating this non-differentiable functional by a differentiable one, a differentiable function P with

$$P(s) \begin{cases} = 0 : s \le 0 \\ > 0 : s > 0 \end{cases}$$

is introduced; with $t(z)$ denoting the time when the cross-section z passes the end of the mould and $t(z,g)$ the time of its complete solidification for a given heat transfer function g, the functional

$$J(g) = \int_{z_1}^{z_2} \left\{ P\left(L_b - \int_{t(z)}^{t(z,g)} v(\tau)\, d\tau \right) + P\left(\int_{t(z)}^{t(z,g)} v(\tau)\, d\tau - L_e \right) \right\} dz$$
(63)

is minimized over all admissible heat transfer functions g. Here, L_b is the beginning of the soft-reduction zone, L_e is its end, z_1 is the cross-section that enters the mould at the time after which the casting speed remains constant again, and z_2 is the cross-section that passes the end of the secondary cooling zone at the time when the casting speed starts to change. The term $P(L_b - \int_{t(z)}^{t(z,g)} v(\tau)\, d\tau)$ penalizes the cross-section z if it solidifies before the soft-reduction zone, the second term penalizes a solidification after L_e. Note that computing J involves solving the direct problem and is hence time-consuming: a single evaluation of the functional (63) requires to solve many direct problems (59)–(62).

Since the practically relevant heat transfer functions are piecewise constant both in space and time (there are only finitely many prescribed times when a change in cooling can be made), the inverse problem finally leads a finite-dimensional nonlinear optimization problem with bounds for the variables, which also serves as some regularization of the problem. However, if the number of degrees of freedom is more than just a few, this regularization alone is not enough to remove the instability. In [GBEM95], the optimization problem was solved with a Quasi-Newton method, i.e., with an iterative method considered as a regularization method with an appropriate stopping rule, where for the efficient computation of the search direction the adjoint method was used. Figure 7 shows an example where the casting speed is increased from 1.8 to 3.2 m/min within about 10 minutes. At the beginning of the process a rather low cooling rate suffices to keep the crater end in the soft-reduction zone. After the acceleration of the strand speed, only a significantly higher cooling rate allows to hold the crater end in the desired region.

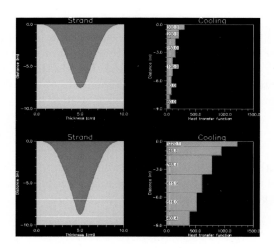

Fig. 7. The strand and computed cooling strategies

The next question that arose in the practical problem was how the heat transfer function g actually depends on the water flow rate - in the following denoted by q - which is the quantity which can be used to control the process in practice, while g is only a mathematical quantity not directly available in practice. While in the first phase of the project, g was considered as $g = g(z + \int_0^t v(\tau)d\tau, t)$ and values from the literature were used, this turned out not to be accurate enough for practical purposes. Hence, we were asked to consider the problem of actually determining the connection between g and q, again an inverse problem about which we report in more detail in [CE00]: The heat transfer function g in the boundary condition (62) is set up as

$$g = g(q) \tag{64}$$

with the water flow rate having the form

$$q = q(z + \int_0^t v(\tau)d\tau, t),$$

and the main focus is laid on the determination of the dependence of g on q in (64). As data, temperature measurements on the strand surface at different cross-sections are available. Since the goal is to estimate a function that appears in the boundary condition of a nonlinear parabolic equation from boundary measurements at a later time, this inverse problem can be understood as a combination of a sideways and backwards heat equation and hence has to be expected to be severely ill-posed.

If we denote by u_i^* the temperature measured at the i-th cross-section and by $u_i(g)$ those predicted by (59)–(62) (with the boundary condition modified according to (64)) for given g at the measurement point, our problem can be formulated as finding $g \in V$ with

$$\|F(q)\|^2 \to \min \quad \text{over } V, \tag{65}$$

where F is defined by

$$F(g) := \begin{pmatrix} u_1(g) - u_1^* \\ \cdot \\ u_N(g) - u_N^* \end{pmatrix}$$

and V is a set of functions one considers admissible for g.

Now, the traditional approach in the engineering literature is to represent g as

$$g(q) := a \cdot q^b$$

such that only two positive parameters a and b are left to be determined. However, for real temperature data, i.e., in the presence of data noise, this exponential ansatz is not appropriate, since it turns out that the parameters a and b obtained in (65) are highly sensitive to noise and the data set used. Even the use of Tikhonov regularization does not improve the results significantly, there are just too few parameters to match real data both in a reasonably accurate and stable way. This instability is in fact not primarily associated with the ill-posedness of the inverse problems, but results from the attempt to match experimental data with just two parameters.

Hence, in the project reported about in [CE00], we increased the number of parameters by subdividing the q-interval into M subintervals and to model g as a piecewise cubic spline in q. But this increase in the number of parameters now gives rise to possible instability due to the ill-posed nature of the inverse problem. Because of this and also since the data were very unevenly distributed over the water flow rates, i.e., in the q-space, a mere minimization of the functional in (65) turned out to be problematic. In order to compensate resulting negative effects, more smoothness is enforced by solving the regularized problem

$$\|F(g)\|^2 + \alpha\|g''\|^2 \to \min, \quad g \in \{\text{cubic splines}\}.$$

Figure 8, taken from [CE00], indicates that satisfactory results can be obtained

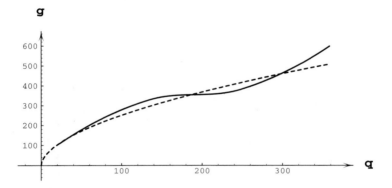

Fig. 8. Exact (dotted line) and computed heat transfer functions

by this method. For this test using simulated data with simulated noise (independent pseudo-random numbers normally distributed in the interval $[-5, 5]$ were added to the simulated data), $M = 4$, $N = 40$ and $\alpha = 14.1$ were chosen. The same remarks as in the first example apply concerning the relation between mathematical theory and practice: Reasonable values for the regularization parameter were determined by numerical experiments on carefully (i.e., theory-based!) chosen examples. The method also worked reasonably when applied to actual data, meaning that the results were reproducible for different data sets.

We close this section by mentioning that similar considerations concerning the proper modeling of functions are also of relevance for the material parameters of steel used in the direct problem (59)–(62). Usually, the thermal conductivity $k(u)$, the density $\rho(u)$ and the specific heat $c(u)$ are either interpolated between given discrete temperature values or adjusted to them via tuning of some parameters in physical laws taken from literature. Alternatively, one could make the full functional dependence of the material parameters on the temperature u subject of a separate inverse problem. For recent work on the identification of nonlinearities we refer to [KE02], [Küg03].

3.3 Inverse Problems in Quantitative Finance

One of the fastest growing fields in applied mathematics is computational finance. The modeling of the fair price of financial derivatives can by now be considered classic (see [Wil98]). All these by now very refined models contain parameters like the volatility of the underlying assets, and in recent years, the problem of identifying such parameters from observed data (e.g., prices of some derivative products observed in the market)) became .....

To make this presentation reasonably self-contained, we review some basic facts about mathematical finance: A financial derivative is a contract where payment is derived from some underlying benchmark like a stock, bond, interest rate or exchange rate. For instance, a European call option on a stock gives its owner the right (but not the obligation) to buy a specified amount of the underlying stock at a given date (the maturity T) for a given price (the strike K). Similarly, a European put option gives the right to sell, while for American options, the strike cannot be done only at, but also before a specified date. The correct pricing of such financial instruments requires mathematical models which reasonably describe the stochastic processes of the underlying, efficient methods for the calculation of fair prices, which are usually described by a partial differential equation model (see [BES03] for one approach), and robust estimates for the coefficients of these PDEs. The prices in the financial markets, where the derivatives are traded, then serve as data for the identification of these coefficients.

In the classical Black Scholes model, see [BS77], the spot price S of an equity follows a random walk

$$dS(t) = \mu S(t)dt + \sigma S dW,$$

where dW denotes the increment of a standard Wiener process and σ is the so-called volatility. For the pricing of options on this underlying, the Black Scholes trick constructs a portfolio whose value evolves risk-free such that the same return r as that of a risk-free cash account has to be expected. Based on these considerations, the value V of a European option can then be shown to satisfy the (by now famous) Black Scholes equation

$$\frac{\partial V}{\partial t} + \frac{1}{2}\sigma^2 S^2 \frac{\partial^2 V}{\partial S^2} + rS\frac{\partial V}{\partial S} - rV = 0. \tag{66}$$

In order to completely describe the (direct) problem of calculating V for a given σ, the parabolic differential equation (66) backwards in time is augmented by an end condition and boundary conditions at zero and at infinity. For the easiest case of European call or put (together called "vanilla")) options, characterized by the terminal conditions

$$V(T) = \max(S - K, 0) \quad \text{or} \quad V(T) = \max(K - S, 0),$$

analytic solutions of (66) are available. Once the volatility is known, a wide range of derivatives with the same underlying S can be priced by the use of equivalents to (66). Hence it is important to find the proper parameter σ.

Given the contract rules of a European vanilla option and its price, the volatility can even be directly calculated from the analytical solution formula of (66). Using this implied volatility in (66) then yields exactly the given option price. However, the market typically observes that the such determined implied volatility depends on the exercise price K of the option, which is unreasonable, and therefore cannot be constant. Based on a duality argument from [Dup94], where this functional correlation of σ can also be expressed as a dependence on the spot price S, one possible generalization of the Black Scholes model is to put up the volatility in (66) as a deterministic function of S, i.e., $\sigma = \sigma(S)$, then denoted as local volatility. Other approaches are based on modeling σ also as a stochastic process ("stochastic volatility"), see [Shr97].

Since the volatility is not directly observable in the markets, one has to face the inverse problem of identifying σ from available data such as the prices of liquidly traded derivatives. Option prices can be observed for different maturities T and strike prices K, i.e., the data take the form $V_{\text{market}}(T_i, K_j)$. Such inverse problems have e.g. been considered in [AFHS97], [CCE00], [Cre03a], [Cre03b], [BI99], [JSH99], [LO97], [LY01]. In [Egg01], the case of data available only for single maturity T_0 has been studied. With $V(T_0, K_j)$ denoting the solution of the direct problem (66) for different K_j and given local volatility, the output least squares problem is formulated as

$$\sum_{j}^{n} (V(T_0, K_j) - V_{\text{market}}(T_0, K_j))^2 \to \min \quad \text{over } \Sigma. \tag{67}$$

Here, Σ denotes the set of piecewise cubic interpolating splines used for the representation of the volatility. As long as there is no noise in the data, the approach (67) works quite well as indicated by Figure 9: The blue dotted line

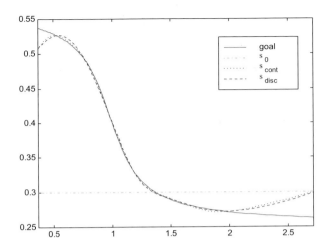

Fig. 9. Simulation results for the volatility

shows the volatility as a function of K recovered from $n = 10$ market prices of a European vanilla option (simulated by using the red volatility function). As expected, the identification is satisfactory as long as we are not too deep in the money or to deep out of the money (for a call option, these terms refer to an asset price above or below the strike price, respectively). At the extreme ends, the option prices do not contain much information and therefore the solution is mainly determined by the initial guess (green). For comparison, also continuous market data have been used. The violet line demonstrates that the error introduced by only using discrete data points (and interpolating between them) is negligible.

However, in practice, there will always be noise in the data, at least in the order of magnitude of the bid-offer spreads in the option prices. Then, the situation becomes unstable without regularization as shown in Figure 10.

For (quite small) noise levels of 0.1 and 0.5 percent, the pure minimization of the functional in (67) leads to oscillating results though the degrees of freedom are kept low in the spline representation. In order to reduce these oscillations, Tikhonov regularization is used in [Egg01], where the penalty term

$$\alpha(\|\sigma'\|^2 + \|\sigma''\|^2),$$

with $\|\cdot\|$ denoting the L^2-norm, is added to the objective functional. Figure 11 now illustrates the influence of the regularization parameter: If α is chosen too small, the main emphasis is still laid on the approximation of the noisy data

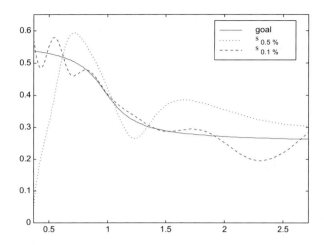

Fig. 10. Simulation results for the volatility without regularization.

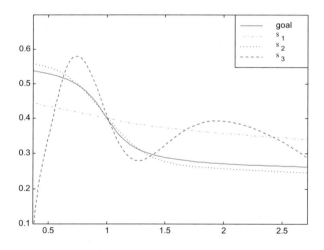

Fig. 11. Simulation results for the volatility with Tikhonov regularization.

only possible by means of an oscillating volatility (blue). On the other hand, for a regularization parameter which is too large, variations in the volatility are penalized too much (green), too little information is used from the data. However, if parameter choice strategies like the ones described in the preceding sections, based on the noise level obtained e.g. from the bid-offer spread, are used, a stable and accurate solution can be obtained (violet). In [Egg01], the theory of Tikhonov regularization including results about convergence rates is actually applied to this inverse problem, so that the gap between theory and practice is not as big as in the two problems from iron and steel industry described above. Similar techniques apply also to interest rates models. E.g., for a one-factor model, the underlying differential equation looks like

$$\frac{\partial V}{\partial t} + \frac{1}{2} W^2 \frac{\partial^2 V}{\partial r^2} + (u - \lambda W) \frac{\partial V}{\partial r} - rV = 0,$$

see e.g. [Wil98], [Reb98]. A special case is the Hull-White interest rate model, where the "short rate" r behaves according to

$$dr(t) = (a(t) - b(t)r(t))dt + \sigma(t)dW.$$

For such models, the term "model calibration" is used for the inverse problem of identifying some or all of the parameters $a(t)$, $b(t)$ and $\sigma(t)$ from market prices of swaps, caps, floors, and swaptions. Given the fast-moving field of mathematical finance, it is not surprising that these methods have already found their way into practice: e.g, the pricing software UnRisk (see [Unr]) uses discrete versions of bounded variation regularization, compare to (25) to obtain robust interest rate model parameters. These parameters can then be used to price more complex instruments like callable bonds or callable reverse floaters.

References

[ARS00] R.G. Airapetyan, A.G. Ramm, A.B. Smirnova, Continuous methods for solving nonlinear ill-posed problems, Fields Institute Communications **25**, 111–137 (2000)

[And86] R.S. Anderssen, The linear functional strategy for improperly posed problems, in: J. Cannon, U. Hornung, eds., Inverse Problems, Birkhäuser, Basel, 11–30 (1986)

[AFHS97] M. Avellaneda, C. Friedman, R. Holmes, D. Samperi, Calibrating Volatility Surfaces Via Relative-Entropy Minimization, Applied Mathematical Finance **4**, 37–64 (1997)

[BG67] G. Backus, F. Gilbert, Numerical applications of a formalism for geophysical inverse problems, Geophys. J. R. Astron. Soc. **13**, 247–276 (1967)

[Bak92] A.B. Bakushinskii, The problem of the convergence of the iteratively regularized Gauss-Newton method, Comput. Math. Math. Phys. **32**, 1353–1359 (1992)

[BB01] T. Banks, K.L. Bihari, Modeling and estimating uncertainty in parameter estimation, Inverse Problems **17**, 1–17 (2001)

[BK89] H. Banks, K. Kunisch, Parameter Estimation Techniques for Distributed Systems, Birkhäuser, Boston (1989)

[BBC85] J. Beck, B. Blackwell, C.S. Clair, Inverse Heat Conduction: Ill–Posed Problems; Wiley, New York (1985)

[BB98] M. Bertero, P. Boccacci, Introduction to Inverse Problems in Imaging, Inst. of Physics Publ., Bristol (1998)

[BEGNS94] A. Binder, H. W. Engl, C. W. Groetsch, A. Neubauer and O. Scherzer, Weakly closed nonlinear operators and parameter identification in parabolic equations by Tikhonov regularization, Appl. Anal. **55**, 215–234 (1994)

[BES03] A. Binder, H.W. Engl, A. Schatz, Advanced numerical techniques for pricing financial derivatives, Derivatives Week Vol. XII **17**, 6–7 (2003)

[BS77] F. Black, M. Scholes, The pricing of options and corporate liabilities, J. of Political Economy **81**, 637–659 (1973)

[BNS97] B. Blaschke, A. Neubauer, O. Scherzer, On convergence rates for the iteratively regularized Gauss-Newton method, IMA Journal of Numerical Analysis **17**, 421–436 (1997)

[BI99] I. Bouchouev, V. Isakov, Uniqueness, stability and numerical methods for the inverse problem that arises in financial markets, Inverse Problems **15**, R95 - R116 (1999)

[Bra87] H. Brakhage, On ill-posed problems and the method of conjugate gradients, in [EG87], 165–175

[BU97] R.M. Brown, G.A. Uhlmann, Uniqueness in the inverse conductivity problem for non-smooth conductivities in two dimensions, Commun. Part. Diff. Equ. **22**, 1009–27 (1997)

[Bur03] M. Burger, A framework for the construction of level set methods for shape optimization and reconstruction, Interfaces and Free Boundaries **5**, 301–329 (2003)

[Bur01] M. Burger, A level set method for inverse problems, Inverse Problems **17**, 1327–1356 (2001)

[BE00] M. Burger, H.W. Engl, Training neural networks with noisy data as an ill–posed problem, Adv. Comp. Math. **13**, 335-354 (2000)

[CE00] C. Carthel, H.W. Engl, Identification of heat transfer functions in continuous casting of steel by regularization, J. Inv. Ill-Posed Problems **8**, 677–693 (2000)

[CCE00] C. Chiarella, M. Craddock, N. El-Hassan, The Calibration of Stock Option Pricing Models Using Inverse Problem Methodology, QFRQ Research Papers, UTS Sydney (2000)

[CER90] D. Colton, R. Ewing, and W. Rundell, Inverse Problems in Partial Differential Equations, SIAM, Philadelphia (1990)

[CK92] D. Colton, R. Kress, Inverse Acoustic and Electromagnetic Scattering Theory, Springer, Berlin (1992)

[Cre03a] S. Crepey, Calibration of the local volatility in a trinomial tree using Tikhonov regularization, Inverse Problems **19**, 91–127 (2003)

[Cre03b] S. Crepey, Calibration of the local volatility in a generalized Black-Scholes model using Tikhonov regularization, SIAM J. Math. Anal. **354**, 1183–1206 (2003)

[DES98] P. Deuflhard, H.W. Engl, O. Scherzer, A convergence analysis of iterative methods for the solution of nonlinear ill-posed problems under affinely invariant conditions, Inverse Problems 14, 1081–1106 (1998)

[DZSKFES98] H. Druckenthaner, H. Zeisel, A. Schiefermüller, G. Kolb, A. Ferstl, H. Engl, A. Schatz, Online simulation of the blast furnace, Advanced Steel, 58–61 (1997-1998)

[Dup94] B. Dupire, Pricing with a smile, RISK **7**, 18–20 (1994)

[Egg01] H. Egger, Identification of Volatility Smiles in the Black Scholes Equation via Tikhonov Regularization, Diploma Thesis, University of Linz, Austria (2001)

[Egg93] , P.P.B. Eggermont, Maximum entropy regularization for Fredholm integral equations of the first kind, SIAM J. Math. Anal. **24**, 1557–1576 (1993)

[Eng82] H.W. Engl, On least squares collocation for solving linear integral equations of the first kind with noisy right-hand side, Boll. Geodesia Sc. Aff. **41**, 291–313 (1982)

[Eng93] H.W. Engl, Regularization methods for the stable solution of inverse problems, Surveys on Mathematics for Industry **3** 71–143 (1993)

[EG88] H.W. Engl, H. Gfrerer, A posteriori parameter choice for general regularization methods for solving linear ill-posed problems, Appl. Numer. Math. **4**, 395–417 (1988)

[EG94] H.W. Engl, W. Grever, Using the L-curve for determining optimal regularization parameters, Numerische Mathematik 69, 25-31 (1994)

[EG87] H.W. Engl, C. W. Groetsch, Inverse and Ill-Posed Problems, Academic Press, Orlando, (1987)

[EHN96] H.W. Engl, M. Hanke, A. Neubauer, Regularization of Inverse Problems, Kluwer Academic Publishers, Dordrecht (1996) (Paperback edition 2000)

[EKN89] H.W. Engl, K. Kunisch and A. Neubauer, Convergence rates for Tikhonov regularization of nonlinear ill-posed problems, Inverse Problems **5**, 523-540 (1989)

[EL93] H.W. Engl, G. Landl, Convergence rates for maximum entropy regularization, SIAM J. Numer. Anal. **30**, 1509–1536 (1993)

[EL96] H.W. Engl, T. Langthaler, Maximum entropy regularization of nonlinear ill-posed problems, in: V. Lakshmikantham, ed., Proceedings of the First World Congress of Nonlinear Analysts, Vol. I, de Gruyter, Berlin, 513–525 (1996)

[EL88] H.W. Engl, T. Langthaler, Control of the solidification front by secondary cooling in continuous casting of steel, in: H.W. Engl, H. Wacker and W. Zulehner, eds. Case Studies in Industrial Mathematics Teubner, Stuttgart, 51–77 (1988)

[EL01] H.W. Engl, A. Leitao, A Mann iterative regularization method for elliptic Cauchy problems, Numer. Func. Anal. Optim. **22**, 861–884 (2001)

[ELR96a] H. W. Engl, A. K. Louis and W. Rundell (eds.), Inverse Problems in Geophysics, SIAM, Philadelphia, (1996)

[ELR96b] H.W. Engl, A.K. Louis, W. Rundell (eds.), Inverse Problems in Medical Imaging and Non-Destructive Testing, Springer, Vienna, New York (1996)

[EN88] H.W. Engl, A. Neubauer, Convergence rates for Tikhonov regularization in finite dimensional subspaces of Hilbert scales, Proc. Amer. Math. Soc. **102**, 587–592 (1988)

[ER95] H. W. Engl, W. Rundell (eds.), Inverse Problems in Diffusion Processes, SIAM, Philadelphia (1995)

[ES00] H.W. Engl, O. Scherzer, Convergence rate results for iterative methods for solving nonlinear ill-posed problems, in: D. Colton, H.W. Engl, A.K. Louis, J. McLaughlin, W.F. Rundell (eds.), Surveys on Solution Methods for Inverse Problems, Springer, Vienna/New York, 7-34 (2000)

[EW85] H.W. Engl, A. Wakolbinger, Continuity properties of the extension of a locally Lipschitz continuous map to the space of probability measures, Monatshefte f. Math. **100**, 85–103 (1985)

[Gre98] W. Grever, A nonlinear parabolic initial boundary value problem modeling the continuous casting of steel with variable casting speed, ZAMM **78**, 109–119 (1998)

[GBEM95] W. Grever, A. Binder, H.W. Engl, K. Mörwald, Optimal cooling strategies in continuous casting of steel with variable casting speed, Inverse Problems in Engineering **2**, 289–300 (1995)

[Gro72] C.W. Groetsch, A note on segmenting Mann iterates, J. Math. Anal. Appl. **40**, 369–372 (1972)

[Gro77] C.W. Groetsch, Generalized Inverses of Linear Operators: Representation and Approximation, Dekker, New York (1977)

[Gro93] C.W. Groetsch, Inverse Problems in the Mathematical Sciences, Vieweg, Braunschweig (1993)

[GN88] C.W. Groetsch, A. Neubauer, Convergence of a general projection method for an operator equation of the first kind, Houston J. Math. **14**, 201–208 (1988)

[Han97] M. Hanke, A regularizing Levenberg-Marquardt scheme with applications to inverse groundwater filtration problems, Inverse Problems **13**, 79–95 (1997)

[Han95] M. Hanke, Conjugate Gradient Type Methods for Ill-Posed Problems, Longman Scientific & Technical, Harlow, Essex, (1995)

[HH93] M. Hanke, P.C. Hansen, Regularization methods for large-scale problems, Surveys Math. Indust. **3**, 253-315 (1993)

[HNS95] M. Hanke, A. Neubauer, O. Scherzer, A convergence analysis of the Landweber iteration for nonlinear ill-posed problems, Numerische Mathematik **72**, 21–37 (1995)

[HO93] P.C. Hansen, D.P. O'Leary, The use of the L-curve in the regularization of discrete ill-posed problems, SIAM J. Sci. Comput. **14**, 1487–1503 (1993)

[HN88] J. Haslinger, P. Neittaanmäki, Finite Element Approximation for Optimal Shape Design: Theory and Applications, John Wiley & Sons Ltd., Chichester (1988)

[Hof99] B. Hofmann, Mathematik inverser Probleme, Teubner, Stuttgart (1999)

[Hoh97] T. Hohage, Logarithmic convergence rates of the iteratively regularized Gauss-Newton method for an inverse potential and an inverse scattering problem, Inverse Problems **13**, 1279 – 1299 (1997)

[IN99] D. Isaacson, J.C. Newell, Electrical Impedance Tomography, SIAM Review **41**, 85–101 (1999)

[Isa98] V. Isakov, Inverse Problems in Partial Differential Equations, Springer, Berlin, New York (1998)

[JSH99] N. Jackson, E. Süli, S. Howison, Computation of Deterministic Volatility Surfaces, Journal of Computational Finance **2**, (1998/99)

[Kal01] B. Kaltenbacher, On the regularizing properties of a full multigrid method for ill-posed problems, Inverse Problems **17**, 767–788 (2001)

[KNR02] B. Kaltenbacher, A. Neubauer, A.G. Ramm, Convergence rates of the continuous regularized Gauss-Newton method, J. Inv. Ill-Posed Problems **10**, 261–280 (2002)

[Kin92] J. King, Multilevel algorithms for ill-posed problems, Numerische Mathematik **61**, 311–334 (1992)

[Kir96] A. Kirsch, An Introduction to the Mathematical Theory of Inverse Problems, Springer, New York (1996)

[KSB88] A. Kirsch, B. Schomburg and G. Berendt, The Backus-Gilbert method, Inverse Problems **4**, 771–783 (1988)

[KS85] C. Kravaris, J.H. Seinfeld, Identification of parameters in distributed parameter systems by regularization, SIAM J. Control Optim. **23**, 217–241 (1985)

[Küg03] P. Kügler, A Derivative Free Landweber Iteration for Parameter Identification in Elliptic Partial Differential Equations with Application to the Manufacture of Car Windshields, PhD-Thesis, Johannes Kepler Universität Linz, Austria, (2003)

[Küg03] P. Kügler, Identification of a temperature dependent heat conductivity from single boundary measurements, SIAM Journal on Numerical Analysis, accepted

[KE02] P. Kügler, H.W. Engl, Identification of a temperature dependent heat conductivity by Tikhonov regularization, Journal of Inverse and Ill-posed Problems **10**, 67–90 (2002)

[KL03] P. Kügler, A. Leitao, Mean value iterations for nonlinear elliptic Cauchy problems, Numerische Mathematik, accepted for publication

[LO97] R. Lagnado, S. Osher, A Technique for Calibrating Derivative Security Pricing Models: Numerical Solution of an Inverse Problem, JCF **1**, (1997)

[LA96] , G. Landl, R.S. Anderssen, Non-negative differentially constrained entropy-like regularization, Inverse Problems **12**, 35–53 (1996)

[LS03] A. Leitao, O. Scherzer, On the relation between constraint regularization, level sets and shape optimization, Inverse Problems **19**, L1–L11 (2003)

[LY98] A. Leonov, A. Yagola, Special regularizing methods for ill–posed problems with sourcewise represented solutions, Inverse Problems **14**, 1539–1550 (1998)

[LY01] J. Lishang, T. Youshan, Identifying the volatility of underlying assets from option prices, Inverse Problems **17**, 137–155 (2001)

[Lou87] A.K. Louis, Convergence of the conjugate gradient method for compact operators, in [EG87], 177–183

[Lou899] A.K. Louis, Inverse und schlecht gestellte Probleme, Teubner, Stuttgart (1989)

[LM90] A.K. Louis, P. Maass, A mollifier method for linear operator equations of the first kind, Inverse Problems **6** 427–440 (1990)

[Man53] W.R. Mann, Mean value methods in iteration, Proc. Am. Math. Soc. **4**, 506–510 (1953)

[Nas76] M. Z. Nashed, Generalized Inverses and Applications, Academic Press, New York (1976)

[NS98] M.Z. Nashed, O. Scherzer, Least squares and bounded variation regularization with nondifferentiable functional, Num. Funct. Anal. and Optimiz. **19**, 873–901 (1998)

[Nat77] F. Natterer, Regularisierung schlecht gestellter Probleme durch Projektionsverfahren, Numer. Math. **28**, 329–341 (1977)

[Nat86] F. Natterer, The Mathematics of Computerized Tomography, Teubner, Stuttgart (1986)

[Neu89] A. Neubauer, Tikhonov regularization for non-linear ill-posed problems: optimal convergence rates and finite-dimensional approximation, Inverse Problems **5**, 541–557 (1989)

[NS90] A. Neubauer, O. Scherzer, Finite-dimensional approximation of Tikhonov regularized solutions of non-linear ill-posed problems, Numer. Funct. Anal. and Optimiz. **11**, 85–99 (1990)

[OF02] S. Osher, R.P. Fedkiw, Level Set Methods and Dynamic Implicit Surfaces, Springer, New York (2002)

[OS01] S. Osher, F. Santosa, Level set methods for optimization problems involving geometry and constraints I. Frequencies of a two-density inhomogeneous drum, J. Comp. Phys. **171**, 272–288 (2001)

[OG69] C. Outlaw, C.W. Groetsch, Averaging iteration in a Banach space, Bull. Am. Math. Soc. **75**, 430–432 (1969)

[PV90] R. Plato, G. Vainikko, On the regularization of projection methods for solving ill-posed problems, Numer. Math. **57**, 63–79 (1990)

[RSW00] E. Radmoser, O. Scherzer, J. Weickert, Scale-space properties of nonstationary iterative regularization methods, Journal of Visual Communication and Image Representation **11**, 96–114 (2000)

[RW98] E. Radmoser, R. Wincor, Determining the inner contour of a furnace from temperature measurements, Industrial Mathematics Institute, Johannes Kepler Universität Linz, Technical Report **12** (1998)

[Ram99] R. Ramlau, A modified Landweber-method for inverse problems, Num. Funct. Anal. Opt. **20**, 79–98 (1999)

[Ram86] A.G. Ramm, Scattering by Obstacles, Reidel, Dordrecht, (1986)

[Rau84] T. Raus, Residue principle for ill-posed problems, Acta et Comment. Univers. Tartuensis **672**, 16–26 (1984)

[Reb98] R. Rebonato, Interest-Rate Option Models, 2nd edition, Wiley, New York (1998)

[Rud94] L.I. Rudin, S. Osher, Total variation based restoration with free local constraints, Proc. of IEEE Int. Conf. on Image Processing, Austin, Tx., 31–35 (1994)

[San96] F. Santosa, A level-set approach for inverse problems involving obstacles, ESAIM: Control, Optimization and Calculus of Variations **1**, 17–33 (1996)

[Sch00] A. Schatz, Lump Ore, Pellets, and Dead Men - Mathematical Modelling and Numerical Simulation of the COREX Reduction Shaft, PhD-Thesis, University of Linz, Austria (2000)

[Shr97] S. Shreve, Stochastic Calculus and Finance, www.cs.cmu.edu/~chal/shreve.html (1997)

[Sch98] O. Scherzer, A modified Landweber iteration for solving parameter estimation problems, Applied Mathematics and Optimization **38**, 45–68 (1998)

[Sch98b] O. Scherzer, An iterative multi level algorithm for solving nonlinear ill-posed problems, Numerische Mathematik **80**, 579–600 (1998)

[Sch95] O. Scherzer, Convergence criteria of iterative methods based on Landweber iteration for solving nonlinear problems, Journal of Mathematical Analysis and Applications **194**, 911–933 (1995)

[Sch02] O. Scherzer, Explicit versus implicit relative error regularization on the space of functions of bounded variation, Contemporary Methematics **313**, 171–198 (2002)

[Sch03] O. Scherzer, Scale space methods for denoising and inverse problems, Advances in Imaging and Electron Physics, accepted for publication

[SEK93] O. Scherzer, H.W. Engl, K. Kunisch, Optimal a-posteriori parameter choice for Tikhonov regularization for solving nonlinear ill-posed problems, SIAM J. Numer. Anal. **30**, 1796–1838 (1993)

[SG01] O. Scherzer, C.W. Groetsch, Scale-Space and Morphology in Computer Vision, Springer Lecture Notes in Computer Science **2106**, Springer, Berlin, 317 – 325 (2001)

[Sch85] E. Schock, Approximate solution of ill-posed equations: arbitrarily slow convergence vs. superconvergence, in: G. Hämmerlin and K. Hoffmann, eds., Constructive Methods for the Practical Treatment of Integral Equations, Birkhäuser, Basel, 234-243 (1985)

[Set99] J.A. Sethian, Level Set Methods and Fast Marching Methods, Cambridge University Press, 2nd edition, Cambridge (1999)

[TMO98] M. Tanaka, T. Matsumoto, S. Oida, Identification of unknown boundary shape of rotationally symmetric body in steady heat conduction via BEM and filter theories, in: M. Tanaka and G.S. Dulikravich eds., Inverse Problems in Engineering Mechanics, Elsevier Science B.V., Tokyo, 121–130 (1998)

[Tau94] U. Tautenhahn, On the asymptotic regularization of nonlinear ill-posed problems, Inverse Problems **10**, 1405–1418 (1994)

[Unr] UnRisk Documentation for the UnRisk PRICING ENGINE for Mathematica, www.unriskderivatives.com

[Vas87] V.V. Vasin, Iterative methods for the approximate solution of ill–posed problems with a priori informations and their applications, in: [EG87], 211–229 (1987)

[Vas92] V.V. Vasin, Ill–posed problems and iterative approximation of fixed points of pseudo–contractive mappings, in: A.N. Tikhonov, A.S. Leonov, A.I. Prilepko, I.A. Vasin, V.A. Vatutin, A.G. Yagola, eds., Ill-Posed Problems in Natural Sciences, VSP, Utrecht, 1992, 214–223, (1992)

[Vas98] V.V. Vasin, On the convergence of gradient-type methods for nonlinear equations, Doklady Mathematics **57**, 173–175 (1998) translated from Doklady Akademii Nauk **359**, 7–9 (1998)

[VA95] V.V. Vasin, A.L. Ageev, Ill–Posed Problems with A-Priori Information, VSP, Utrecht (1995)

[Vog96] C.R. Vogel, Non-convergence of the L-curve regularization parameter selection method, Inverse Problems **12**, 535–547 (1996)

[Wah90] G. Wahba, Spline Models for Observational Data, SIAM, Philadelphia (1990)

[Wil98] P. Wilmott, Derivatives - The Theory and Practice of Financial Engineering, Wiley, New York, (1998)

Numerical Methods for the Simulation of Incompressible Viscous Flow: An Introduction

Roland Glowinski[1], Tsorng-Whay Pan[1], and Lorenzo Hector Juarez V.[2]
Edward Dean[1]

[1] Department of Mathematics, University of Houston, Houston, Texas 77204-3008, USA
[2] Departamento de Matemáticas, Universidad Autónoma Metropolitana-Iztapalapa, Iztaplapa, D. F. 09340, MEXICO

Dedicated to Jacques–Louis Lions (1928–2001)

Introduction

The Navier–Stokes equations have been known for more than a century and they still provide the most commonly used mathematical model to describe and study the motion of viscous fluids, including phenomena as complicated as turbulent flow. One can only marvel at the fact that these equations accurately describe phenomena whose length scales (resp., time scale) range from fractions of a millimeter (resp., of a second) to thousands of kilometers (resp., several years). Indeed, the Navier–Stokes equations have been validated by numerous comparisons between analytical or computational results and experimental measurements; some of these comparisons are reported in Canuto et al. [1], Lesieur [2], Guyon et al. [3], and Glowinski [4].

These notes do not have the pretension to cover the full field of finite element methods for the Navier–Stokes equations; they are organized in sections as follows:

1. The Navier–Stokes equations for incompressible viscous flow
2. Some operator splitting methods for initial value problems and applications to the Navier–Stokes equations
3. Iterative solution of the advection–diffusion sub–problems and the wave-like equation method for the advection sub–problems
4. Iterative solution of the Stokes type sub–problems
5. Finite element approximation of the Navier–Stokes equations
6. Numerical results

1 The Navier–Stokes Equations for Incompressible Viscous Flow

1.1 Model

Let Ω be an *open* and *connected* region (i.e. a domain) of $\mathbb{R}^d$ ($d = 2$ or 3) filled with a fluid. The generic point of $\mathbb{R}^d$ will be denoted by $x = \{x_i\}_{i=1}^d$ while dx will denote $dx_1 dx_2$ and $dx_1 dx_2 dx_3$ for $d = 2$ and $d = 3$, respectively.

Derivations of the Navier–Stokes equations may be found in, e.g., Prager [5], Batchelor [6], Guyon et al. [3], Chorin et al. [7], and Glowinski [4]. Here skipping the details of derivation, we have the following so-called *momentum equation*

$$\frac{\partial \mathbf{u}}{\partial t} + (\mathbf{u} \cdot \boldsymbol{\nabla})\mathbf{u} - \nu \Delta \mathbf{u} + \boldsymbol{\nabla} p = \mathbf{f} \quad in \quad \Omega \times (0, T), \tag{1}$$

and the *continuity equation*

$$\boldsymbol{\nabla} \cdot \mathbf{u} = 0 \quad in \quad \Omega \times (0, T) \tag{2}$$

for unsteady, isothermal flows of incompressible, viscous, Newtonian fluids. In (1), (2) (and in the following),

1. $\mathbf{u} = \{u_i\}_{i=1}^d$ is the *velocity* and p is the *pressure*;
2. $\nu (> 0)$ is the *(kinematic) viscosity* coefficient;
3. $\boldsymbol{\nabla} = \{\frac{\partial}{\partial x_i}\}_{i=1}^d$, $\Delta = \sum_{i=1}^d \frac{\partial^2}{\partial x_i^2}$, $\mathbf{u} \cdot \mathbf{v} = \sum_{i=1}^d u_i v_i$, $\forall \ \mathbf{u} = \{u_i\}_{i=1}^d$, $\mathbf{v} = \{v_i\}_{i=1}^d$,

$$\boldsymbol{\nabla} \mathbf{u} : \boldsymbol{\nabla} \mathbf{v} = \sum_{i=1}^d \sum_{j=1}^d \frac{\partial u_i}{\partial x_j} \frac{\partial v_i}{\partial x_j}, \ \forall \ \mathbf{u}, \mathbf{v}, \quad |\mathbf{v}|^2 = \mathbf{v} \cdot \mathbf{v}, |\boldsymbol{\nabla} \mathbf{v}|^2 = \boldsymbol{\nabla} \mathbf{v} : \boldsymbol{\nabla} \mathbf{v};$$

4. $\boldsymbol{\nabla} \cdot \mathbf{v} = \sum_{i=1}^d \frac{\partial v_i}{\partial x_i}, \ \forall \ \mathbf{v}, \quad (\mathbf{v} \cdot \boldsymbol{\nabla})\mathbf{w} = \{\sum_{j=1}^d v_j \frac{\partial w_i}{\partial x_j}\}_{i=1}^d, \ \forall \ \mathbf{v}, \mathbf{w};$
5. $\mathbf{f} = \{f_i\}_{i=1}^d$ is a density of external forces.

Let Γ be the boundary of Ω (here we suppose that Ω is bounded) and let $\mathbf{n}$ be the unit outward normal vector at Γ. Relations (1), (2) are not sufficient to define a flow; we have to consider further conditions, such as the *initial condition*

$$\mathbf{u}(x, 0) = \mathbf{u}_0(x) \quad (with \ \boldsymbol{\nabla} \cdot \mathbf{u}_0 = 0), \tag{3}$$

and the *boundary condition*

$$\mathbf{u} = \mathbf{g} \quad on \quad \Gamma \times (0, T) \quad (with \int_\Gamma \mathbf{g} \cdot \mathbf{n} \, d\Gamma = 0). \tag{4}$$

The boundary condition (4) is of Dirichlet type; more complicated boundary conditions are described in, e.g., Glowinski [8], Bristeau et al. [9], and Pironneau [10], among them, the following mixed boundary condition which occurs often in applications

$$\mathbf{u} = \mathbf{g}_0 \quad on \quad \Gamma_0 \times (0,T), \quad \nu \frac{\partial \mathbf{u}}{\partial \mathbf{n}} - \mathbf{n}p = \mathbf{g}_1 \quad on \quad \Gamma_1 \times (0,T), \tag{5}$$

with $\frac{\partial \mathbf{u}}{\partial \mathbf{n}} = \{\frac{\partial u_i}{\partial \mathbf{n}}\}_{i=1}^d \ (= \{\nabla u_i \cdot \mathbf{n}\}_{i=1}^d)$; in (5), Γ_0 and Γ_1 are two subsets of Γ satisfying $\Gamma_0 \cap \Gamma_1 = \emptyset$, closure of $\Gamma_0 \cup \Gamma_1 = \Gamma$. Another mixed boundary condition is

$$\mathbf{u} = \mathbf{g}_0 \quad on \quad \Gamma_0 \times (0,T), \quad \boldsymbol{\sigma}\mathbf{n} = \mathbf{g}_1 \quad on \quad \Gamma_1 \times (0,T), \tag{6}$$

with the (stress) tensor $\boldsymbol{\sigma} = 2\nu \mathbf{D}(\mathbf{u}) - p\mathbf{I}$ and $2\mathbf{D}(\mathbf{u}) = \nabla \mathbf{u} + (\nabla \mathbf{u})^t$. The mixed boundary condition (5) is less physical than (6), but like (6), it is quite useful to implement *downstream boundary conditions* for flow in unbounded regions.

Remark 1. The Dirichlet conditions in (4), (5) and (6) are called *no-slip conditions* if $\mathbf{g} = \mathbf{0}$ (resp., $\mathbf{g}_0 = \mathbf{0}$) on Γ (resp., Γ_0) if Γ (resp., Γ_0) is not moving. □

Remark 2. The decrease in the popularity of the solution methods for the Navier-Stokes equations based on the *stream function–vorticity* formulation has been seen in these last years. We see two main reasons for this trend:

(i) These methods are really convenient for *two–dimensional* flow. The generalization to three–dimensional flow, although possible, leads to complicated formulations.

(ii) The treatment of the boundary conditions is more delicate than with the velocity–pressure formulation, particularly for flow in *multi–connected* regions.

In this note, we will not discuss the stream function–vorticity formulation for the Navier-Stokes equations. □

Remark 3. As today, it is not known if the *time dependent* Navier-Stokes equations modeling the *unsteady flow* of three-dimensional incompressible viscous fluids have a unique solution. For those readers who may be surprised that some decisive indications - in one direction or the other - have not been obtained via laboratory or computational experiments we would like to make the following comments:

(i) The Navier-Stokes equations are just mathematical models (obtained after idealization) for some real life phenomena. Mathematical modeling cannot reflect the full complexity of a laboratory experimentation; indeed, it is practically impossible to reproduce *exactly* a given experiment in order to validate its results by those of another one. We also have to remember that at large Reynolds numbers (the interesting case) small perturbations in the data can imply very large differences in the ensuing results.

(ii) Unlike their two-dimensional counterparts, three-dimensional viscous flows at large Reynolds numbers are not routine yet when it comes to numerical simulation. They require a lot of computer resources in time and memory. In order to explore the uniqueness issue it will be necessary to define significant test problems and store in a large data base the results obtained by solution methods using different type of space and time discretizations. We anticipate that this program will take place in the near future and that parallel computing will play an important role in this endeavor. □

Remark 4. The *mathematical theory* of the Navier-Stokes equations for incompressible viscous fluids has inspired many investigators. The first rigorous mathematical results were obtained by J. Leray who proved (in Leray [11]) the *existence* of solutions when the flow region Ω is the full space $\mathbb{R}^d$ with $d = 2$ or 3. The Leray's results were extended to flow regions with boundaries by Leray himself [12] in 1934 and by E. Hopf [13] in 1951. The methods and tools developed by the above two authors have proved to be very useful to the solution of many problems in mechanics, physics, etc., modeled by linear or nonlinear partial differential equations, many of these problems being outside the field of fluid mechanics. The results of J. Leray and E. Hopf have been improved and generalized by several authors (see Leray [14] for an historical account), one of the most remarkable milestones in that direction being the proof by J.L. Lions and G. Prodi [15] in 1959 that if the flow region is *two-dimensional* (i.e. $\Omega \subset \mathbb{R}^2$), then the *time-dependent* Navier-Stokes equations have a *unique* solution. The proof of these existence and uniqueness results and of many others (on the *regularity* of the solutions, for example) can be found in the books by, e.g., J.L. Lions (Chapter 10 of [16]), (Chapter 1 of [17]), Ladysenskaya [18], Temam (Chapters 2 and 3 of [19]), Tartar [20], Kreiss and Lorenz (Chapters 9 and 10 of [21]), and P.L. Lions (Chapters 2 and 3 of [22]). Regarding the *Handbook of Numerical Analysis* an important source of results (and of methods to obtain them) is the Chapter 1 of Marion and Temam [23]. The above list is far from complete, and the books and articles mentioned above contain bibliographical references worth consulting. □

1.2 Variational Formulations of the Navier-Stokes Equations

We return now to the Navier-Stokes equations for incompressible Newtonian viscous flow. When using $\boldsymbol{\sigma} = 2\nu \mathrm{d}(\mathbf{u}) - p i$, we have then

$$\frac{\partial \mathbf{u}}{\partial t} + (\mathbf{u} \cdot \boldsymbol{\nabla})\mathbf{u} - \boldsymbol{\nabla} \cdot \boldsymbol{\sigma} = \mathbf{f} \ in \ \Omega \times (0, T), \tag{7}$$

$$\boldsymbol{\nabla} \cdot \mathbf{u} = 0 \ in \ \Omega \times (0, T), \tag{8}$$

$$\mathbf{u}(0) = \mathbf{u}_0 \ (with \ \boldsymbol{\nabla} \cdot \mathbf{u}_0 = 0), \tag{9}$$

that we complete by the following boundary conditions

$$\mathbf{u} = \mathbf{g}_0 \ on \ \Gamma_0 \times (0, T), \quad \boldsymbol{\sigma}\mathbf{n} = \mathbf{g}_1 \ on \ \Gamma_1 \times (0, T); \tag{10}$$

with Γ_0, Γ_1 as in Section 1.1.

We define now the functional space V_0 by

$$V_0 = \{\mathbf{v} \mid \mathbf{v} \in (H^1(\Omega))^d, \mathbf{v} = 0 \ on \ \Gamma_0\}. \tag{11}$$

Space V_0 is a Hilbert space for the scalar product and norm defined by

$$(\mathbf{v}, \mathbf{w})_{V_0} = \sum_{i=1}^d (v_i, w_i)_{H^1(\Omega)}, \quad \forall \mathbf{v} = \{v_i\}_{i=1}^d, \mathbf{w} = \{w_i\}_{i=1}^d \in V_0,$$

$$||\mathbf{v}||_{V_0} = (\sum_{i=1}^d ||v_i||^2_{H^1(\Omega)})^{1/2}, \quad \forall \mathbf{v} = \{v_i\}_{i=1}^d \in V_0,$$

respectively. In the particular case where $\Gamma_0 \neq \emptyset$ (with $\int_{\Gamma_0} d\Gamma > 0$) and Ω is bounded, we can use over V_0 the scalar product and norm defined by

$$\{\mathbf{v}, \mathbf{w}\} \longrightarrow \sum_{i=1}^d \sum_{j=1}^d \int_\Omega \frac{\partial v_i}{\partial x_j} \frac{\partial w_i}{\partial x_j} dx = \sum_{i=1}^d \int_\Omega \boldsymbol{\nabla} v_i \cdot \boldsymbol{\nabla} w_i \ dx,$$

$$\mathbf{v} \longrightarrow (\sum_{i=1}^d \int_\Omega |\boldsymbol{\nabla} v_i|^2 \ dx)^{1/2},$$

respectively. Suppose that $\mathbf{R}$ and $\mathbf{S}$ are two $d \times d$ tensors so that $\mathbf{R} = \{r_{ij}\}$, $\mathbf{S} = \{s_{ij}\}$; from now on we shall use the notation $\mathbf{R} : \mathbf{S}$ for $\sum_{i=1}^d \sum_{j=1}^d r_{ij} s_{ij}$. With this notation the above V_0-scalar product and norm can be written as

$$\int_\Omega \boldsymbol{\nabla}\mathbf{v}:\boldsymbol{\nabla}\mathbf{w} \ dx \quad and \quad (\int_\Omega \boldsymbol{\nabla}\mathbf{v}:\boldsymbol{\nabla}\mathbf{v} \ dx)^{1/2},$$

respectively. For simplicity, we shall use in the sequel the notation $|\boldsymbol{\nabla}\mathbf{v}|^2$ for $\boldsymbol{\nabla}\mathbf{v}:\boldsymbol{\nabla}\mathbf{v}$. We suppose now that the functions occurring in the system (7)-(9) are sufficiently smooth; taking the $\mathbb{R}^d$-dot product of both sides of (7) with $\mathbf{v}$, an arbitrary element of V_0, and then integrating over Ω we obtain from Green's formula that for almost any t on $(0, T)$ we have

$$\begin{cases} \int_\Omega \frac{\partial \mathbf{u}(t)}{\partial t} \cdot \mathbf{v} \ dx + \int_\Omega (\mathbf{u}(t) \cdot \boldsymbol{\nabla})\mathbf{u}(t) \cdot \mathbf{v} \ dx + 2\nu \int_\Omega \mathrm{d}(\mathbf{u}(t)) : \mathrm{d}(\mathbf{v}) \ dx \\ - \int_\Omega p(t)\boldsymbol{\nabla} \cdot \mathbf{v} \ dx = \int_\Omega \mathbf{f}(t) \cdot \mathbf{v} \ dx + \int_{\Gamma_1} \mathbf{g}_1(t) \cdot \mathbf{v} \ d\Gamma, \forall \mathbf{v} \in V_0, \end{cases} \tag{12}$$

to be completed by (8), (9) and

$$\mathbf{u} = \mathbf{g}_0 \ on \ \Gamma_0 \times (0, T). \tag{13}$$

The "Neumann" condition $\boldsymbol{\sigma}\mathbf{n} = \mathbf{g}_1$ on $\Gamma_1 \times (0, T)$ is *automatically* enforced by the formulation (12), which is known as a *variational formulation* of the momentum equation (7). Actually, it can be shown that relation (12) *implies* the momentum equation (7) *and* the "Neumann" condition $\boldsymbol{\sigma}\mathbf{n} = \mathbf{g}_1$ on $\Gamma_1 \times (0, T)$.

Suppose now that instead of (10) the boundary conditions are given by

$$\mathbf{u} = \mathbf{g}_0 \ on \ \Gamma_0 \times (0, T), \quad \nu\frac{\partial \mathbf{u}}{\partial n} - p\mathbf{n} = \mathbf{g}_1 \ on \ \Gamma_1 \times (0, T). \tag{14}$$

Multiplying both sides of (1) by $\mathbf{v} \in V_0$, integrating over Ω and using Green's formula we obtain this time that for almost any $t \in (0, T)$ we have

$$\begin{cases} \displaystyle\int_\Omega \frac{\partial \mathbf{u}}{\partial t}(t) \cdot \mathbf{v} \, dx + \int_\Omega (\mathbf{u}(t) \cdot \boldsymbol{\nabla})\mathbf{u}(t) \cdot \mathbf{v} \, dx + \nu\int_\Omega \boldsymbol{\nabla}\mathbf{u}(t) : \boldsymbol{\nabla}\mathbf{v} \, dx \\ \displaystyle -\int_\Omega p(t)\boldsymbol{\nabla}\cdot \mathbf{v} \, dx = \int_\Omega \mathbf{f}(t) \cdot \mathbf{v} \, dx + \int_{\Gamma_1} \mathbf{g}_1(t) \cdot \mathbf{v} d\Gamma, \forall \mathbf{v} \in V_0. \end{cases} \tag{15}$$

Conversely, the variational formulation (15) implies the momentum equation (1) and the generalized Neumann condition $\nu\frac{\partial \mathbf{u}}{\partial n} - p\mathbf{n} = \mathbf{g}_1$ on $\Gamma_1 \times (0, T)$.

The variational formulations (12) and (15) of the momentum equation will play a fundamental role in the *finite element* approximation of the Navier-Stokes problems (7)-(10) and (1)-(3), (5), respectively. We shall return to this issue. Actually, for the finite element approximations of the above problems, we shall take advantage of the fact that the *incompressibility condition* $\boldsymbol{\nabla}\cdot\mathbf{u} = 0$ is *equivalent* to

$$\int_\Omega q\boldsymbol{\nabla}\cdot \mathbf{u} \, dx = 0, \forall q \in L^2(\Omega). \tag{16}$$

2 Operator Splitting Methods for Initial Value Problems: Application to the Navier–Stokes Equations

Solving the above Navier-Stokes equations is a non-trivial task for the following reasons:

(i) the momentum equation is *nonlinear*;
(ii) the incompressibility condition $\boldsymbol{\nabla}\cdot\mathbf{u} = 0$;
(iii) solving the Navier-Stokes equations amounts to solve a *system* of partial differential equations ($d + 1$ if $\Omega \subset \mathbb{R}^d$) coupled through the nonlinear term $(\mathbf{u}\cdot\boldsymbol{\nabla})\mathbf{u}$, the incompressibility condition $\boldsymbol{\nabla}\cdot\mathbf{u} = 0$, and sometimes through the viscous term and the boundary conditions (as it is the case in (5) and (6)).

In the following subsections we will only focus on *time discretization* by *operator-splitting* schemes, mainly the θ–scheme and the Marchuk–Yanenko scheme (the details of other well-known operator-splitting schemes, such as Peaceman–Rachford, Douglas–Rachford, alternating directions etc., can be found in, e.g., Glowinski [4]), which will partly overcome the above difficulties; in particular, we will be able to decouple the difficulties associated to the *nonlinearity* with those associated to the *incompressibility condition*.

2.1 A Family of Initial Value Problems

We consider the following *initial value problem*:

$$\frac{d\varphi}{dt} + A(\varphi, t) = 0, \quad \varphi(0) = \varphi_0, \tag{17}$$

where, for a given t, A is an operator (possibly nonlinear, and even multivalued) from a Hilbert space H into itself and where $\varphi_0 \in H$.

Suppose now that operator A has the following *nontrivial decomposition* $A = A_1 + A_2$ (by *nontrivial* we mean that A_1 and A_2 are individually simpler than A). It is then quite natural to integrate the initial value problem (17) by numerical methods taking advantage of the decomposition property, $A = A_1 + A_2$; such a goal can be achieved by the *operator splitting schemes*. discussed in the following subsections.

2.2 A θ–scheme

This scheme, introduced in Glowinski [32] and [33], is a variation of schemes discussed in Strang [34], Beale and Majda [35], Leveque and Oliger [36]; it is discussed with further details in Glowinski and Le Tallec [31]. The θ–scheme to be described below is in fact a variant of the Peaceman–Rachford scheme.

Let θ be a number of the open interval $(0, \frac{1}{2})$ (in practice $\theta \in (0, \frac{1}{3})$); the θ-scheme applied to the solution of the initial value problem (17), when $A = A_1 + A_2$, is described as follows:

$$\varphi^0 = \varphi_0; \tag{18}$$

then for $n \geq 0$, φ^n being known, we compute $\varphi^{n+\theta}, \varphi^{n+1-\theta}$ and φ^{n+1} as follows:

$$\frac{\varphi^{n+\theta} - \varphi^n}{\theta \Delta t} + A_1\left(\varphi^{n+\theta}, (n+\theta)\Delta t\right) + A_2(\varphi^n, n\Delta t) = 0, \tag{19}$$

$$\frac{\varphi^{n+1-\theta} - \varphi^{n+\theta}}{(1-2\theta)\Delta t} + A_1\left(\varphi^{n+\theta}, (n+\theta)\Delta t\right) + A_2\left(\varphi^{n+1-\theta}, (n+1-\theta)\Delta t\right)$$
$$= 0, \tag{20}$$

$$\frac{\varphi^{n+1} - \varphi^{n+1-\theta}}{\theta \Delta t} + A_1\left(\varphi^{n+1}, (n+1)\Delta t\right) + A_2\left(\varphi^{n+1-\theta}, (n+1-\theta)\Delta t\right)$$
$$= 0. \tag{21}$$

We consider now the simple situation where $H = \mathbb{R}^N$, $\varphi_0 \in \mathbb{R}^N$, where A is an $N \times N$ matrix, *symmetric, positive definite* and *independent* of t. The solution of the corresponding *autonomous* system (17) is then

$$\varphi(t) = e^{-At}\varphi_0. \tag{22}$$

If one projects (22) over a vector basis of $\mathbb{R}^N$, consisting of *eigenvectors* of A, we obtain - with obvious notation -

$$\varphi_i(t) = e^{-\lambda_i t}\varphi_{0i}, i = 1, ...N, \tag{23}$$

where $0 < \lambda_1 \leq \lambda_2 ... \leq \lambda_N$ denote the *eigenvalues* of A.

In order to apply scheme (18)–(21), we consider the following decomposition of matrix A

$$A = \alpha A + \beta A, \tag{24}$$

with $\alpha + \beta = 1, 0 < \alpha, \beta < 1$. Applying (18)–(21) with $A_1 = \alpha A, A_2 = \beta A$ yields

$$\varphi^{n+1} = (I + \alpha\theta\Delta tA)^{-2}(I - \beta\theta\Delta tA)^2(I + \beta\theta'\Delta tA)^{-1}(I - \alpha\theta'\Delta tA)\varphi^n, \tag{25}$$

where $\theta' = 1 - 2\theta$, which implies

$$\varphi_i^n = \frac{(1 - \beta\theta\Delta t\lambda_i)^{2n}(1 - \alpha\theta'\Delta t\lambda_i)^n}{(1 + \alpha\theta\Delta t\lambda_i)^{2n}(1 + \beta\theta'\Delta t\lambda_i)^n} \varphi_{0i}, \forall i = 1, ...N. \tag{26}$$

Consider now the rational function R_1 defined by

$$R_1(\xi) = \frac{(1 - \beta\theta\xi)^2(1 - \alpha\theta'\xi)}{(1 + \alpha\theta\xi)^2(1 + \beta\theta'\xi)}. \tag{27}$$

Since

$$\lim_{\xi \to +\infty} |R_1(\xi)| = \beta/\alpha, \tag{28}$$

we should prescribe the condition

$$\alpha > \beta \tag{29}$$

which is a *necessary* one for the *stiff A-stability* of the θ–scheme (18)–(21). To obtain the *unconditional stability* we need to have

$$|R_1(\xi)| \leq 1, \forall \xi \in \mathbb{R}_+;$$

actually, a closer inspection of the function R_1 would show that

$$|R_1(\xi)| < 1, \forall \xi > 0, \forall \theta \in [\frac{1}{4}, \frac{1}{2}), \forall \alpha, \beta \ so \ that \ 0 < \beta < \alpha < 1, \alpha + \beta = 1, \tag{30}$$

which implies the unconditional stability of scheme (18)–(21) with respect to Δt (the *lower bound* $\frac{1}{4}$ in (30) is *not optimal* for θ, but we shall be satisfied with it since, as we shall see below, the "optimal" value of θ is $1 - \frac{1}{\sqrt{2}} = 0.292893219... > \frac{1}{4}$).

Concerning now the accuracy of scheme (18)–(21), we can show that in the neighborhood of $\xi = 0, R_1$ satisfies:

$$R_1(\xi) = 1 - \xi + \frac{\xi^2}{2}[1 + (\beta - \alpha)(2\theta^2 - 4\theta + 1)] + O(1)\xi^3. \tag{31}$$

Comparing (31) to the expansion of $e^{-\xi}$

$$e^{-\xi} = 1 - \xi + \frac{\xi^2}{2} - \frac{\xi^3}{6} + O(1)\xi^4, \tag{32}$$

we obtain that scheme (18)–(21) is *second order accurate if and only if*

$$\alpha = \beta(= \frac{1}{2} \ from \ \alpha + \beta = 1), \tag{33}$$

and/or

$$\theta = 1 - 1/\sqrt{2} = .292893219...; \tag{34}$$

scheme (18)–(21) is *first order accurate* if neither (33) nor (34) holds. If one takes $\alpha = \beta = \frac{1}{2}$ it follows from (26) and (27) that scheme (18)–(21) is *unconditionally stable*, $\forall \theta \in (0, \frac{1}{2})$; however, we have (from (28))

$$\lim_{\xi \to +\infty} |R_1(\xi)| = 1, \tag{35}$$

implying that in that particular case scheme (18)–(21) *is not stiff A-stable*. Relations (23) show that the larger λ_i, the faster $\varphi_i(t)$ converges to zero as $t \to +\infty$; considering now the discrete analogue of (23), namely (26) we observe that for large values of $\lambda_i \Delta t$ we have $R_1(\lambda_i \Delta t) \sim 1$, implying that, in (26), φ_i^n converges slowly to zero as $n \to +\infty$; from this property (which is also shared by the *Peaceman–Rachford scheme scheme*, the *Douglas–Rachford scheme scheme*, and the *Crank–Nicolson scheme*) we can expect scheme (18)–(21) with $\alpha = \beta = \frac{1}{2}$ and $\theta \in (0, \frac{1}{2})$ to be not well suited (unless Δt is very small) to simulate fast transient phenomena and to capture efficiently the possible *steady state* solutions of (17) (i.e. the solutions of $A(\varphi, +\infty) = 0$), if operator A is *stiff* (the notion of *stiffness* is defined in, e.g., Crouzeix and Mignot [30] (pages 86 to 88)).

Let us consider now the case where α and β have been chosen so that *we have the same matrix for all the partial steps of the θ–scheme*; in that case α, β, θ have to satisfy

$$\alpha\theta = \beta(1 - 2\theta), \tag{36}$$

which implies

$$\alpha = (1 - 2\theta)/(1 - \theta), \ \beta = \theta/(1 - \theta). \tag{37}$$

Combining (29) and (37) yields

$$0 < \theta < 1/3; \tag{38}$$

for $\theta = 1/3$, (37) implies $\alpha = \beta = 1/2$, a situation which has been discussed already.

If $0 < \theta < 1/3$ and if α and β are given by (37) we have

$$\lim_{\xi \to +\infty} |R_1(\xi)| = \beta/\alpha = \theta/(1 - 2\theta) < 1. \tag{39}$$

Indeed, we can prove that if $\theta^* \leq \theta \leq 1/3$ (with $\theta^* = .087385580...$) and if α and β are given by (37), then scheme (18)-(21) is *unconditionally stable*; moreover if $\theta^* < \theta < 1/3$ (with α and β still given by (37)), property (39) implies that scheme (18)-(21) is *stiff A-stable* and has therefore good asymptotic properties as $n \to +\infty$, *making it well suited to compute steady state solutions.*

If $\theta = 1 - 1/\sqrt{2}$ (resp., $\theta = 1/4$) we have $\alpha = 2 - \sqrt{2}, \beta = \sqrt{2} - 1, \beta/\alpha = 1/\sqrt{2}$ (resp., $\alpha = 2/3, \beta = 1/3, \beta/\alpha = 1/2$).

Remark 5. We consider the case where in (17) we have

$$A(\varphi, t) = B(\varphi) - f(t) \ with \ B = B_1 + B_2. \tag{40}$$

In order to decide how to decompose f when applying the θ-scheme (18)-(21) to the solution of the initial value problem

$$\begin{cases} \dfrac{d\varphi}{dt} + B(\varphi) = f, \\ \varphi(0) = \varphi_0, \end{cases} \tag{41}$$

we suppose that

$$f = f_1 + f_2 \tag{42}$$

with $f_1 = \alpha f, f_2 = \beta f, 0 \leq \alpha, \ \beta \leq 1, \ \alpha + \beta = 1$, and we assume that $B = 0$, for simplicity.

Applying scheme (18)-(21) to the solution of

$$\frac{d\varphi}{dt} = f, \ \varphi(0) = \varphi_0, \tag{43}$$

we obtain (with $\theta' = 1 - 2\theta$)

$$\varphi^0 = \varphi_0, \tag{44}$$

and for $n \geq 0$,

$$\frac{\varphi^{n+\theta} - \varphi^n}{\theta \Delta t} = \alpha f\left((n + \theta)\Delta t\right) + \beta f(n\Delta t). \tag{45}$$

$$\frac{\varphi^{n+1-\theta} - \varphi^{n+\theta}}{\theta' \Delta t} = \alpha f\left((n + \theta)\Delta t\right) + \beta f\left((n + 1 - \theta)\Delta t\right), \tag{46}$$

$$\frac{\varphi^{n+1} - \varphi^{n+1-\theta}}{\theta \Delta t} = \alpha f\left((n + 1)\Delta t\right) + \beta f\left((n + 1 - \theta)\Delta t\right), \tag{47}$$

which imply that

$$\begin{cases} \varphi^n = \varphi_0 + \Delta t \sum_{q=0}^{n-1} \{\beta\theta f(q\Delta t) + \alpha(1 - \theta)f\left((q + \theta)\Delta t\right) + \\ \beta(1 - \theta)f\left((q + 1 - \theta)\Delta t\right) + \alpha\theta f\left((q + 1)\Delta t\right)\}. \end{cases} \tag{48}$$

Since $\beta\theta + \alpha(1 - \theta) + \beta(1 - \theta) + \alpha\theta = 1$, the numerical integration rule which, in (48), approximates $\int_{q\Delta t}^{(q+1)\Delta t} f(t)dt$, is *first-order accurate*, at least; actually, it is *second-order accurate*, if and only if

$$\alpha(1 - \theta)\theta + \beta(1 - \theta)^2 + \alpha\theta = \frac{1}{2},$$

or equivalently

$$(\beta - \alpha)(2\theta^2 - 4\theta + 1) = 0. \tag{49}$$

Not surprisingly, we recover from (49) conditions (33) and (34), namely scheme (44)-(47) is *second-order accurate if and only if*

$$\alpha = \beta = \frac{1}{2} \tag{50}$$

and/or

$$\theta = 1 - 1/\sqrt{2}. \tag{51}$$

Assuming that (51) holds, we can wonder if there are values of α and β for which scheme (44)-(47) is *third-order accurate*; this will be the case if and only if the numerical integration rule in (48) is *exact for second degree polynomials*, i.e. if and only if

$$\alpha(1 - \theta)\theta^2 + \beta(1 - \theta)^3 + \alpha\theta = \frac{1}{3} \tag{52}$$

with $\theta = 1 - 1/\sqrt{2}$ in (52). Taking $\beta = 1 - \alpha$ into account, it follows from (52) that

$$\alpha(2\theta^2 - 4\theta + 1) = (1 - \theta)^3 - 1/3.$$

which implies in turn, since (51) holds, that

$$0 = \frac{3 - 2\sqrt{2}}{6\sqrt{2}} \tag{53}$$

which makes no sense. Strictly speaking, therefore, if $\theta = 1 - 1/\sqrt{2}$ scheme (44)-(47) is never third-order accurate, $\forall \alpha, \beta$, so that $0 \leq \alpha, \beta \leq 1, \alpha + \beta = 1$. However, since $\dfrac{3 - 2\sqrt{2}}{6\sqrt{2}} \simeq 2 \times 10^{-2}$ we can say that (52) is "almost" verified, implying that scheme (44)-(47) is "not far" from being third-order accurate if $\theta = 1 - 1/\sqrt{2}$. Similarly, if $\alpha = \beta = 1/2$, we can prove that there is no value of θ in $(0,1/2)$ so that scheme (44)-(47) is third-order accurate.

From the above results, we suggest to proceed as follows when applying the θ-scheme (18)-(21) to the solution of the initial value problem (40):

1) If $\theta \neq 1 - 1/\sqrt{2}$, use

$$\varphi^0 = \varphi_0, \tag{54}$$

and for $n \geq 0$

$$\frac{\varphi^{n+\theta} - \varphi^n}{\theta \Delta t} + B_1(\varphi^{n+\theta}) + B_2(\varphi^n) = \frac{1}{2}(f^{n+\theta} + f^n), \tag{55}$$

$$\frac{\varphi^{n+1-\theta} - \varphi^{n+\theta}}{(1 - 2\theta)\Delta t} + B_1(\varphi^{n+\theta}) + B_2(\varphi^{n+1-\theta}) = \frac{1}{2}(f^{n+\theta} + f^{n+1-\theta}), \tag{56}$$

$$\frac{\varphi^{n+1} - \varphi^{n+1-\theta}}{\theta \Delta t} + B_1(\varphi^{n+1}) + B_2(\varphi^{n+1-\theta}) = \frac{1}{2}(f^{n+1} + f^{n+1-\theta}). \tag{57}$$

2) If $\theta = 1 - 1/\sqrt{2}$ we can still use scheme (54)-(57), but simpler choices are provided by

$$\varphi^0 = \varphi_0, \tag{58}$$

and for $n \geq 0$

$$\frac{\varphi^{n+\theta} - \varphi^n}{\theta \Delta t} + B_1(\varphi^{n+\theta}) + B_2(\varphi^n) = f^{n+\theta}, \tag{59}$$

$$\frac{\varphi^{n+1-\theta} - \varphi^{n+\theta}}{(1 - 2\theta)\Delta t} + B_1(\varphi^{n+\theta}) + B_2(\varphi^{n+1-\theta}) = f^{n+\theta}, \tag{60}$$

$$\frac{\varphi^{n+1} - \varphi^{n+1-\theta}}{\theta \Delta t} + B_1(\varphi^{n+1}) + B_2(\varphi^{n+1-\theta}) = f^{n+1} \tag{61}$$

(which corresponds to $\{\alpha, \beta\} = \{1, 0\}$) and by

$$\varphi^0 = \varphi_0, \tag{62}$$

and for $n \geq 0$

$$\frac{\varphi^{n+\theta} - \varphi^n}{\theta \Delta t} + B_1(\varphi^{n+\theta}) + B_2(\varphi^n) = f^n, \tag{63}$$

$$\frac{\varphi^{n+1-\theta} - \varphi^{n+\theta}}{(1-2\theta)\Delta t} + B_1(\varphi^{n+\theta}) + B_2(\varphi^{n+1-\theta}) = f^{n+1-\theta}, \tag{64}$$

$$\frac{\varphi^{n+1} - \varphi^{n+1-\theta}}{\theta \Delta t} + B_1(\varphi^{n+1}) + B_2(\varphi^{n+1-\theta}) = f^{n+1-\theta} \tag{65}$$

(which corresponds to $\{\alpha, \beta\} = \{0, 1\}$).   □

2.3 Fractional–step scheme à la Marchuk–Yanenko

Among the many operator–splitting methods which can be employed to solve (17), we also advocate (following, e.g., Marchuk [29]) the very simple one below; it is only first order accurate, but its low order accuracy is compensated by easy implementation, less cost in computation, good stability, and robustness properties. We consider the initial value problem (17) with $A = A_1 + A_2$ where A_1 and A_2 are linear and independent of t; we have then (at least formally)

$$\varphi(t) = e^{-(A_1+A_2)t}\varphi_0. \tag{66}$$

We consider a time discretization step $\Delta t(> 0)$ and denote $(n+\alpha)\Delta t$ by $t^{n+\alpha}$. Then from (66) we have

$$\varphi(t^{n+1}) = e^{-(A_1+A_2)\Delta t}\varphi(t^n). \tag{67}$$

Now we suppose that A_1 and A_2 do not commute. We have then

$$e^{-(A_1+A_2)\Delta t} = e^{-A_2\Delta t}e^{-A_1\Delta t} + O(\Delta t^2). \tag{68}$$

Relation (68) leads to the following first order scheme for the solution of problem (17):

$$\varphi^0 = \varphi_0, \tag{69}$$

for $n \geq 0, \varphi^n$ being known, we compute $\varphi^{n+1/2}, \varphi^{n+1}$ via the solution of two initial value problems below:

$$d\varphi/dt + A_1\varphi = 0 \ on \ (t^n, t^{n+1}), \ \varphi(t^n) = \varphi^n; \ \varphi^{n+1/2} = \varphi(t^{n+1}), \tag{70}$$

$$d\varphi/dt + A_2\varphi = 0 \ on \ (t^n, t^{n+1}), \varphi(t^n) = \varphi^{n+1/2}; \ \varphi^{n+1} = \varphi(t^{n+1}), \tag{71}$$

We consider again the simple situation where $H = \mathbb{R}^N$, $\varphi_0 \in \mathbb{R}^N$, where A is an $N \times N$ matrix, *symmetric, positive definite* and *independent* of t. Applying (70), (71) with $A_1 = \alpha A, A_2 = \beta A$ satisfying $\alpha + \beta = 1, 0 < \alpha, \beta < 1$ and backward Euler method yields

$$\frac{\varphi^{n+1/2} - \varphi^n}{\Delta t} + \alpha A \varphi^{n+1/2} = 0, \tag{72}$$

$$\frac{\varphi^{n+1} - \varphi^{n+1/2}}{\Delta t} + \beta A \varphi^{n+1} = 0, \tag{73}$$

and

$$\varphi^{n+1} = (I + \beta \Delta t A)^{-1} (I + \alpha \Delta t A)^{-1} \varphi^n$$

which implies

$$\varphi_i^n = \frac{\varphi_{0i}}{(1 + \beta \Delta t \lambda_i)^n (1 + \alpha \Delta t \lambda_i)^n}, \forall i = 1, ... N.$$

Hence $\varphi_i^n \to 0$ as $n \to \infty$ for all α, β satisfying $\alpha + \beta = 1$, $0 < \alpha, \beta < 1$. So the scheme is unconditionally stable. Consider now the rational function R_2 defined by

$$R_2(\xi) = (1 + \beta \xi)^{-1} (1 + \alpha \xi)^{-1}.$$

We have

$$R_2(\xi) = 1 - \xi + \xi^2 (\alpha\beta + \beta^2 + \alpha^2) + O(1)\xi^3.$$

Comparing the above expansion of $R_2(\xi)$ to the expansion of $e^{-\xi}$ in (32), we obtain that the Marchuk–Yanenko scheme is first order accurate (due to the way we approximate problem (17) by the two problems (70) and (71)) and unconditionally stable, at least for the above simple case under consideration.

Remark 6. A second order scheme can be obtained by *symmetrization* (see, e.g., Dean and Glowinski [100] and Dean, Glowinski and Pan [101] for the application of *symmetrized* splitting schemes to the solution of the Navier-Stokes equations). $\square$

Generalizing (72) and (73) to problem (2.1) (with $A = A_1 + A_2$) leads us to:

$$\varphi^0 = \varphi_0, \tag{74}$$

for $n \geq 0, \varphi^n$ being known, we compute $\varphi^{n+1/2}, \varphi^{n+1}$ as follows:

$$\frac{\varphi^{n+1/2} - \varphi^n}{\Delta t} + A_1(\varphi^{n+1/2}, t^{n+1}) = 0, \tag{75}$$

$$\frac{\varphi^{n+1} - \varphi^{n+1/2}}{\Delta t} + A_2(\varphi^{n+1}, t^{n+1}) = 0, \tag{76}$$

Remark 7. We consider again the case where in (17) we have $A(\varphi, t) = -f(t)$. As before, we suppose that, that $f = f_1 + f_2$ with $f_1 = \alpha f$, $f_2 = \beta f$ and $0 \leq \alpha$, $\beta \leq 1$, $\alpha + \beta = 1$.

Applying scheme (74)-(76) to the solution of

$$\frac{d\varphi}{dt} = f, \ \varphi(0) = \varphi_0,$$

we obtain

$$\varphi^0 = \varphi_0, \tag{77}$$

and for $n \geq 0$,

$$\frac{\varphi^{n+1/2} - \varphi^n}{\Delta t} = \alpha f(t^{n+1}), \tag{78}$$

$$\frac{\varphi^{n+1} - \varphi^{n+1/2}}{\Delta t} = \beta f(t^{n+1}). \tag{79}$$

which imply that

$$\varphi^n = \varphi_0 + \Delta t \sum_{q=1}^{n} f(t^q).$$

Hence the above scheme is first order accurate if f' is continuous. $\quad\square$

2.4 Application to the Navier-Stokes equations

We discuss now the application of the time discretization schemes described in the above sections to the solution of the *time-dependent Navier-Stokes equations* (1)-(3), (5).

Actually, we shall consider application of the θ-scheme, since it is the one which gives the best results regarding accuracy and convergence to steady-state solutions. We obtain then the following time discretization scheme (with $0 < \alpha < 1, 0 < \beta < 1$ and $\alpha + \beta = 1$) :

$$\mathbf{u}^0 = \mathbf{u}_0; \tag{80}$$

then for $n \geq 0, \mathbf{u}^n$ being known, we compute $\mathbf{u}^{n+\theta}, \mathbf{u}^{n+1-\theta}$ and $\mathbf{u}^{n+1}$ via the solution of

$$\frac{\mathbf{u}^{n+\theta} - \mathbf{u}^n}{\theta \Delta t} - \alpha \nu \Delta \mathbf{u}^{n+\theta} + \nabla p^{n+\theta} = \mathbf{f}^{n+\theta} + \beta \nu \Delta \mathbf{u}^n - (\mathbf{u}^n \cdot \nabla)\mathbf{u}^n$$

$$\text{in } \Omega, \tag{81}$$

$$\nabla \cdot \mathbf{u}^{n+\theta} = 0 \text{ in } \Omega, \tag{82}$$

$$\mathbf{u}^{n+\theta} = \mathbf{g}_0^{n+\theta} \text{ on } \Gamma_0, \quad \alpha \nu \frac{\partial \mathbf{u}^{n+\theta}}{\partial n} - \mathbf{n}p^{n+\theta} = \mathbf{g}_1^{n+\theta} - \beta \nu \frac{\partial \mathbf{u}^n}{\partial n} \text{ on } \Gamma_1, \tag{83}$$

and then, of

$$\frac{\mathbf{u}^{n+1-\theta} - \mathbf{u}^{n+\theta}}{(1 - 2\theta)\Delta t} - \beta\nu\Delta\mathbf{u}^{n+1-\theta} + (\mathbf{u}^{n+1-\theta} \cdot \boldsymbol{\nabla})\mathbf{u}^{n+1-\theta} = \mathbf{f}^{n+\theta} + \alpha\nu\Delta\mathbf{u}^{n+\theta}$$

$$-\boldsymbol{\nabla}p^{n+\theta} \ in \ \Omega, \tag{84}$$

$$\mathbf{u}^{n+1-\theta} = \mathbf{g}_0^{n+1-\theta} \ on \ \Gamma_0, \quad \beta\nu\frac{\partial\mathbf{u}^{n+1-\theta}}{\partial n} = \mathbf{g}_1^{n+\theta} + \mathbf{n}p^{n+\theta}$$

$$-\alpha\nu\frac{\partial\mathbf{u}^{n+\theta}}{\partial n} \ on \ \Gamma_1, \tag{85}$$

and finally, of

$$\frac{\mathbf{u}^{n+1} - \mathbf{u}^{n+1-\theta}}{\theta\Delta t} - \alpha\nu\Delta\mathbf{u}^{n+1} + \boldsymbol{\nabla}p^{n+1} = \mathbf{f}^{n+1} + \beta\nu\Delta\mathbf{u}^{n+1-\theta}$$

$$-(\mathbf{u}^{n+1-\theta} \cdot \boldsymbol{\nabla})\mathbf{u}^{n+1-\theta} \ in \ \Omega, \tag{86}$$

$$\boldsymbol{\nabla} \cdot \mathbf{u}^{n+1} = 0 \ in \ \Omega, \tag{87}$$

$$\mathbf{u}^{n+1} = \mathbf{g}_0^{n+1} \ on \ \Gamma_0, \quad \alpha\nu\frac{\partial\mathbf{u}^{n+1}}{\partial n} - \mathbf{n}p^{n+1} = \mathbf{g}_1^{n+1}$$

$$-\beta\nu\frac{\partial\mathbf{u}^{n+1-\theta}}{\partial n} \ on \ \Gamma_1; \tag{88}$$

the choice of α and β will be discussed below. We observe that using the θ-scheme we have been able to *decouple* the nonlinearity and the incompressibility in the Navier-Stokes equations (1)-(3), (5). We observe also that $\mathbf{u}^{n+\theta}$ and $\mathbf{u}^{n+1}$ are obtained from the solution of *linear* problems very close to the *Stokes problem*

$$\begin{cases} -\nu\Delta\mathbf{u} + \boldsymbol{\nabla}p = \mathbf{f} \ in \ \Omega, \\ \boldsymbol{\nabla} \cdot \mathbf{u} = 0 \ in \ \Omega, \\ \mathbf{u} = \mathbf{g}_0 \ on \ \Gamma_0, \quad \nu\frac{\partial\mathbf{u}}{\partial n} - \mathbf{n}p = \mathbf{g}_1 \ on \ \Gamma_1. \end{cases} \tag{89}$$

In Sections 3 and 4, we shall describe the specific treatment of the subproblems encountered at each step of scheme (80)-(88). Concerning now the choice of α and β, we advocate the one given by (37); with such a choice many computer subprograms are *common* to both the linear and nonlinear subproblems, saving therefore quite a substantial amount of core memory. Concerning θ, numerical experiments show that $\theta = 1 - 1/\sqrt{2}$ seems to produce the best results, even in those situations where the Reynolds number is large.

Remark 8. Numerical experiments show that there is practically no loss in accuracy and stability by replacing $(\mathbf{u}^{n+1-\theta} \cdot \boldsymbol{\nabla})\mathbf{u}^{n+1-\theta}$ by $(\mathbf{u}^{n+\theta} \cdot \boldsymbol{\nabla})\mathbf{u}^{n+1-\theta}$ in (84). This observation has important practical consequences since the following problem

$$\begin{cases} \dfrac{\mathbf{u}^{n+1-\theta} - \mathbf{u}^{n+\theta}}{(1-2\theta)\Delta t} - \beta\nu\Delta\mathbf{u}^{n+1-\theta} + (\mathbf{u}^{n+\theta}\cdot\boldsymbol{\nabla})\mathbf{u}^{n+1-\theta} = \\ \mathbf{f}^{n+\theta} + \alpha\nu\Delta\mathbf{u}^{n+\theta} - \boldsymbol{\nabla}p^{n+\theta} \ in \ \Omega, \\ \mathbf{u}^{n+1-\theta} = \mathbf{g}_0^{n+1-\theta} \ on \ \Gamma_0, \beta\nu\dfrac{\partial\mathbf{u}^{n+1-\theta}}{\partial n} = \mathbf{g}_1^{n+\theta} + \mathbf{n}p^{n+\theta} - \alpha\nu\dfrac{\partial\mathbf{u}^{n+\theta}}{\partial n} \ on \ \Gamma_1, \end{cases}$$

$$(90)$$

being linear, is easier to solve than the nonlinear problem (84). □

Remark 9. Operator splitting methods have always been popular tools for the numerical simulation of incompressible viscous flow. To be more precise, the so-called *projection methods*, which have been used for more than thirty years now, for solving the Navier-Stokes equations can be viewed as operator splitting methods. The projection methods can also be viewed as *predictor-corrector* schemes, where a predicted value (not necessarily divergence-free) of the approximate solution at time $(n+1)\Delta t$ is projected in the $L^2(\Omega)$-sense over an appropriate space of divergence-free functions. We will discuss a projection method obtained by the scheme à la Marchuk-Yanenko in Section 5.3. To our knowledge, projection methods for solving the Navier-Stokes equations have been introduced by Chorin [38] and [39] and Temam [40, 41]; the original projection methods contained several drawbacks, concerning particularly the quality of the approximate pressure at low Reynolds numbers, but, fortunately, these flaws have been essentially eliminated in the modern projection methods. A concise, but fairly complete introduction to projection schemes can be found in Quarteroni and Valli [42] (Section 13.5), a more detailed one being Marion and Temam [23] (Chapter 3). □

3 Classical and Variational Formulations of the Advection-Diffusion Subproblems Associated with the Operator Splitting Schemes

At each full step of scheme (80)-(88) we have to solve a *nonlinear elliptic system* of the following type (with Ω, Γ, Γ_0 and Γ_1 as in Section 1):

$$\begin{cases} \alpha\mathbf{u} - \nu\Delta\mathbf{u} + (\mathbf{u}\cdot\boldsymbol{\nabla})\mathbf{u} = \mathbf{f} \ in \ \Omega, \\ \mathbf{u} = \mathbf{g}_0 \ on \ \Gamma_0, \ \nu\dfrac{\partial\mathbf{u}}{\partial n} = \mathbf{g}_1 \ on \ \Gamma_1, \end{cases} \tag{91}$$

where α and ν are two positive constants, and $\mathbf{f}, \mathbf{g}_0$ and $\mathbf{g}_1$ are three given functions, defined on Ω, Γ_0 and Γ_1, respectively. We shall not discuss here the existence and uniqueness of solution for problem (91), which can be found in, e.g., Glowinski [4] (Section 15). We consider now the following functional spaces of *Sobolev* type:

$$H^1(\Omega) = \{\varphi | \varphi \in L^2(\Omega), \frac{\partial \varphi}{\partial x_i} \in L^2(\Omega), \forall i = 1, ...d\}, \tag{92}$$

$$H_0^1(\Omega) = \{\varphi | \varphi \in H^1(\Omega), \ \varphi = 0 \ on \ \Gamma\}, \tag{93}$$

$$V_0 = \{\mathbf{v} | \mathbf{v} \in (H^1(\Omega))^d, \ \mathbf{v} = \mathbf{0} \ on \ \Gamma_0\}, \tag{94}$$

$$V_g = \{\mathbf{v} | \mathbf{v} \in (H^1(\Omega))^d, \ \mathbf{v} = \mathbf{g}_0 \ on \ \Gamma_0\}; \tag{95}$$

if $\mathbf{g}_0$ is sufficiently smooth, then space V_g is nonempty.

Using *Green's formula* we can prove that for sufficiently smooth functions $\mathbf{u}$ and $\mathbf{v}$ belonging to $(H^1(\Omega))^d$ and V_0, respectively, we have

$$\int_{\Gamma_1} \frac{\partial \mathbf{u}}{\partial n} \cdot \mathbf{v} d\Gamma = \int_\Omega \boldsymbol{\nabla} \mathbf{u} : \boldsymbol{\nabla} \mathbf{v} \, dx + \int_\Omega \Delta \mathbf{u} \cdot \mathbf{v} \, dx. \tag{96}$$

Taking now the dot-product with $\mathbf{v}$ of both sides of the first equation (91), using (96) and taking the boundary conditions in (91) into account we obtain that if $\mathbf{u}$ is a solution of problem (91) belonging to V_g, it is also a solution of the following *nonlinear variational problem*:

$$\begin{cases} \mathbf{u} \in V_g; \ \forall \mathbf{v} \in V_0 \ we \ have \\ \alpha \int_\Omega \mathbf{u} \cdot \mathbf{v} \, dx + \nu \int_\Omega \boldsymbol{\nabla} \mathbf{u} : \boldsymbol{\nabla} \mathbf{v} \, dx + \int_\Omega (\mathbf{u} \cdot \boldsymbol{\nabla}) \mathbf{u} \cdot \mathbf{v} \, dx = \int_\Omega \mathbf{f} \cdot \mathbf{v} \, dx \\ \hspace{8cm} + \int_{\Gamma_1} \mathbf{g}_1 \cdot \mathbf{v} d\Gamma. \end{cases} \tag{97}$$

Actually, the reciprocal property is true and (97) implies (91). Problem (91), (97) is not equivalent to a problem of the *Calculus of Variations*, since $\mathbf{v} \rightarrow (\mathbf{v} \cdot \boldsymbol{\nabla})\mathbf{v}$ is not the differential of a functional of $\mathbf{v}$; using, however, a convenient *least squares formulation* we shall be able to solve the above advection-diffusion problem by iterative methods from *Nonlinear Programming*, such as *conjugate gradient algorithms*.

3.1 Least–Squares Formulation of (91), (97)

Let $\mathbf{v} \in V_g$; to $\mathbf{v}$ we associate the solution $\mathbf{y} = \mathbf{y}(\mathbf{v}) \in V_0$ of

$$\begin{cases} \alpha \mathbf{y} - \nu \Delta \mathbf{y} = \alpha \mathbf{v} - \nu \Delta \mathbf{v} + (\mathbf{v} \cdot \boldsymbol{\nabla})\mathbf{v} - \mathbf{f} \ in \ \Omega, \\ \mathbf{y} = \mathbf{0} \ on \ \Gamma_0, \ \nu \dfrac{\partial \mathbf{y}}{\partial n} = \nu \dfrac{\partial \mathbf{v}}{\partial n} - \mathbf{g}_1 \ on \ \Gamma_1. \end{cases} \tag{98}$$

We observe that $\mathbf{y}$ is obtained from $\mathbf{v}$ via the solution of d uncoupled linear elliptic problems (one for each component of $\mathbf{y}$); using (96), it is easily shown that (98) is equivalent to the linear variational problem

$$\begin{cases} \mathbf{y} \in V_0; \ \forall \mathbf{z} \in V_0 \ we \ have \\ \alpha \int_\Omega \mathbf{y} \cdot \mathbf{z} \, dx + \nu \int_\Omega \nabla \mathbf{y} : \nabla \mathbf{z} \, dx = \alpha \int_\Omega \mathbf{v} \cdot \mathbf{z} \, dx + \nu \int_\Omega \nabla \mathbf{v} : \nabla \mathbf{z} \, dx \\ + \int_\Omega (\mathbf{v} \cdot \nabla) \mathbf{v} \cdot \mathbf{z} \, dx - \int_\Omega \mathbf{f} \cdot \mathbf{z} \, dx - \int_{\Gamma_1} \mathbf{g}_1 \cdot \mathbf{z} d\Gamma, \end{cases} \qquad (99)$$

which has a unique solution.

Suppose now that $\mathbf{v}$ is a solution of the nonlinear problem (91), (97); the corresponding $\mathbf{y}$ (obtained from the solution of (98), (99)) is clearly $\mathbf{y} = \mathbf{0}$; from this observation, it is quite natural to introduce the following (nonlinear) least–squares formulation of (91), (97):

$$\begin{cases} find \ \mathbf{u} \in V_g \ such \ that \\ J(\mathbf{u}) \le J(\mathbf{v}), \ \forall \mathbf{v} \in V_g, \end{cases} \qquad (100)$$

where the functional $J : (H^1(\Omega))^d \to \mathbb{R}$ is defined by

$$J(\mathbf{v}) = \frac{1}{2} \int_\Omega \{\alpha |\mathbf{y}|^2 + \nu |\nabla \mathbf{y}|^2\} \, dx \qquad (101)$$

with $\mathbf{y}$ defined from $\mathbf{v}$ by (98), (99). Observe that if $\mathbf{u}$ is a solution of (100), such that $J(\mathbf{u}) = 0$, then it is also a solution of (91), (97).

3.2 Conjugate Gradient Methods for the Solution of Minimization Problems in Hilbert Spaces

The main goal of this subsection is to discuss the *iterative solution of min-imization problems in Hilbert spaces by conjugate gradient algorithms*. For years, our main sources of information concerning conjugate gradient algorithms have been Daniel [43] and Polak [44], the first reference in particular since it is also concerned with infinite dimensional problems.

Conjugate gradient algorithms have been introduced by M. Hestenes and E. Stiefel in the early fifties for the solution of finite dimensional linear systems associated with *symmetric and positive definite* matrices (see Hestenes and Stiefel [45] for details). Since then, these methods have enjoyed considerable generalizations and have motivated a very large number of publications. The interested reader may find abundant information on these methods and their implementation in, e.g., the review articles Freund, Golub and Nachtigal [46], Nocedal [47] and in the monographs of Kelley [48] (Chapter 2), Saad [49] (see also the references therein, and Golub and O'Leary [50] for a historical account).

Conjugate Gradient Solution of Linear Variational Problems in Hilbert Spaces

We shall discuss first the *conjugate gradient solution* of the *linear variational problems in Hilbert spaces*. We consider:

(i) V is a *real Hilbert space* for the scalar product $(\cdot, \cdot)$ and the associated norm $|| \cdot ||$;

(ii) $a(\cdot, \cdot)$ is a *bilinear* functional from $V \times V \to \mathbb{R}$, *continuous* and V-*elliptic* (i.e., $\exists\, \alpha > 0$ such that $a(v, v) \geq \alpha ||v||^2,\ \forall v \in V$);

(iii) L is *linear* and *continuous* over V.

In this section we make the following additional assumption on the bilinear functional $a(\cdot, \cdot)$:

$$\begin{cases} the\ bilinear\ functional\ a(\cdot, \cdot)\ is\ symmetric, \\ i.e.,\ \ a(v, w) = a(w, v),\ \forall v, w \in V. \end{cases} \tag{102}$$

If the symmetry property (102) holds, then the linear variational problem

$$\begin{cases} u \in V, \\ a(u, v) = L(v),\ \forall v \in V, \end{cases} \tag{103}$$

has a *unique* solution by the Lax-Milgram theorem, which is also the solution of the *minimization* problem

$$\begin{cases} u \in V, \\ J(u) \leq J(v),\ \forall v \in V, \end{cases} \tag{104}$$

with

$$J(v) = \frac{1}{2}a(v, v) - L(v),\ \forall v \in V. \tag{105}$$

Here is a typical example of an above like linear variational problem:

Example 3.1: We consider here the variational formulation of the the homogeneous Dirichlet problem, namely

$$\begin{cases} u \in H_0^1(\Omega), \\ \displaystyle\int_\Omega \boldsymbol{\nabla} u \cdot \boldsymbol{\nabla} v\, dx = \int_\Omega fv\, dx,\ \forall v \in H_0^1(\Omega), \end{cases}$$

with $f \in L^2(\Omega)$. In this example, we have $V = H_0^1(\Omega)$,

$$a(u, v) = \int_\Omega \boldsymbol{\nabla} u \cdot \boldsymbol{\nabla} v\, dx,\quad L(v) = \int_\Omega fv\, dx$$

and

$$J(v) = \frac{1}{2}\int_\Omega |\boldsymbol{\nabla} v|^2\, dx - \int_\Omega fv\, dx.$$

Description of the conjugate gradient algorithm.

In order to solve problem (103), (104) we propose the following conjugate gradient algorithm.

Step 0: Initialization

$$u^0 \in V \text{ is given;} \tag{106}$$

solve

$$\begin{cases} g^0 \in V, \\ (g^0, v) = a(u^0, v) - L(v), \ \forall v \in V, \end{cases} \tag{107}$$

and set

$$w^0 = g^0. \square \tag{108}$$

For $n \geq 0$, assuming that u^n, g^n, w^n are known with $g^n \neq 0$ and $w^n \neq 0$, compute u^{n+1}, g^{n+1}, w^{n+1} as follows
Step 1: Steepest descent
Compute

$$\rho_n = ||g^n||^2 / a(w^n, w^n) \tag{109}$$

and set

$$u^{n+1} = u^n - \rho_n w^n. \tag{110}$$

Step 2: Testing the convergence and construction of the new descent direction
Solve

$$\begin{cases} g^{n+1} \in V, \\ (g^{n+1}, v) = (g^n, v) - \rho_n a(w^n, v), \ \forall v \in V. \end{cases} \tag{111}$$

If $||g^{n+1}||/||g^0|| \leq \varepsilon$ take $u = u^{n+1}$; else, compute

$$\gamma_n = ||g^{n+1}||^2 / ||g^n||^2 \tag{112}$$

and update w^n by

$$w^{n+1} = g^{n+1} + \gamma_n w^n. \tag{113}$$

Do $n = n + 1$ and return to (109). $\square$

Despite its apparent simplicity, algorithm (106)-(113) is *one of the most powerful tools* of Scientific Computing; it is currently used to solve very complicated problems from Science and Engineering which may involve many millions of unknowns. Large scale application of the above algorithm will be found in several parts of these notes.

Convergence of algorithm (106) ... (113)

Before discussing the convergence of algorithm (106)-(113), it can be shown by using the Riesz theorem that problem (103), (104) is equivalent to

$$Au = l, \tag{114}$$

where $l \in V, A \in \mathcal{L}(V, V)$ and verify

$$L(v) = (l, v), \ \forall v \in V \ and \ a(v, w) = (Av, w), \ \forall v, w \in V;$$

operator A is an *automorphism* of V (*symmetric* since $a(\cdot, \cdot)$ is symmetric). Incidentally, we have

$$\alpha ||v||^2 \le a(v, v) \le ||A||\,||v||^2, \ \forall v \in V; \tag{115}$$

in (115), the best constant α (i.e., the largest one) is given by $1/||A^{-1}||$.

Concerning the *convergence* of algorithm (106)-(113), we are going to prove the following:

Theorem 1. *Suppose that $\varepsilon = 0$ in algorithm (106)-(113); we have then*

$$\lim_{n \to +\infty} ||u^n - u|| = 0, \ \forall u^0 \in V, \tag{116}$$

where u is the solution of problem (103), (104).

PROOF: For clarity, the proof has been divided in two parts.
Orthogonality properties: First, we are going to show that the following *orthogonality* properties hold, as long as we can iterate (i.e., as long as g^n and w^n are different from 0 in (106)-(113)):

$$(g^i, g^j) = 0, \ \forall i, j, i \ne j, \tag{117}$$
$$(g^i, w^j) = 0, \ \forall i, j, i > j, \tag{118}$$
$$a(w^i, w^j) = 0, \ \forall i, j, i \ne j. \tag{119}$$

We are going to proceed by *induction*, assuming first that relations (117)-(119) hold up to n; let us show that they also hold up to $n + 1$. We start with (117): We have, from (111) and from (113) (with n replaced by $n - 1$)

$$\begin{aligned} (g^{n+1}, g^n) &= ||g^n||^2 - \rho_n a(w^n, g^n) \\ &= ||g^n||^2 - \rho_n a(w^n, w^n - \gamma_{n-1} w^{n-1}); \end{aligned}$$

using (119) (true up to n) and (109) we obtain

$$(g^{n+1}, g^n) = ||g^n||^2 - \rho_n a(w^n, w^n) = 0.$$

Similarly, we have for $j < n$

$$\begin{cases} (g^{n+1}, g^j) = (g^n, g^j) - \rho_n a(w^n, g^j) \\ \qquad = (g^n, g^j) - \rho_n a(w^n, w^j - \gamma_{j-1} w^{j-1}) = 0. \end{cases}$$

We have thus shown that if (117) holds up to n, it also holds up to $n + 1$. $\square$

We consider now the relations (118); operating as above we have

$$\begin{aligned} (g^{n+1}, w^n) &= (g^n, w^n) - \rho_n a(w^n, w^n) \\ &= (g^n, g^n + \gamma_{n-1} w^{n-1}) - \rho_n a(w^n, w^n) \\ &= ||g^n||^2 - \rho_n a(w^n, w^n) = 0, \end{aligned}$$

and for $j < n$

$$(g^{n+1}, w^j) = (g^n, w^j) - \rho_n a(w^n, w^j) = 0.$$

We have shown, here also, that if (118) holds up to n, it holds up to $n+1$. $\square$

Proving similar results for (119) is slightly more complicated; however, using the relations in algorithm (106)-(113) and the fact that (117), (118) (resp., (119)) hold up to $n + 1$ (resp., n) we have

$$\begin{aligned} a(w^{n+1}, w^n) &= a(w^n, w^{n+1}) = \rho_n^{-1}[(g^n, w^{n+1}) - (g^{n+1}, w^{n+1})] \\ &= \rho_n^{-1}[(g^n, g^{n+1} + \gamma_n w^n) - (g^{n+1}, g^{n+1} + \gamma_n w^n)] \\ &= \rho_n^{-1}[\gamma_n(g^n, w^n) - ||g^{n+1}||^2] \\ &= \rho_n^{-1}[\gamma_n(g^n, g^n + \gamma_{n-1} w^{n-1}) - ||g^{n+1}||^2] \\ &= \rho_n^{-1}[\gamma_n ||g^n||^2 - ||g^{n+1}||^2] = 0, \end{aligned}$$

and then for $j < n$

$$\begin{aligned} a(w^{n+1}, w^j) &= a(g^{n+1} + \gamma_n w^n, w^j) = a(g^{n+1}, w^j) \\ &= a(w^j, g^{n+1}) \\ &= \rho_j^{-1}[(g^j, g^{n+1}) - (g^{j+1}, g^{n+1})] = 0; \end{aligned}$$

the above relations imply that (119) hold up to $n + 1$ if it holds up to n. $\square$

To complete the proof of (117)-(119) it suffices to show that these relations also hold for $i = 1$ and $j = 0$. Using the fact that $w^0 = g^0$, we have

$$(g^1, g^0) = ||g^0||^2 - \rho_0 a(w^0, g^0) = ||g^0||^2 - \rho_0 a(w^0, w^0) = 0,$$
$$(g^1, w^0) = 0.$$

Concerning now $a(w^1, w^0)$, we have

$$\begin{aligned} a(w^1, w^0) &= a(w^0, w^1) = \rho_0^{-1}[(g^0, w^1) - (g^1, w^1)] \\ &= \rho_0^{-1}[(g^0, g^1 + \gamma_0 w^0) - (g^1, g^1 + \gamma_0 w^0)] \\ &= \rho_0^{-1}[\gamma_0(g^0, w^0) - ||g^1||^2] \\ &= \rho_0^{-1}[\gamma_0 ||g^0||^2 - ||g^1||^2] = 0, \end{aligned}$$

which completes the proof of relations (117)-(119). $\square$

Convergence: We can easily show (by induction, again) that

$$(g^{n+1}, v) = a(u^{n+1}, v) - L(v), \quad \forall v \in V.$$

If $g^{n+1} = 0$ in algorithm (106)-(113), we have therefore $u^{n+1} = u$ (since problem (103) has a unique solution). Suppose now that $w^{n+1} = 0$; it follows from (113) that

$$g^{n+1} + \gamma_n w^n = 0$$

which implies in turn (from (118)) that

$$||g^{n+1}||^2 + \gamma_n(g^{n+1}, w^n) = ||g^{n+1}||^2 = 0;$$

we have thus $u^{n+1} = u$.

Suppose now that we have $g^n \neq 0$ and $w^n \neq 0$, $\forall n \geq 0$; in order to show that $\lim_{n \to +\infty} u^n = u$ we consider the difference $J(u^n) - J(u^{n+1})$; we clearly have (Taylor's expansion)

$$
\begin{aligned}
J(u^{n+1}) &= J(u^n - \rho_n w^n) = J(u^n) - \rho_n(J'(u^n), w^n) + \tfrac{1}{2}\rho_n^2 a(w^n, w^n) \\
&= J(u^n) - \rho_n[a(u^n, w^n) - L(w^n)] + \tfrac{1}{2}\rho_n^2 a(w^n, w^n) \\
&= J(u^n) - \rho_n(g^n, w^n) + \tfrac{1}{2}\rho_n^2 a(w^n, w^n) \\
&= J(u^n) - \rho_n(g^n, g^n + \gamma_{n-1}w^{n-1}) + \tfrac{1}{2}\rho_n^2 a(w^n, w^n) \\
&= J(u^n) - \rho_n||g^n||^2 + \tfrac{1}{2}\rho_n^2 a(w^n, w^n)
\end{aligned}
$$

which implies that

$$J(u^n) - J(u^{n+1}) = \rho_n||g^n||^2 - \frac{1}{2}\rho_n^2 a(w^n, w^n) = \frac{1}{2}||g^n||^4/a(w^n, w^n), \quad \forall n \geq 0. \tag{120}$$

It follows from (120) that the sequence $\{J(u^n)\}_{n \geq 0}$ is a decreasing one; since it is bounded from below by $J(u)$, it converges to some limit ($\geq J(u)$) which implies that

$$\lim_{n \to +\infty} [J(u^n) - J(u^{n+1})] = 0.$$

We have thus shown (from (120)) that

$$\lim_{n \to +\infty} ||g^n||^4/a(w^n, w^n) = 0. \tag{121}$$

Since $g^n = w^n - \gamma_{n-1}w^{n-1}$, we have (from (119)) that

$$a(g^n, g^n) = a(w^n, w^n) + \gamma_{n-1}^2 a(w^{n-1}, w^{n-1}) \geq a(w^n, w^n) > 0; \tag{122}$$

we also have, from (115),

$$a(g^n, g^n) \leq ||A|| \, ||g^n||^2. \tag{123}$$

Combining (121), (122), (123) yields $\lim_{n \to +\infty} ||g^n|| = 0$, which implies in turn (since $g^n = Au^n - l$, $\forall n \geq 0$)

$$\lim_{n \to +\infty} u^n = A^{-1}l = u,$$

which completes the proof of the theorem.

Remark 10. Suppose that V is *finite dimensional* with $dim V = d$; in that case *we have convergence in d iterations at most.* Suppose that it is not the case, then $\{g^0, g^1, \ldots g^d\}$ will be a system of $d+1$ vectors of V, *linearly independent* since all different from zero and mutually orthogonal (from (117)). Since this is impossible there exists $N \le d$ such that $g^N = 0$, which implies in turn that $u^N = u$. $\square$

Remark 11. The above proof of Theorem 1 is a variant of the classical one used to prove, in finite dimension, the *finite termination* property discussed in Remark 10; these proofs completely rely on the orthogonality properties (117)-(119). Computer implementations (necessarily finite-dimensional) of algorithm (106)-(113) will suffer from the effects of round-off errors, one of the effects being precisely the loss of the above orthogonality properties; we can wonder, therefore, about the convergence properties of algorithm (106)-(113) in practice. Actually they are quite good, in general, despite the fact that the finite termination is lost, strictly speaking. This good behavior of algorithm (106)-(113) is a direct consequence of the following estimate of its speed of convergence (proved in, e.g., Daniel [43]):

$$a(u^n - u, u^n - u) \le 4a(u^0 - u, u^0 - u) \left(\frac{\sqrt{\nu_a} - 1}{\sqrt{\nu_a} + 1} \right)^{2n}, \ \forall n \ge 1, \qquad (124)$$

where, in (124), the *condition number* ν_a of the bilinear functional $a(\cdot, \cdot)$ is defined by

$$\nu_a = \sup_{v \in S} a(v, v) / \inf_{v \in S} a(v, v), \qquad (125)$$

with $S = \{v | v \in V, ||v|| = 1\}$ (we can easily show that $\nu_a = ||A|| ||A^{-1}||$, operator A being this element of $\mathcal{L}(V, V)$ such that $a(v, w) = (Av, w), \ \forall v, w \in V$). We observe that the closer ν_a is to 1, the faster is the speed of convergence. For problems of large dimension the convergence behavior associated with (124) is much more important than the hypothetical finite termination property mentioned above. $\square$

Using the following *equivalence* relations between the norms $||v||$ and $\sqrt{a(v, v)}$

$$||A^{-1}||^{-1} ||v||^2 \le a(v, v) \le ||A|| ||v||^2, \ \forall v \in V,$$

we can easily show that (124) implies

$$||u^n - u|| \le 2\sqrt{\nu_a} \left(\frac{\sqrt{\nu_a} - 1}{\sqrt{\nu_a} + 1} \right)^n ||u^0 - u||, \ \forall n \ge 1, \qquad (126)$$

which is less sharp than (124).

Conjugate Gradient Methods for the Solution of Minimization Problems in Hilbert Spaces.

Formulation of the Minimization Problems.

The *minimization problems* to be considered have the following formulation:

$$\begin{cases} u \in V, \\ J(u) \leq J(v), \ \forall v \in V, \end{cases} \tag{127}$$

where:
• V is a Hilbert space for the scalar product $(\cdot, \cdot)$ and the corresponding norm $|| \cdot ||$; we do not assume, here, that V has been identified to its dual space V'.
• $J : V \to \mathbb{R}$ is a *differentiable* functional whose differential is denoted by J' (some authors use the notation ∇J for the differential of J).

Since V has not been necessarily identified to V', it is convenient to introduce the *duality isomorphism* $S : V \to V'$, which is the *unique* operator in *Isom* (V, V') such that

$$< Sv, w >=< Sw, v >= (v, w), \ \forall v, w \in V, \tag{128}$$

where $< \cdot, \cdot >$ denotes the *duality pairing* between V' and V; operator S is *self-adjoint* and *strongly-elliptic* over V since (128) implies

$$< Sv, v >= ||v||^2, \ \forall v \in V. \tag{129}$$

Actually, in addition to (129), relation (128) implies

$$||f||_*^2 =< f, S^{-1}f >, \ \forall f \in V' \tag{130}$$

(where the dual norm $|| \cdot ||_*$ is defined - classically - by

$$||f||_* = \sup_{v \in \Sigma} | < f, v > | \ with \ \Sigma = \{v | v \in V, ||v|| = 1\}),$$

and

$$(f, g)_* =< f, S^{-1}g >, \ \forall f, g \in V', \tag{131}$$

where $(\cdot, \cdot)_*$ denotes the scalar product in V', compatible with the norm $|| \cdot ||_*$.

Concerning now the *differentiability* of J, we shall assume that J is either *Fréchet-differentiable* or *Gâteaux-differentiable*. We recall (see, e.g., Zeidler [51] (Chapter 4)) that J is *Fréchet-differentiable* over V if, $\forall v \in V$, there exists $J'(v) \in V'$, the derivative of J at v, such that

$$J(v + w) - J(v) =< J'(v), w > +||w||\varepsilon(v, w), \tag{132}$$

with $\lim_{w \to 0} \varepsilon(v, w) = 0$. Similarly, (see again Zeidler [51], loc. cit.), J is *Gâteaux-differentiable* over V, if, $\forall v, w \in V$, there exists $J'(v) \in V'$ such that

$$J(v + tw) - J(v) = t < J'(v), w > +t\varepsilon(t, v, w). \tag{133}$$

with $\lim_{t \to 0} \varepsilon(t, v, w,) = 0$. It is quite obvious that the Fréchet-differentiability of J implies its continuity and its Gâteaux-differentiability.

Back to (127), if we suppose that the minimization problem has a solution u, it *necessarily* verifies

$$J'(u) = 0. \tag{134}$$

Proving (134) is fairly obvious, but owing to the importance of this result, we feel obliged to prove it. Observe, therefore, that (127) implies

$$\frac{J(u + tv) - J(u)}{t} \geq 0, \ \forall v \in V \ and \ \forall t > 0; \tag{135}$$

taking the limit in (135), as $t \to 0_+$, we obtain, from (133),

$$< J'(u), v >\geq 0, \forall v \in V,$$

which clearly implies (replace v by $-v$)

$$< J'(u), v >= 0, \forall v \in V. \tag{136}$$

Finally, to show (134) take $v = S^{-1} J'(u)$ in (136) and use relation (130).

As already mentioned, the optimal condition (134) is *sufficient* if J is *convex*; furthermore if J is *strictly convex*, i.e.

$$\begin{cases} J(tv + (1 - t)w) < tJ(v) + (1 - t)J(w), \\ \forall t \in (0, 1), \ \forall v, w \in V, v \neq w, \end{cases} \tag{137}$$

then *existence implies uniqueness*.

We shall conclude this section by mentioning typical conditions which imply the existence of a solution to the minimization problem (127); these conditions are

$$\lim_{||v|| \to +\infty} J(v) = +\infty, \tag{138}$$

$$J \ is \ weakly \ lower \ semi - continuous \ over \ V; \tag{139}$$

condition (139) means that:

$$If \ \lim_{n \to +\infty} v_n = v \ weakly \ in \ V, \ then \ \lim_{n \to +\infty} \inf J(v_n) \geq J(v).$$

Showing that (138), (139) implies the existence of a solution to problem (127) is fairly easy.

Remark 12. If J is convex and differentiable over V, condition (139) is automatically satisfied. To show this result we observe that from the convexity of J we have (by definition)

$$J((1-t)v + tw) \leq tJ(w) + (1-t)J(v), \ \forall v, w \in V, \ \forall t \in (0,1],$$

which can be rewritten as

$$\frac{J(v + t(w-v)) - J(v)}{t} \leq J(w) - J(v), \ \forall v, w \in V, \ \forall t \in (0,1]. \tag{140}$$

Taking the limit in (140), as $t \to 0_+$ we obtain (from (133))

$$J(w) - J(v) \geq\, < J'(v), w - v >, \ \forall v, w \in V. \tag{141}$$

Condition (141) is in fact a celebrated *characterization* of the convexity of differentiable functionals (as shown in, e.g., Ekeland and Temam [52]). Consider now a sequence $\{v_n\}_{n \geq 0}$ in V such that $\lim_{n \to +\infty} v_n = v$ weakly in V; we have, from (141),

$$J(v_n) - J(v) \geq\, < J'(v), v_n - v >, \ \forall n \geq 0,$$

which implies at the limit, as $n \to +\infty$,

$$\lim_{n \to +\infty} \inf J(v_n) \geq J(v),$$

which shows the weak lower semi-continuity of J.   □

Description of Conjugate Gradient Algorithm for the Solution of Problem (127).

In order to solve problem (127) we shall use the following conjugate gradient type algorithms:

Step 0: Initialization

$$u^0 \in V \ is \ given; \tag{142}$$

solve

$$\begin{cases} g^0 \in V, \\ (g^0, v) =\, < J'(u^0), v >, \forall v \in V, \end{cases} \tag{143}$$

and set

$$w^0 = g^0. \square \tag{144}$$

Then for $n \geq 0$, assuming that u^n, g^n, w^n are known, compute $u^{n+1}, g^{n+1}, w^{n+1}$ as follows:

Step 1: Steepest Descent
Solve

$$\begin{cases} \rho_n \in \mathbb{R}, \\ J(u^n - \rho_n w^n) \le J(u^n - \rho w^n), \; \forall \rho \in \mathbb{R} \end{cases} \tag{145}$$

and set

$$u^{n+1} = u^n - \rho_n w^n. \tag{146}$$

Step 2: Testing the convergence and construction of the new descent direction
Solve

$$\begin{cases} g^{n+1} \in V, \\ (g^{n+1}, v) = < J'(u^{n+1}), v >, \forall v \in V; \end{cases} \tag{147}$$

if $||g^{n+1}||/||g^0|| \le \varepsilon$ *take* $u = u^{n+1}$; *else, compute either*

$$\gamma_n = ||g^{n+1}||^2/||g^n||^2 \; (Fletcher - Reeves \; update) \tag{148}$$

or

$$\gamma_n = (g^{n+1} - g^n, g^{n+1})/||g^n||^2 \; (Polak - Ribière \; update) \tag{149}$$

and then

$$w^{n+1} = g^{n+1} + \gamma_n w^n. \tag{150}$$

Do $n = n+1$ *and return to (145).* $\square$

Remark 13. Suppose that the functional J in (127) is given by (105). We can easily show that

$$< J'(v), w >= a(v, w) - L(w), \; \forall v, w \in V \tag{151}$$

and that algorithm (142)-(150) applied to the minimization of J yields

$$\rho_n = (g^n, w^n)/a(w^n, w^n) \; in \; (145). \tag{152}$$

Consider now algorithm (106)-(113): the orthogonality conditions (117)-(119) imply that

$$\rho_n = ||g^n||^2/a(w^n, w^n) = (g^n, w^n)/a(w^n, w^n) \; in \; (109), \tag{153}$$

$$\gamma_n = ||g^{n+1}||^2/||g^n||^2 = (g^{n+1}, g^{n+1} - g^n)/||g^n||^2 \; in \; (112). \tag{154}$$

It follows from (151)-(154) that algorithms (106)-(113) and (142)-(150) *coincide if J is given by (105).* $\square$

The convergence properties of the Fletcher-Reeves and Polak-Ribière conjugate gradient algorithms have inspired many investigators; let us mention among others Daniel [43], Ortega and Rheinboldt [53, 54, 55], Polak [44] (Chapter 6), Avriel [56] (Chapter 10), Powell [57, 58], Girault and Raviart [59] (Chapter 4) and also two recent references, namely Nocedal [47] and Hiriart-Urruty and Lemarechal [60] (Chapter 2). We found the last two references particularly interesting, since they contain a large number of further references on conjugate gradient algorithms, and also very detailed advices and recipes on the practical implementation of these algorithms, based on three decades of theoretical investigations and computer experiments.

Application to the Advection-Diffusion Problem (91)

In order to solve problem (91) by the least-squares/conjugate gradient techniques discussed in previous subsections, we need to equip V_0 and V_g with an appropriate Hilbertian structure; we chose as scalar product on V_0 and V_g

$$\{\mathbf{v}, \mathbf{w}\} \rightarrow \int_\Omega (\alpha \mathbf{v} \cdot \mathbf{w} + \nu \boldsymbol{\nabla} \mathbf{v} : \boldsymbol{\nabla} \mathbf{w})\, dx,$$

the corresponding norm being, obviously,

$$\mathbf{v} \rightarrow (\int_\Omega (\alpha |\mathbf{v}|^2 + \nu |\boldsymbol{\nabla} \mathbf{v}|^2)\, dx)^{\frac{1}{2}}.$$

To apply the Fletcher-Reeves algorithm (142)-(150) to the solution of the problem (91), (100) we need, in principle, to take as unknown $\tilde{\mathbf{u}} = \mathbf{u} - \tilde{\mathbf{g}}_0$ with some $\tilde{\mathbf{g}}_0 \in V_g$ so that $\mathbf{g}_0 = \tilde{\mathbf{g}}_0|_{\Gamma_0}$, in order to transform the problem (91), (100) into an equivalent problem in V_0; actually, this is not necessary and we can proceed directly with algorithm (142)-(150). We obtain then:

$$\mathbf{u}^0 \in V_g \ \textit{is given}; \tag{155}$$

solve

$$\begin{cases} \mathbf{g}^0 \in V_0, \\ \int_\Omega (\alpha \mathbf{g}^0 \cdot \mathbf{z} + \nu \boldsymbol{\nabla} \mathbf{g}^0 : \boldsymbol{\nabla} \mathbf{z}) dx = < J'(\mathbf{u}^0), \mathbf{z} >, \quad \forall \mathbf{z} \in V_0, \end{cases} \tag{156}$$

and set

$$\mathbf{w}^0 = \mathbf{g}^0. \tag{157}$$

For $n \geq 0$, assuming that $\mathbf{u}^n, \mathbf{g}^n, \mathbf{w}^n$ are known, we obtain $\mathbf{u}^{n+1}, \mathbf{g}^{n+1}, \mathbf{w}^{n+1}$ by:
Step 1: Steepest Descent
Solve

$$\begin{cases} \rho_n \in \mathbb{R}, \\ J(\mathbf{u}^n - \rho_n \mathbf{w}^n) \leq J(\mathbf{u}^n - \rho \mathbf{w}^n), \ \forall \rho \in \mathbb{R} \end{cases} \tag{158}$$

and set

$$\mathbf{u}^{n+1} = \mathbf{u}^n - \rho_n \mathbf{w}^n. \tag{159}$$

Step 2: Testing the convergence and construction of the new descent direction
Solve

$$\begin{cases} \mathbf{g}^{n+1} \in V_0, \\ \int_\Omega (\alpha \mathbf{g}^{n+1} \cdot \mathbf{z} + \nu \nabla \mathbf{g}^{n+1} : \nabla \mathbf{z}) dx = <J'(\mathbf{u}^{n+1}), \mathbf{z}>, \ \forall \mathbf{z} \in V_0; \end{cases} \tag{160}$$

if $\int_\Omega (\alpha |\mathbf{g}^{n+1}|^2 + \nu |\nabla \mathbf{g}^{n+1}|^2) \, dx / \int_\Omega (\alpha |\mathbf{g}^0|^2 + \nu |\nabla \mathbf{g}^0|^2) \, dx \leq \varepsilon^2, \ take \ \mathbf{u} =$
$\mathbf{u}^{n+1};$ *else, compute*

$$\gamma_n = \int_\Omega (\alpha |\mathbf{g}^{n+1}|^2 + \nu |\nabla \mathbf{g}^{n+1}|^2) \, dx / \int_\Omega (\alpha |\mathbf{g}^n|^2 + \nu |\nabla \mathbf{g}^n|^2) \, dx \tag{161}$$

and then

$$\mathbf{w}^{n+1} = \mathbf{g}^{n+1} + \gamma_n \mathbf{w}^n. \tag{162}$$

Do $n = n + 1$ *and return to (158).* $\square$

Calculations of J' and ρ_n when solving the non-linear problem (91), (100).

To compute $J'(\mathbf{u}^n)$ at each iteration in algorithm (155)-(162) when solving the non-linear problem (91), (100), let us follow again the definition (133). For $\mathbf{v} \in V_g$ and $\mathbf{w} \in V_0$, we have

$$\begin{cases} \mathbf{y}(t) \in V_0; \ \forall \mathbf{z} \in V_0 \ we \ have \\ \alpha \int_\Omega \mathbf{y}(t) \cdot \mathbf{z} \, dx + \nu \int_\Omega \nabla \mathbf{y}(t) : \nabla \mathbf{z} \, dx \\ = \alpha \int_\Omega (\mathbf{v} + t\mathbf{w}) \cdot \mathbf{z} \, dx + \nu \int_\Omega \nabla (\mathbf{v} + t\mathbf{w}) : \nabla \mathbf{z} \, dx \\ + \int_\Omega ((\mathbf{v} + t\mathbf{w}) \cdot \nabla)(\mathbf{v} + t\mathbf{w}) \cdot \mathbf{z} \, dx - \int_\Omega \mathbf{f} \cdot \mathbf{z} \, dx - \int_{\Gamma_1} \mathbf{g}_1 \cdot \mathbf{z} \, d\Gamma, \end{cases} \tag{163}$$

and

$$\begin{cases} \mathbf{y} \in V_0; \ \forall \mathbf{z} \in V_0 \ we \ have \\ \alpha \int_\Omega \mathbf{y} \cdot \mathbf{z} \, dx + \nu \int_\Omega \nabla \mathbf{y} : \nabla \mathbf{z} \, dx = \alpha \int_\Omega \mathbf{v} \cdot \mathbf{z} \, dx + \nu \int_\Omega \nabla \mathbf{v} : \nabla \mathbf{z} \, dx \\ + \int_\Omega (\mathbf{v} \cdot \nabla) \mathbf{v} \cdot \mathbf{z} \, dx - \int_\Omega \mathbf{f} \cdot \mathbf{z} \, dx - \int_{\Gamma_1} \mathbf{g}_1 \cdot \mathbf{z} \, d\Gamma. \end{cases} \tag{164}$$

Clearly we have $\mathbf{y}(t) \simeq \mathbf{y} + t \, \delta\mathbf{y}(t)$ where $\delta\mathbf{y}(t)$ is the solution of

$$\begin{cases} \delta\mathbf{y}(t) \in V_0; \ \forall \mathbf{z} \in V_0 \ \textit{we have} \\ \alpha \int_{\Omega} \delta\mathbf{y}(t) \cdot \mathbf{z} \, dx + \nu \int_{\Omega} \boldsymbol{\nabla}\delta\mathbf{y}(t) : \boldsymbol{\nabla}\mathbf{z} \, dx = \alpha \int_{\Omega} \mathbf{w} \cdot \mathbf{z} \, dx + \nu \int_{\Omega} \boldsymbol{\nabla}\mathbf{w} : \boldsymbol{\nabla}\mathbf{z} \, dx \\ \quad + \int_{\Omega} (\mathbf{v} \cdot \boldsymbol{\nabla})\mathbf{w} \cdot \mathbf{z} \, dx + \int_{\Omega} (\mathbf{w} \cdot \boldsymbol{\nabla})\mathbf{v} \cdot \mathbf{z} \, dx + t \int_{\Omega} (\mathbf{w} \cdot \boldsymbol{\nabla})\mathbf{w} \cdot \mathbf{z} \, dx \end{cases} \tag{165}$$

Hence we have the difference

$$\begin{aligned} J(\mathbf{v} + t\mathbf{w}) - J(\mathbf{v}) &= \frac{1}{2} \int_{\Omega} (\alpha|\mathbf{y} + t \, \delta\mathbf{y}(t)|^2 + \nu|\boldsymbol{\nabla}(\mathbf{y} + t \, \delta\mathbf{y}(t))|^2) \, dx \\ &\quad - \frac{1}{2} \int_{\Omega} (\alpha|\mathbf{y}|^2 + \nu|\boldsymbol{\nabla}\mathbf{y}|^2) \, dx \\ &= t \int_{\Omega} (\alpha\mathbf{y} \cdot \delta\mathbf{y}(t) + \nu\boldsymbol{\nabla}\mathbf{y} : \boldsymbol{\nabla}\delta\mathbf{y}(t)) \, dx \\ &\quad + \frac{t^2}{2} \int_{\Omega} (\alpha|\delta\mathbf{y}(t)|^2 + \nu|\boldsymbol{\nabla}\delta\mathbf{y}(t)|^2) \, dx \end{aligned} \tag{166}$$

and

$$< J'(\mathbf{v}), \mathbf{w} > = \lim_{t \to 0} \frac{J(\mathbf{v} + t\mathbf{w}) - J(\mathbf{v})}{t} = \int_{\Omega} (\alpha\mathbf{y} \cdot \delta\mathbf{y}(0) + \nu\boldsymbol{\nabla}\mathbf{y} : \boldsymbol{\nabla}\delta\mathbf{y}(0)) \, dx. \tag{167}$$

Since $\mathbf{y} \in V_0$, we set $\mathbf{z} = \mathbf{y}$ and $t = 0$ in (165) and obtain

$$\begin{aligned} < J'(\mathbf{v}), \mathbf{w} > = \alpha \int_{\Omega} \mathbf{y} \cdot \mathbf{w} \, dx + \nu \int_{\Omega} \boldsymbol{\nabla}\mathbf{y} : \boldsymbol{\nabla}\mathbf{w} \, dx + \int_{\Omega} (\mathbf{v} \cdot \boldsymbol{\nabla})\mathbf{w} \cdot \mathbf{y} \, dx \\ + \int_{\Omega} (\mathbf{w} \cdot \boldsymbol{\nabla})\mathbf{v} \cdot \mathbf{y} \, dx. \end{aligned} \tag{168}$$

Therefore $< J'(\mathbf{v}), \mathbf{w} >$ has a purely integral representation, which is of major importance in view of finite element implementation of algorithm (155)-(162).

Another problem of practical importance is the calculation of ρ_n in (158) when solving the non-linear problem (91), (100). Let $\mathbf{y}^n(\rho)$ be the solution of (98), (99) with $\mathbf{v} = \mathbf{u}^n - \rho\mathbf{w}^n$, then we have $\mathbf{y}^n(0) = \mathbf{y}^n$ and $\mathbf{y}^n(\rho_n) = \mathbf{y}^{n+1}$ and

$$\mathbf{y}^n(\rho) = \mathbf{y}^n - \rho\mathbf{y}_1^n + \rho^2\mathbf{y}_2^n, \tag{169}$$

where $\mathbf{y}_1^n$ and $\mathbf{y}_2^n$ are the solution of

$$\begin{cases} \mathbf{y}_1^n \in V_0; \ \forall \mathbf{z} \in V_0 \ \textit{we have} \\ \int_{\Omega} (\alpha\mathbf{y}_1^n \cdot \mathbf{z} + \nu\boldsymbol{\nabla}\mathbf{y}_1^n : \boldsymbol{\nabla}\mathbf{z}) dx = \alpha \int_{\Omega} \mathbf{w}^n \cdot \mathbf{z} \, dx + \nu \int_{\Omega} \boldsymbol{\nabla}\mathbf{w}^n : \boldsymbol{\nabla}\mathbf{z} \, dx \\ \quad + \int_{\Omega} (\mathbf{u}^n \cdot \boldsymbol{\nabla})\mathbf{w}^n \cdot \mathbf{z} \, dx + \int_{\Omega} (\mathbf{w}^n \cdot \boldsymbol{\nabla})\mathbf{u}^n \cdot \mathbf{z} \, dx, \end{cases} \tag{170}$$

$$\begin{cases} \mathbf{y}_2^n \in V_0; \ \forall \mathbf{z} \in V_0 \ we \ have \\ \displaystyle\int_\Omega (\alpha \mathbf{y}_2^n \cdot \mathbf{z} + \nu \boldsymbol{\nabla} \mathbf{y}_2^n : \boldsymbol{\nabla}\mathbf{z}) dx = \int_\Omega (\mathbf{w}^n \cdot \boldsymbol{\nabla})\mathbf{w}^n \cdot \mathbf{z} \, dx, \end{cases} \tag{171}$$

respectively. Since

$$J(\mathbf{u}^n - \rho \mathbf{w}^n) = \frac{1}{2}\int_\Omega \{\alpha |\mathbf{y}^n(\rho)|^2 + \nu |\boldsymbol{\nabla}\mathbf{y}^n(\rho)|^2\} \, dx, \tag{172}$$

the function $j_n(\rho) = J(\mathbf{u}^n - \rho \mathbf{w}^n)$ is a *quartic polynomial* in ρ; ρ_n is therefore a solution of the *cubic* equation

$$j_n'(\rho) = 0. \tag{173}$$

We shall use the standard *Newton's method* to compute ρ_n from (173), starting from $\rho = 0$. The resulting algorithm is as follows:

$$\rho^0 = 0, \tag{174}$$

and for $k \geq 0$, ρ^k being known,

$$\rho^{k+1} = \rho^k - j_n'(\rho^k)/j_n''(\rho^k). \tag{175}$$

Conjugate gradient algorithm for the non-linear problem (91), (100).

From the above results, algorithm (155)-(162) can be written as follows:

$$\mathbf{u}^0 \in V_g \ is \ given; \tag{176}$$

solve

$$\begin{cases} \mathbf{y}^0 \in V_0; \ \forall \mathbf{z} \in V_0 \ we \ have \\ \displaystyle\alpha \int_\Omega \mathbf{y}^0 \cdot \mathbf{z} \, dx + \nu \int_\Omega \boldsymbol{\nabla}\mathbf{y}^0 : \boldsymbol{\nabla}\mathbf{z} \, dx = \alpha \int_\Omega \mathbf{u}^0 \cdot \mathbf{z} \, dx + \nu \int_\Omega \boldsymbol{\nabla}\mathbf{u}^0 : \boldsymbol{\nabla}\mathbf{z} \, dx \\ \displaystyle + \int_\Omega (\mathbf{u}^0 \cdot \boldsymbol{\nabla})\mathbf{u}^0 \cdot \mathbf{z} \, dx - \int_\Omega \mathbf{f} \cdot \mathbf{z} \, dx - \int_{\Gamma_1} \mathbf{g}_1 \cdot \mathbf{z} \, d\Gamma, \end{cases} \tag{177}$$

and

$$\begin{cases} \mathbf{g}^0 \in V_0; \ \forall \mathbf{z} \in V_0 \ we \ have \\ \displaystyle\int_\Omega (\alpha \mathbf{g}^0 \cdot \mathbf{z} + \nu \boldsymbol{\nabla}\mathbf{g}^0 : \boldsymbol{\nabla}\mathbf{z}) dx = \alpha \int_\Omega \mathbf{y}^0 \cdot \mathbf{z} \, dx + \nu \int_\Omega \boldsymbol{\nabla}\mathbf{y}^0 : \boldsymbol{\nabla}\mathbf{z} \, dx \\ \displaystyle + \int_\Omega (\mathbf{u}^0 \cdot \boldsymbol{\nabla})\mathbf{z} \cdot \mathbf{y}^0 \, dx + \int_\Omega (\mathbf{z} \cdot \boldsymbol{\nabla})\mathbf{u}^0 \cdot \mathbf{y}^0 \, dx; \end{cases} \tag{178}$$

and set

$$\mathbf{w}^0 = \mathbf{g}^0. \tag{179}$$

For $n \geq 0$, assuming that $\mathbf{u}^n$, $\mathbf{y}^n$, $\mathbf{g}^n$, $\mathbf{w}^n$ are known, we obtain $\mathbf{u}^{n+1}$, $\mathbf{y}^{n+1}$, $\mathbf{g}^{n+1}$, $\mathbf{w}^{n+1}$ by:
Solve

$$\begin{cases} \mathbf{y}_1^n \in V_0; \ \forall \mathbf{z} \in V_0 \text{ we have} \\ \displaystyle\int_\Omega (\alpha \mathbf{y}_1^n \cdot \mathbf{z} + \nu \nabla \mathbf{y}_1^n : \nabla \mathbf{z}) dx = \alpha \int_\Omega \mathbf{w}^n \cdot \mathbf{z} \, dx + \nu \int_\Omega \nabla \mathbf{w}^n : \nabla \mathbf{z} \, dx \\ \displaystyle + \int_\Omega (\mathbf{u}^n \cdot \nabla) \mathbf{w}^n \cdot \mathbf{z} \, dx + \int_\Omega (\mathbf{w}^n \cdot \nabla) \mathbf{u}^n \cdot \mathbf{z} \, dx, \end{cases} \tag{180}$$

$$\begin{cases} \mathbf{y}_2^n \in V_0; \ \forall \mathbf{z} \in V_0 \text{ we have} \\ \displaystyle\int_\Omega (\alpha \mathbf{y}_2^n \cdot \mathbf{z} + \nu \nabla \mathbf{y}_2^n : \nabla \mathbf{z}) dx = \int_\Omega (\mathbf{w}^n \cdot \nabla) \mathbf{w}^n \cdot \mathbf{z} \, dx. \end{cases} \tag{181}$$

Define

$$\mathbf{y}^n(\rho) = \mathbf{y}^n - \rho \mathbf{y}_1^n + \rho^2 \mathbf{y}_2^n, \tag{182}$$

$$j_n(\rho) = \frac{1}{2} \int_\Omega \{\alpha |\mathbf{y}^n(\rho)|^2 + \nu |\nabla \mathbf{y}^n(\rho)|^2\} \, dx, \tag{183}$$

and solve the cubic equation

$$j_n^{'}(\rho_n) = 0; \tag{184}$$

we have then

$$\mathbf{u}^{n+1} = \mathbf{u}^n - \rho_n \mathbf{w}^n, \tag{185}$$
$$\mathbf{y}^{n+1} = \mathbf{y}^n(\rho_n). \tag{186}$$

Solve

$$\begin{cases} \mathbf{g}^{n+1} \in V_0; \ \forall \mathbf{z} \in V_0 \text{ we have} \\ \displaystyle\int_\Omega (\alpha \mathbf{g}^{n+1} \cdot \mathbf{z} + \nu \nabla \mathbf{g}^{n+1} : \nabla \mathbf{z}) dx = \alpha \int_\Omega \mathbf{y}^{n+1} \cdot \mathbf{z} \, dx + \nu \int_\Omega \nabla \mathbf{y}^{n+1} : \nabla \mathbf{z} \, dx \\ \displaystyle + \int_\Omega (\mathbf{u}^{n+1} \cdot \nabla) \mathbf{z} \cdot \mathbf{y}^{n+1} \, dx + \int_\Omega (\mathbf{z} \cdot \nabla) \mathbf{u}^{n+1} \cdot \mathbf{y}^{n+1} \, dx. \end{cases} \tag{187}$$

If $\displaystyle\int_\Omega (\alpha |\mathbf{g}^{n+1}|^2 + \nu |\nabla \mathbf{g}^{n+1}|^2) \, dx / \int_\Omega (\alpha |\mathbf{g}^0|^2 + \nu |\nabla \mathbf{g}^0|^2) \, dx \leq \varepsilon^2$, take $\mathbf{u} = \mathbf{u}^{n+1}$; else, compute

$$\gamma_n = \int_\Omega (\alpha |\mathbf{g}^{n+1}|^2 + \nu |\nabla \mathbf{g}^{n+1}|^2)\,dx \Big/ \int_\Omega (\alpha |\mathbf{g}^n|^2 + \nu |\nabla \mathbf{g}^n|^2)\,dx \qquad (188)$$

and then

$$\mathbf{w}^{n+1} = \mathbf{g}^{n+1} + \gamma_n \mathbf{w}^n. \qquad (189)$$

Do $n = n + 1$ and return to (180). □

Remark 14. The linearized advection–diffusion problem (90) can be also solved by a least–squares conjugate gradient method close to the one discussed in this section, but cheaper since the linearity of (90), we have to solve only 2 elliptic systems associated to $\alpha I - \nu \triangle$ per iteration. An interesting alternative is clearly to use a preconditioned GMRES algorithm to solve (90), with $\alpha I - \nu \triangle$ as preconditioner. □

Remark 15. We observe that each iteration of algorithm (176)-(189) requires the solution of three systems of mixed Dirichlet-Neumann boundary value problem associated with the elliptic operator $\alpha I - \nu \triangle$. This number is optimal for a nonlinear problem, since the solution of a linear problem by a least-squares conjugate gradient method requires the solution at each iteration of two linear systems associated with the preconditioning operator.

Another important issue concerning algorithm (176)-(189) is their stopping criterion; we have used

$$J(\mathbf{u}^n)/J(\mathbf{u}^0) \le \varepsilon \qquad (190)$$

with ε of the order of 10^{-6}. □

3.3 Solution of Advection Subproblem

Now we would like to consider the *pure advection problem*

$$\begin{cases} \dfrac{\partial \mathbf{u}}{\partial t} + (\mathbf{V} \cdot \nabla)\mathbf{u} = \mathbf{0} \ in \ \Omega \times (t^n, t^{n+1}), \\ \mathbf{u}(t^n) = \mathbf{u}_0, \ \mathbf{u} = \mathbf{g} \ on \ \Gamma_- \times (t^n, t^{n+1}). \end{cases} \qquad (191)$$

with $\nabla \cdot \mathbf{V} = 0$ and $\partial \mathbf{V}/\partial t = \mathbf{0}$ in $\Omega \times (t^n, t^{n+1})$, $\Gamma_- = \{\mathbf{x} \mid \mathbf{x} \in \Gamma, \mathbf{V}(\mathbf{x}) \cdot \mathbf{n}(\mathbf{x}) < 0\}$ and $\partial \mathbf{g}/\partial t = 0$ on $\Gamma_- \times (t^n, t^{n+1})$. The above problem can be obtained when applying to the Navier-Stokes equations an operator–splitting scheme à la Marchuk-Yanenko to be discussed later.

Solving the *pure advection problem* is a more delicate issue. Clearly, problem (191) can be solved by a *method of characteristics* (see, e.g., Pironneau [10] and Glowinski and Pironneau [103] and the references therein). An easy to implement alternative to the method of characteristics is provided by the *wave-like equation* method below. After translation and dilation on the time axis, each component of $\mathbf{u}$ is a solution of a *transport equation* of the following type:

$$\begin{cases} \dfrac{\partial \varphi}{\partial t} + \mathbf{V} \cdot \boldsymbol{\nabla} \varphi = 0 \ in \ \Omega \times (0,1), \\ \varphi(0) = \varphi_0, \ \varphi = g \ on \ \Gamma_- \times (0,1). \end{cases} \tag{192}$$

Let us follow Dean, Glowinski and Pan [101] to discuss the solution of the transport problem (192). Since each component of $\mathbf{u}$, in equation (191) verifies a transport equation such as (192), we shall focus on the solution of this last equation. The properties $\boldsymbol{\nabla} \cdot \mathbf{V} = 0$ and $\partial \mathbf{V}/\partial t = \mathbf{0}$ on $\Omega \times (0,1)$ (we also have $\partial g/\partial t = 0$ on $\Gamma_- \times (0,1)$) imply that problem (192) is "equivalent" to the (formally) well-posed problem:

$$\begin{cases} \dfrac{\partial^2 \varphi}{\partial t^2} - \boldsymbol{\nabla} \cdot ((\mathbf{V} \cdot \boldsymbol{\nabla} \varphi)\mathbf{V}) = 0 \ in \ \Omega \times (0,1), \\ \varphi(0) = \varphi_0, \ \dfrac{\partial \varphi}{\partial t}(0) = -\mathbf{V} \cdot \boldsymbol{\nabla} \varphi_0, \\ \varphi = g \ on \ \Gamma_- \times (0,1), \ \mathbf{V} \cdot \mathbf{n} \, (\dfrac{\partial \varphi}{\partial t} + \mathbf{V} \cdot \boldsymbol{\nabla} \varphi) = 0 \ on \ (\Gamma \setminus \overline{\Gamma}_-) \times (0,1). \end{cases} \tag{193}$$

Solving the *wave-like equation* (193) by a classical finite element/time stepping method is quite easy since a variational formulation of (193) is given by

$$\begin{cases} \displaystyle\int_\Omega \dfrac{\partial^2 \varphi}{\partial t^2} v \, d\mathbf{x} + \int_\Omega (\mathbf{V} \cdot \boldsymbol{\nabla} \varphi)(\mathbf{V} \cdot \boldsymbol{\nabla} v) \, d\mathbf{x} \\ \qquad + \displaystyle\int_{\Gamma \setminus \overline{\Gamma}_-} \mathbf{V} \cdot \mathbf{n} \varphi_t v \, d\Gamma = 0, \ \forall v \in W_0, \ a.e. \ on \ (0,T), \\ \varphi(0) = \varphi_0, \ \dfrac{\partial \varphi}{\partial t}(0) = -\mathbf{V} \cdot \boldsymbol{\nabla} \varphi_0, \\ \varphi = g \ on \ on \ \Gamma_- \times (0,1), \end{cases} \tag{194}$$

with the test function space W_0 defined by

$$W_0 = \{v | v \in H^1(\Omega), \ v = 0 \ on \ \Gamma_-\}.$$

We observe that $\mathbf{V} \cdot \mathbf{n} \geq 0$ on $\Gamma \setminus \Gamma_-$, implying that the boundary term in (194) is *dissipative*.

Let H_h^1 be a C^0 - conforming *finite element* subspace of $H^1(\Omega)$ as discussed in, e.g., Ciarlet [24, 81]. We define W_{0h} by $W_{0h} = H_h^1 \cap W_0$; we suppose that $\lim_{h \to 0} W_{0h} = W_0$ in the usual finite element sense (see Ciarlet [24, 81]). Next, we define $\tau_1 > 0$ by $\tau_1 = \Delta t/Q_1$, where Q_1 is a positive integer and we discretize problem (194) by

$$\varphi^0 = \varphi_{0h}(\simeq \varphi_0), \tag{195}$$

$$\begin{cases} \displaystyle\int_\Omega (\varphi^{-1} - \varphi^1)v \, d\mathbf{x} = 2\tau_1 \int_\Omega (\mathbf{V}_h \cdot \boldsymbol{\nabla} \varphi^0)v \, d\mathbf{x}, \ \forall v \in W_{0h}, \\ \varphi^{-1} - \varphi^1 \in W_{0h}, \end{cases} \tag{196}$$

and for $q = 0, \ldots, Q_1 - 1$,

$$\begin{cases} \varphi^{q+1} \in H_h^1, \ \varphi^{q+1} = g_h \ \text{on} \ \Gamma_-, \\ \displaystyle\int_\Omega \frac{\varphi^{q+1} + \varphi^{q-1} - 2\varphi^q}{\tau_1^2} v \, d\mathbf{x} + \int_\Omega (\mathbf{V}_h \cdot \boldsymbol{\nabla}\varphi^q)(\mathbf{V}_h \cdot \boldsymbol{\nabla}v) \, d\mathbf{x} \\ + \displaystyle\int_{\Gamma \backslash \Gamma_-} \mathbf{V}_h \cdot \mathbf{n}(\frac{\varphi^{q+1} - \varphi^{q-1}}{2\tau_1}) v \, d\Gamma = 0, \ \forall v \in W_{0h}, \end{cases} \qquad (197)$$

where, in (196) and (197), $\mathbf{V}_h$ and g_h approximate $\mathbf{V}$ and g respectively. Scheme (195)-(197) is a centered scheme which is formally second-order accurate with respect to space and time discretizations. To be stable, scheme (195)-(197) has to verify a condition such as

$$\tau_1 \le ch, \qquad (198)$$

with c of the order of $1/\|\mathbf{V}\|$. If one chooses an appropriate numerical integration method to compute the first and third integrals in (197), the above scheme becomes *explicit*, i.e. φ^{q+1} is obtained via the solution of a linear system with a *diagonal* matrix.

Remark 16. Scheme (195)-(197) does not introduce *numerical dissipation*, unlike the *upwinding* schemes commonly used to solve transport problems like (191) and (192). □

Remark 17. Since the wave equation in (193) is, for arbitrary data, a model for simultaneous transport phenomena in the directions $\mathbf{V}$ and $-\mathbf{V}$, both playing the same role, one has to be aware that the initial condition and the boundary conditions have to be treated very accurately in order to keep at a small level the transport phenomenon taking place in the $-\mathbf{V}$ direction, which is here a numerical artifact. □

Remark 18. In order to show that this projection/wave-like equation method is closely related to the Chorin's projection method [39]. Let us consider the homogeneous boundary condition, $\mathbf{u}|_\Gamma = \mathbf{0}$ for all t and set $Q_1 = 1$ in (195)-(197). Then we have the following scheme:

$$\mathbf{u}^0 = \mathbf{u}_{0h} \ is \ given; \qquad (199)$$

for $n \ge 0$, $\mathbf{u}^n$ being known,

$$\begin{cases} \displaystyle\int_\Omega \frac{\mathbf{u}^{n+1/3} - \mathbf{u}^n}{\Delta t} \cdot \mathbf{v} \, d\mathbf{x} - \int_\Omega p^{n+1/3} \boldsymbol{\nabla} \cdot \mathbf{v} \, d\mathbf{x} = 0, \ \forall \mathbf{v} \in V_{0h}, \\ \displaystyle\int_\Omega q \boldsymbol{\nabla} \cdot \mathbf{u}^{n+1/3} \, d\mathbf{x} = 0, \ \forall q \in L_h^2; \\ \mathbf{u}^{n+1/3} \in V_{0h}, p^{n+1/3} \in L_{0h}^2, \end{cases} \qquad (200)$$

$$\begin{cases} \int_\Omega \dfrac{\mathbf{u}^{n+2/3} - \mathbf{u}^{n+1/3}}{\Delta t} \cdot \mathbf{v}\, dx + \int_\Omega (\mathbf{u}^{n+1/3} \cdot \boldsymbol{\nabla})\mathbf{u}^{n+1/3} \cdot \mathbf{v}\, dx \\[2mm] = -\dfrac{\Delta t}{2} \int_\Omega (\mathbf{u}^{n+1/3} \cdot \boldsymbol{\nabla})\mathbf{u}^{n+1/3}(\mathbf{u}^{n+1/3} \cdot \boldsymbol{\nabla})\mathbf{v}\, dx, \\[2mm] \forall \mathbf{v} \in V_{0h}; \mathbf{u}^{n+2/3} \in V_{0h}. \end{cases} \tag{201}$$

$$\int_\Omega \frac{\mathbf{u}^{n+1} - \mathbf{u}^{n+2/3}}{\Delta t} \cdot \mathbf{v}\, dx + \nu \int_\Omega \boldsymbol{\nabla}\mathbf{u}^{n+1} : \boldsymbol{\nabla}\mathbf{v}\, dx = 0, \ \forall \mathbf{v} \in V_{0h}; \mathbf{u}^{n+1} \in V_{0h}. \tag{202}$$

The difference between the above scheme and the Chorin's projection method is a right-hand-side term in (201) which is a naturally built-in diffusion term only acting in the direction of streamlines. This extra term is also close to the one introduced in streamline-diffusion methods (see, e.g.,Johnson [104]). □

4 Iterative Solution of the Stokes Type Sub–problem

At each full time step of scheme (80)-(88), we have to solve twice the following *generalized Stokes problem*:

$$\begin{cases} \alpha\mathbf{u} - \nu\Delta\mathbf{u} + \nabla p = \mathbf{f} \ in \ \Omega, \\ \nabla \cdot \mathbf{u} = 0 \ in \ \Omega, \\ \mathbf{u} = \mathbf{g}_0 \ on \ \Gamma_0, \ \nu\dfrac{\partial\mathbf{u}}{\partial n} - \mathbf{n}p = \mathbf{g}_1 \ on \ \Gamma_1 \end{cases} \tag{203}$$

with α and ν two positive parameters. Our main goal in this section is to discuss *iterative methods* for the solution of the *generalized Stokes problem* (203).

4.1 Mathematical Properties of the Generalized Stokes Problem

We suppose that in (203), Ω is a *bounded* domain of $\mathbb{R}^d$ (with $d = 2$ or 3, in practice), $\alpha \geq 0$, $\nu > 0$, $\Gamma_0 \cap \Gamma_1 = \emptyset$, $\overline{\Gamma_0 \cup \Gamma_1} = \Gamma$; we suppose also that $\mathbf{f} \in (L^2(\Omega))^d$, $\mathbf{g}_0 = \tilde{\mathbf{g}}_0|_{\Gamma_0}$ with $\tilde{\mathbf{g}}_0 \in (H^1(\Omega))^d$, $\mathbf{g}_1 \in (L^2(\Gamma_1))^d$. If (203) has a solution $\{\mathbf{u}, p\}$ belonging to $(H^1(\Omega))^d \times L^2(\Omega)$, this solution verifies clearly

$$\begin{cases} \mathbf{u} \in V_{g_0}, p \in L^2(\Omega), \\ \displaystyle\int_\Omega (\alpha\mathbf{u} \cdot \mathbf{v} + \nu\nabla\mathbf{u} : \nabla\mathbf{v})dx - \int_\Omega p\nabla \cdot \mathbf{v}dx = \\ \displaystyle\int_\Omega \mathbf{f} \cdot \mathbf{v}dx + \int_{\Gamma_1} \mathbf{g}_1 \cdot \mathbf{v}d\Gamma, \ \forall \mathbf{v} \in V_0, \\ \nabla \cdot \mathbf{u} = 0, \end{cases} \tag{204}$$

where

$$V_0 = \{\mathbf{v}|\mathbf{v} \in (H^1(\Omega))^d, \ \mathbf{v} = \mathbf{0} \ on \ \Gamma_0\}, \qquad (205)$$

$$V_{g_0} = \{\mathbf{v}|\mathbf{v} \in (H^1(\Omega))^d, \ \mathbf{v} = \mathbf{g}_0 \ on \ \Gamma_0\}. \qquad (206)$$

Actually, things would be no more complicated if, in (204), one replaces the linear functional

$$\mathbf{v} \rightarrow \int_\Omega \mathbf{f} \cdot \mathbf{v} dx + \int_{\Gamma_1} \mathbf{g}_1 \cdot \mathbf{v} d\Gamma$$

by $L : (H^1(\Omega))^d \rightarrow \mathbb{R}$, defined as follows

$$L(\mathbf{v}) = \int_\Omega \mathbf{f}_0 \cdot \mathbf{v} dx + \sum_{i=1}^d \int_\Omega \mathbf{f}_i \cdot \frac{\partial \mathbf{v}}{\partial x_i} dx + \int_{\Gamma_1} \mathbf{g}_1 \cdot \mathbf{v} d\Gamma, \qquad (207)$$

with $\mathbf{f}_i \in (L^2(\Omega))^d$, $\forall i=0, 1, \ldots d$; functional L is clearly *linear* and *continuous* over $(H^1(\Omega))^d$. We have then the following theorem of *uniqueness*:

Theorem 2. *Suppose that the above hypotheses on $\alpha, \nu, L, \mathbf{g}_0, \mathbf{g}_1$ hold and that $\{\mathbf{u}, p\}$ is a solution to*

$$\begin{cases} \mathbf{u} \in V_{g_0}, \ p \in L^2(\Omega), \\ \int_\Omega (\alpha \mathbf{u} \cdot \mathbf{v} + \nu \nabla \mathbf{u} : \nabla \mathbf{v}) dx - \int_\Omega p \nabla \cdot \mathbf{v} dx = L(\mathbf{v}), \ \forall \mathbf{v} \in V_0, \qquad (208) \\ \nabla \cdot \mathbf{u} = 0. \end{cases}$$

Then $\{\mathbf{u}, p\}$ is unique in $V_{g_0} \times L^2(\Omega)$ (resp., in $V_{g_0} \times (L^2(\Omega)/\mathbb{R}))$ if $\int_{\Gamma_i} d\Gamma > 0$, $\forall i = 0, 1$ (resp., if $\Gamma_0 = \Gamma$, i.e., $\Gamma_1 = \emptyset$).

The proof of the above theorem can be found in, e.g., Glowinski [4]. *Existence* results for problem (208) will be discussed in Section 4.3.

Remark 19. If $\alpha > 0$, then Theorem 2 still holds if $\Gamma_1 = \Gamma$. □

4.2 The Stokes Operator

We suppose from now on that in addition to Ω *bounded*, we also have $\nu > 0$ and $\alpha \geq 0$ (resp., $\alpha > 0$) if $\int_{\Gamma_0} d\Gamma > 0$ (resp., $\Gamma_0 = \emptyset$). We call, then, *Stokes operator* the linear operator from $L^2(\Omega)$ into $L^2(\Omega)$ defined by

$$Aq = \nabla \cdot \mathbf{u}_q, \ \forall q \in L^2(\Omega), \qquad (209)$$

where, in (209), $\mathbf{u}_q$ is the *unique* solution (from the Lax-Milgram Theorem (e.g., see Section 14 in Glowinski [4] or Ciarlet [24] (Chapter 1)) of the following *linear variational problem* in V_0 (the space V_0 is defined by (205)):

$$\begin{cases} \mathbf{u}_q \in V_0, \\ \displaystyle\int_\Omega (\alpha \mathbf{u}_q \cdot \mathbf{v} + \nu \nabla \mathbf{u}_q : \nabla \mathbf{v}) dx = \int_\Omega q \nabla \cdot \mathbf{v} dx, \ \forall \mathbf{v} \in V_0. \end{cases} \qquad (210)$$

If function q is sufficiently smooth (say $q \in H^1(\Omega)$), then $\mathbf{u}_q$ and q are related by

$$\begin{cases} \alpha \mathbf{u}_q - \nu \Delta \mathbf{u}_q + \nabla q = \mathbf{0} \ in \ \Omega, \\ \mathbf{u}_q = \mathbf{0} \ on \ \Gamma_0, \ \nu \dfrac{\partial \mathbf{u}_q}{\partial n} - nq = \mathbf{0} \ on \ \Gamma_1 \end{cases} \qquad (211)$$

(use the *divergence theorem* to derive (211) from (210)).

Next, we define the (pressure) space P as follows:

$$P = L_0^2(\Omega)(= \{q | q \in L^2(\Omega), \int_\Omega q dx = 0\}) \ if \ \Gamma_0 = \Gamma, \qquad (212)$$

$$P = L^2(\Omega) \ if \ \int_{\Gamma_1} d\Gamma > 0. \qquad (213)$$

One of the key results of this section is provided by the following:

Theorem 3. *Operator A is a strongly elliptic, symmetric automorphism of P (i.e., is a strongly elliptic, symmetric isomorphism from P onto itself).*

PROOF: See, e.g., Glowinski [4].

Remark 20. The *spectral properties* of the *Stokes operator A*, and of related operators, are thoroughly discussed (in the particular case $\alpha = 0$ and $\Gamma_0 = \Gamma$) in two beautiful papers by Crouzeix [61, 62]; the main motivation of the above two references is to provide a detailed analysis of the convergence properties of some of the iterative methods, for solving (203), to be discussed in the following subsections. □

4.3 Existence Results for the Generalized Stokes Problem (208)

We can complete, now, the *uniqueness* Theorem 2; we have thus

Theorem 4. *Suppose that the pressure space P is defined by (212) or (213); suppose also that*

$$\alpha > 0 \ if \ \Gamma_1 = \Gamma, \qquad (214)$$

$$\int_\Gamma \mathbf{g}_0 \cdot \mathbf{n} d\Gamma = 0 \ if \ \Gamma_0 = \Gamma. \qquad (215)$$

Then the generalized Stokes problem (208) has a unique solution in $V_{g_0} \times P$.

PROOF: Let us consider first the following *linear* variational problem

$$\begin{cases} \mathbf{u}_0 \in V_{g_0}, \\ \int_\Omega (\alpha \mathbf{u}_0 \cdot \mathbf{v} + \nu \nabla \mathbf{u}_0 : \nabla \mathbf{v}) dx = L(\mathbf{v}), \ \forall \mathbf{v} \in V_0; \end{cases} \tag{216}$$

problem (216) has a *unique* solution. Suppose now that problem (208) has a solution $\{\mathbf{u}, p\}$ in $V_{g_0} \times P$ (necessarily *unique*, from Theorem 2) and define $\bar{\mathbf{u}}$ by

$$\bar{\mathbf{u}} = \mathbf{u} - \mathbf{u}_0. \tag{217}$$

By subtraction between (208) and (216), the pair $\{\bar{\mathbf{u}}, p\}$ verifies, necessarily

$$\begin{cases} \bar{\mathbf{u}} \in V_0, \ p \in P, \\ \int_\Omega (\alpha \bar{\mathbf{u}} \cdot \mathbf{v} + \nu \nabla \bar{\mathbf{u}} : \nabla \mathbf{v}) dx = \int_\Omega p \nabla \cdot \mathbf{v} dx, \ \forall \mathbf{v} \in V_0, \\ \nabla \cdot \bar{\mathbf{u}} = -\nabla \cdot \mathbf{u}_0; \end{cases} \tag{218}$$

system (217), (218) is clearly equivalent to the generalized Stokes problem (208). Actually, it follows from (218) and from the results of previous subsection that the pressure p (if it exists) verifies

$$Ap = -\nabla \cdot \mathbf{u}_0. \tag{219}$$

Conversely, if equation (219) has a solution p in P and if $\bar{\mathbf{u}}$ is the corresponding solution of problem (210) (i.e., $\bar{\mathbf{u}} = \mathbf{u}_p$) then, the pair $\{\bar{\mathbf{u}} + \mathbf{u}_0, p\}$ is the unique solution of problem (208) in $V_{g_0} \times P$. Thus, the proof of the theorem will be complete if we can show that equation (219) has a solution in P. Since operator A is, from Theorem 3, an isomorphism from P onto P, equation (219) will have a unique solution in P if we can show that its right hand side $-\nabla \cdot \mathbf{u}_0$ belongs to P. If condition (213) holds, this is obviously the case since $\mathbf{u}_0 \in V_{g_0} \subset (H^1(\Omega))^d$ implies $\nabla \cdot \mathbf{u}_0 \in L^2(\Omega)(= P$, in that case). If condition (212) holds we still have $\nabla \cdot \mathbf{u}_0 \in L^2(\Omega)$, and also, from (215), $\int_\Omega \nabla \cdot \mathbf{u}_0 dx = \int_\Gamma \mathbf{g}_0 \cdot \mathbf{n} d\Gamma = 0$, i.e., $\nabla \cdot \mathbf{u}_0 \in L_0^2(\Omega)(= P$, here). The proof of the theorem is complete.

Remark 21. As we shall see in the following subsections, it is possible to solve the generalized Stokes problem (208), via the iterative solution of equation (219), *without knowing explicitly* operator A; similarly, it will not be necessary to know the vector valued function $\mathbf{u}_0$, to solve (208), via (219). All we shall need, is to be able to compute $Aq + \nabla \cdot \mathbf{u}_0, \ \forall q \in L^2(\Omega)$; this can be done relatively easily since, from (216) and previous subsection, we have

$$Aq + \nabla \cdot \mathbf{u}_0 = \nabla \cdot \mathbf{U}_q,$$

where $\mathbf{U}_q$ is the *unique* solution of

$$\begin{cases} \mathbf{U}_q \in V_{g_0}, \\ \displaystyle\int_\Omega (\alpha \mathbf{U}_q \cdot \mathbf{v} + \nu \nabla \mathbf{U}_q : \nabla \mathbf{v})dx = \int_\Omega q\nabla \cdot \mathbf{v}dx + L(\mathbf{v}), \ \forall \mathbf{v} \in V_0, \end{cases}$$

i.e., of an *elliptic system* for the operator $\alpha I - \nu\Delta$. □

4.4 A Saddle-Point Interpretation of the Generalized Stokes Problem

As we shall see in a moment, any pair $\{\mathbf{u}, p\}$ solution of the generalized Stokes problem (208) can be viewed as a *saddle-point* of a well chosen *Lagrangian functional*, defined over $(H^1(\Omega))^d \times L^2(\Omega)$. This interpretation is not necessary to prove the convergence of the various iterative methods to be discussed in the following subsections; what matters really there are the properties of the Stokes operator A defined in Section 4.2.

Let X and Y be two *non-empty* sets and let f be a mapping from $X \times Y$ into $\overline{\mathbb{R}}$, where $\overline{\mathbb{R}} = \mathbb{R} \cup \{+\infty\} \cup \{-\infty\}$. We suppose that f is *proper*, i.e., there exists at least one pair $\{x, y\} \in X \times Y$ so that $f(x, y)$ is *finite*.

Definition 19.1 A pair $\{a, b\}$ is called a *saddle-point* of the functional f over $X \times Y$ if

$$\begin{cases} \{a, b\} \in X \times Y, \ f(a, b) \in \mathbb{R}, \\ f(a, y) \leq f(a, b) \leq f(x, b), \ \forall \{x, y\} \in X \times Y. \end{cases} \tag{220}$$

We associate with the generalized Stokes problem (208) the *Lagrangian functional*

$$\mathcal{L}(\mathbf{v}, q) = \frac{1}{2}\int_\Omega (\alpha|\mathbf{v}|^2 + \nu|\nabla\mathbf{v}|^2)dx - L(\mathbf{v}) - \int_\Omega q\nabla \cdot \mathbf{v}dx; \tag{221}$$

functional $\mathcal{L}$ is C^∞ on $(H^1(\Omega))^d \times L^2(\Omega)$. We have then the following:

Theorem 5. *Suppose that functional $\mathcal{L}$ has a saddle-point $\{\mathbf{u}, p\}$ over $V_{g_0} \times L^2(\Omega)$, i.e.,*

$$\begin{cases} \{\mathbf{u}, p\} \in V_{g_0} \times L^2(\Omega), \\ \mathcal{L}(\mathbf{u}, q) \leq \mathcal{L}(\mathbf{u}, p) \leq \mathcal{L}(\mathbf{v}, p), \ \forall \{\mathbf{v}, q\} \in V_{g_0} \times L^2(\Omega). \end{cases} \tag{222}$$

Then $\{\mathbf{u}, p\}$ is a solution of the Stokes problem (208). Conversely, any solution of (208) belonging to $V_{g_0} \times L^2(\Omega)$ is a saddle-point of $\mathcal{L}$ over $V_{g_0} \times L^2(\Omega)$.

4.5 A Gradient Method for the Generalized Stokes Problem

It follows from Theorem 5 of the previous subsection that any solution of the generalized Stokes problem (208) is also a saddle–point over $V_{g_0} \times L^2(\Omega)$ of the Lagrangian functional defined by (221); conversely, any saddle-point of $\mathcal{L}$ over

$V_{g_0} \times L^2(\Omega)$ is also a solution of the Stokes problem (208). This *equivalence* property implies, among other things, that it makes sense to attempt solving problem (208) by solving the saddle–point problem (222), by the following Uzawa's algorithm:

$$p^0 \in L^2(\Omega), \ given; \tag{223}$$

for $n \geq 0, p^n \in L^2(\Omega)$ *being known, we obtain* $\mathbf{u}^n$ *and* p^{n+1} *via*

$$\begin{cases} \mathbf{u}^n \in V_{g_0}, \\ \displaystyle\int_\Omega (\alpha \mathbf{u}^n \cdot \mathbf{v} + \nu \nabla \mathbf{u}^n : \nabla \mathbf{v}) dx = \int_\Omega p^n \nabla \cdot \mathbf{v} dx + L(\mathbf{v}), \ \forall \mathbf{v} \in V_0, \end{cases} \tag{224}$$

$$p^{n+1} = p^n - \rho \nabla \cdot \mathbf{u}^n. \tag{225}$$

Remark 22. Problem (224) is a *system* for the elliptic operator $\alpha I - \nu \Delta$. If

$$L(\mathbf{v}) = \int_\Omega \mathbf{f} \cdot \mathbf{v} dx + \int_{\Gamma_1} \mathbf{g}_1 \cdot \mathbf{v} d\Gamma,$$

with, for example, $\mathbf{f} \in (L^2(\Omega))^d$ and $\mathbf{g}_1 \in (L^2(\Gamma_1))^d$, respectively, then problem (224) is *equivalent* to solving in V_{g_0} the elliptic system

$$\begin{cases} \alpha \mathbf{u}^n - \nu \Delta \mathbf{u}^n = \mathbf{f} - \nabla p^n \ in \ \Omega, \\ \mathbf{u}^n = \mathbf{g}_0 \ on \ \Gamma_0, \ \nu \dfrac{\partial \mathbf{u}^n}{\partial n} = \mathbf{g}_1 + \mathbf{n} p^n \ on \ \Gamma_1 \end{cases} \tag{226}$$

(the boundary condition on Γ_1 makes sense only if p^n has a *trace* on Γ_1). □

Concerning the convergence of algorithm (223)-(225) we have then

Theorem 6. *Suppose that the parameter* ρ *in (225) satisfies*

$$0 < \rho < 2\nu/d. \tag{227}$$

We have then the following convergence properties for algorithm (223)-(225):

$$\lim_{n \to +\infty} \mathbf{u}^n = \mathbf{u} \ in \ (H^1(\Omega))^d, \tag{228}$$

$$\lim_{n \to +\infty} p^n = p \ in \ L^2(\Omega), \ if \ P = L^2(\Omega), \tag{229}$$

$$\lim_{n \to +\infty} p^n = p + (\int_\Omega p^0 dx)/meas.(\Omega) \ in \ L^2(\Omega), \ if \ P = L_0^2(\Omega), \tag{230}$$

where, in (228)-(230), $\{\mathbf{u}, p\}$ *is the unique solution of the generalized Stokes problem (208) in* $V_{g_0} \times P$.

The proof of the above theorem can be found in, e.g., Glowinski [4], or Glowinski [8] (Theorem 5.11).

Algorithm (223)-(225) can be interpreted as a gradient method by using operator A introduced in Section 4.2. If $\Gamma_0 = \Gamma$, we suppose for simplicity that, in (223), we take $p^0 \in L_0^2(\Omega)(= P$ in that case), implying (from (225), (215)) that $p^n \in L_0^2(\Omega)$, $\forall n \geq 0$. Proceeding as in Section 4.3, we define $\mathbf{u}_0$ by

$$\begin{cases} \mathbf{u}_0 \in V_{g_0}, \\ \int_\Omega (\alpha \mathbf{u}_0 \cdot \mathbf{v} + \nu \nabla \mathbf{u}_0 : \nabla \mathbf{v}) dx = L(\mathbf{v}), \ \forall \mathbf{v} \in V_0. \end{cases} \tag{231}$$

Subtracting (231) to (224) we obtain

$$\begin{cases} \mathbf{u}^n - \mathbf{u}_0 \in V_0, \\ \int_\Omega [\alpha(\mathbf{u}^n - \mathbf{u}_0) \cdot \mathbf{v} + \nu \nabla(\mathbf{u}^n - \mathbf{u}_0) : \nabla \mathbf{v}] dx = \int_\Omega p^n \nabla \cdot \mathbf{v} dx, \ \forall \mathbf{v} \in V_0, \end{cases}$$

which implies, from the definition of operator A (see Section 4.2) that

$$Ap^n = \nabla \cdot (\mathbf{u}^n - \mathbf{u}_0),$$

i.e.,

$$\nabla \cdot \mathbf{u}^n = Ap^n + \nabla \cdot \mathbf{u}_0,$$

which implies in turn that algorithm (223)-(225) is equivalent to

$$p^0 \in P \ is \ given; \tag{232}$$

then for $n \geq 0, p^n \in P$ being known,

$$p^{n+1} = p^n - \rho(Ap^n + \nabla \cdot \mathbf{u}_0). \tag{233}$$

Algorithm (232)-(233) is clearly a *fixed point* method for solving problem (219), namely

$$Ap = -\nabla \cdot \mathbf{u}_0.$$

We introduce now the functional $J_* : P \to \mathbb{R}$ defined by

$$J_*(q) = \frac{1}{2} \int_\Omega (Aq)q dx + \int_\Omega \nabla \cdot \mathbf{u}_0 q dx, \ \forall q \in P. \tag{234}$$

The differential J_*' of functional J_* is given by

$$J_*'(q) = Aq + \nabla \cdot \mathbf{u}_0, \tag{235}$$

implying that algorithms (223)-(225) and (232)-(233) can also be written as follows:

$$p^0 \in P \text{ is given;} \tag{236}$$

and for $n \geq 0$, $p^n \in P$ *being known*

$$p^{n+1} = p^n - \rho J_*'(p^n); \tag{237}$$

algorithm (236)-(237) is clearly a *gradient algorithm, with constant step* ρ, applied to the solution of the minimization problem

$$\begin{cases} p \in P, \\ J_*(p) \leq J_*(q), \ \forall q \in P. \end{cases} \tag{238}$$

Equation (219) is the *Euler-Lagrange equation* associated to the minimization problem (238) and can also be written as

$$J_*'(p) = 0. \tag{239}$$

Actually, the minimization problem (238) is the *dual problem* associated with the *saddle-point problem*

$$\begin{cases} \{\mathbf{u}, p\} \in V_{g_0} \times P, \\ \mathcal{L}(\mathbf{u}, q) \leq \mathcal{L}(\mathbf{u}, p) \leq \mathcal{L}(\mathbf{v}, p), \ \forall \{\mathbf{v}, q\} \in V_{g_0} \times P, \end{cases} \tag{240}$$

with $\mathcal{L}$ still defined by (221); this follows from Glowinski [4] (Chapter 5, Section 20), namely

Theorem 7. *The minimization problem (238) and the dual problem associated with problem (240) coincide.*

4.6 Conjugate Gradient Algorithms for the Generalized Stokes Problem

To apply the *conjugate gradient algorithm* (106)-(113) to the solution of the minimization problem (219), (238), we first equip the space P with the classical scalar product of $L^2(\Omega)$, namely

$$\{q, q'\} \rightarrow \int_\Omega qq' dx, \ \forall \{q, q'\} \in P \times P, \tag{241}$$

and the corresponding norm and then obtain the following conjugate gradient algorithm, a variant of the *Uzawa's algorithm* (223)-(225):

$$p^0 \in P \text{ is given;} \tag{242}$$

solve

$$\begin{cases} \mathbf{u}^0 \in V_{g_0}; \ \forall \mathbf{v} \in V_0 \text{ we have} \\ \displaystyle\int_\Omega (\alpha \mathbf{u}^0 \cdot \mathbf{v} + \nu \nabla \mathbf{u}^0 : \nabla \mathbf{v}) \, dx = L(\mathbf{v}) + \int_\Omega p^0 \nabla \cdot \mathbf{v} dx, \end{cases} \tag{243}$$

compute

$$g^0 = \nabla \cdot \mathbf{u}^0 \qquad (244)$$

and set

$$w^0 = g^0. \qquad (245)$$

For $n \geq 0$, assuming that p^n, g^n, w^n are known, solve

$$\begin{cases} \bar{\mathbf{u}}^n \in V_0, \\ \displaystyle\int_\Omega (\alpha \bar{\mathbf{u}}^n \cdot \mathbf{v} + \nu \nabla \bar{\mathbf{u}}^n : \nabla \mathbf{v}) \, dx = \int_\Omega w^n \nabla \cdot \mathbf{v} dx, \ \forall \mathbf{v} \in V_0, \end{cases} \qquad (246)$$

compute

$$\bar{g}^n = \nabla \cdot \bar{\mathbf{u}}^n, \qquad (247)$$

and then

$$\rho_n = \int_\Omega |g^n|^2 dx / \int_\Omega \bar{g}^n w^n dx. \qquad (248)$$

Update p^n and g^n by

$$\mathbf{u}^{n+1} = \mathbf{u}^n - \rho_n \bar{\mathbf{u}}^n, \qquad (249)$$

$$p^{n+1} = p^n - \rho_n w^n, \qquad (250)$$

$$g^{n+1} = g^n - \rho_n \bar{g}^n. \qquad (251)$$

If $\|g^{n+1}\|_{L^2(\Omega)} / \|g^0\|_{L^2(\Omega)} \leq \varepsilon$ take $p = p^{n+1}$; else, compute

$$\gamma_n = \|g^{n+1}\|^2_{L^2(\Omega)} / \|g^n\|^2_{L^2(\Omega)} \qquad (252)$$

and update w^n via

$$w^{n+1} = g^{n+1} + \gamma_n w^n. \qquad (253)$$

Do $n = n + 1$ and return to (246). $\square$

The rate of convergence of the above conjugate gradient algorithm (242)-(253) has been studied in Glowinski [4] (Chapter V, Section 21).

Algorithms (223)-(225) and (242)-(253) may be slow in practice, particularly for flow at large Reynolds number where $\alpha \sim 1/\triangle t$ is taken very large (to follow the fast dynamics of such flow) and where ν is very small. To explain this behavior let us recall that, from the definition of operator A (see Section 4.2), we have $\mathbf{u}_q = -(\alpha I - \nu \Delta)^{-1} \nabla q$ via (210) and

$$Aq = -\nabla \cdot (\alpha I - \nu \Delta)^{-1} \nabla q, \ \forall q \in P. \qquad (254)$$

Assuming that $(\alpha I - \nu \Delta)^{-1}$ and ∇ *commute* (which is not strictly true, in general) we obtain, from (254),

$$Aq = -\nabla \cdot \nabla(\alpha I - \nu \Delta)^{-1}q = -\Delta(\alpha I - \nu \Delta)^{-1}q,$$

i.e., $A = -\Delta(\alpha I - \nu \Delta)^{-1}$, which implies in turn that

$$A^{-1} = (\alpha I - \nu \Delta)(-\Delta)^{-1} = \alpha(-\Delta)^{-1} + \nu I. \tag{255}$$

Relation (255) shows that if

$$\nu >> \alpha \tag{256}$$

operator A behaves, essentially, like I/ν, explaining why algorithm (242)-(253) has good convergence properties if condition (256) holds. On the other hand, if

$$\alpha >> \nu, \tag{257}$$

we have $A \simeq -\dfrac{1}{\alpha}\Delta$ and therefore operator A is very far from being a multiple of the identity operator, explaining the very slow convergence of the above algorithms (we can expect operator A to have a condition number of the order of h^{-2}, after space discretization, if (257) holds). In Cahouet and Chabard [63], one has considered the case (*without boundary*) where $\Omega = \mathbb{R}^d$ and justified, in some sense, relation (255), whose derivation was quite heuristical. On the basis of these results, we shall assume that A^{-1} behaves like

$$\nu I, \quad if \ \nu >> \alpha, \tag{258}$$

and

$$\begin{cases} \alpha(-\Delta)^{-1} \ (for \ the \ homogeneous \ Dirichlet \ condition \ on \ \Gamma_1 \\ and \ the \ homogeneous \ Neumann \ condition \ on \ \Gamma_0), \ if \ \alpha >> \nu. \end{cases} \tag{259}$$

Relation (259) implies that *preconditioning is necessary if $\alpha >> \nu$*. In order to have a preconditioning operator whose good properties remain uniform when the ratio α/ν varies from 0 to $+\infty$, we suggest to take as preconditioner (as done in Cahouet and Chabard [63], for the case $\Gamma_0 = \Gamma$, $\Gamma_1 = \emptyset$) the *isomorphism S* from P onto P defined by

$$S^{-1} = \alpha(-\Delta)^{-1} + \nu I; \tag{260}$$

in (260), the *Green operator* $(-\Delta)^{-1}$ is associated to the boundary conditions described in (259). The fact that the preconditioning operator S is defined by its inverse does not create practical problems as shown in the following section.

A preconditioned conjugate gradient algorithm

As already observed above, it follows from the properties of operators A that problems (219), (238) can be solved by *conjugate gradient algorithms* when the Hilbert space P is the usual $L^2(\Omega)$-scalar product, namely

$$\{q, q'\} \rightarrow \int_\Omega qq' dx, \ \forall q, q' \in P. \tag{261}$$

In order to avoid the deterioration of the convergence properties, associated with large values of the ratio α/ν, and to keep the convergence as uniform as possible, we suggested to employ as scalar product on space P the one advocated in Cahouet and Chabard [63], namely

$$\{q, q'\} \rightarrow \int_\Omega (Sq)q' dx, \ \forall q, q' \in P, \quad \text{with operator } S \text{ defined, via } S^{-1}, \text{ by (260)}. \tag{262}$$

Using the scalar product (262) leads to the following conjugate gradient algorithm, a sophisticated variant of algorithm (242)-(253):

$$p^0 \in P \text{ is given;} \tag{263}$$

solve

$$\begin{cases} \mathbf{u}^0 \in V_{g_0}; \ \forall \mathbf{v} \in V_0, \\ \int_\Omega [\alpha \mathbf{u}^0 \cdot \mathbf{v} + \nu \nabla \mathbf{u}^0 : \nabla \mathbf{v}] \, dx = L(\mathbf{v}) + \int_\Omega p^0 \nabla \cdot \mathbf{v} dx, \end{cases} \tag{264}$$

and set

$$r^0 = \nabla \cdot \mathbf{u}^0. \tag{265}$$

Solve now

$$\begin{cases} -\Delta \varphi^0 = r^0 \text{ in } \Omega, \\ \dfrac{\partial \varphi^0}{\partial n} = 0 \text{ on } \Gamma_0, \ \varphi^0 = 0 \text{ on } \Gamma_1, \end{cases} \tag{266}$$

if $\int_{\Gamma_i} d\Gamma > 0, \ \forall i = 0, 1;$ *or*

$$\begin{cases} -\Delta \varphi^0 = r^0 \text{ in } \Omega, \\ \dfrac{\partial \varphi^0}{\partial n} = 0 \text{ on } \Gamma, \ \int_\Omega \varphi^0 dx = 0, \end{cases} \tag{267}$$

if $\Gamma_0 = \Gamma;$ *or*

$$\begin{cases} -\Delta \varphi^0 = r^0 \text{ in } \Omega, \\ \varphi^0 = 0 \text{ on } \Gamma, \end{cases} \tag{268}$$

if $\Gamma_1 = \Gamma$. *Then set*

$$g^0 = \nu r^0 + \alpha \boldsymbol{\varphi}^0, \tag{269}$$

$$w^0 = g^0. \tag{270}$$

Then, for $n \geq 0$, *assuming that* p^n, r^n, g^n, w^n *are known, compute* p^{n+1}, r^{n+1}, g^{n+1}, w^{n+1} *as follows:*
    Solve:

$$\begin{cases} \bar{\mathbf{u}}^n \in V_0; \ \forall \mathbf{v} \in V_0, \\ \displaystyle\int_\Omega [\alpha \bar{\mathbf{u}}^n \cdot \mathbf{v} + \nu \nabla \bar{\mathbf{u}}^n : \nabla \mathbf{v}]\, dx = \int_\Omega w^n \nabla \cdot \mathbf{v} dx, \end{cases} \tag{271}$$

and set

$$\bar{r}^n = \nabla \cdot \bar{\mathbf{u}}^n. \tag{272}$$

Compute

$$\rho_n = \int_\Omega r^n g^n dx \Big/ \int_\Omega \bar{r}^n w^n dx, \tag{273}$$

and then

$$\mathbf{u}^{n+1} = \mathbf{u}^n - \rho_n \bar{\mathbf{u}}^n, \tag{274}$$
$$p^{n+1} = p^n - \rho_n w^n, \tag{275}$$
$$r^{n+1} = r^n - \rho_n \bar{r}^n. \tag{276}$$

Solve, next,

$$\begin{cases} -\Delta \bar{\varphi}^n = \bar{r}^n \ in \ \Omega, \\ \dfrac{\partial \bar{\varphi}^n}{\partial n} = 0 \ on \ \Gamma_0, \ \bar{\varphi}^n = 0 \ on \ \Gamma_1, \end{cases} \tag{277}$$

if $\displaystyle\int_{\Gamma_i} d\Gamma > 0$, $\forall i = 0, 1$; *or*

$$\begin{cases} -\Delta \bar{\varphi}^n = \bar{r}^n \ in \ \Omega, \\ \dfrac{\partial \bar{\varphi}^n}{\partial n} = 0 \ on \ \Gamma, \ \displaystyle\int_\Omega \bar{\varphi}^n dx = 0, \end{cases} \tag{278}$$

if $\Gamma_0 = \Gamma$; *or*

$$\begin{cases} -\Delta \bar{\varphi}^n = \bar{r}^n \ in \ \Omega, \\ \bar{\varphi}^n = 0 \ on \ \Gamma, \end{cases} \tag{279}$$

if $\Gamma_1 = \Gamma$. Then, compute

$$g^{n+1} = g^n - \rho_n(\nu\bar{r}^n + \alpha\bar{\varphi}^n). \tag{280}$$

If $\int_\Omega r^{n+1}g^{n+1}dx / \int_\Omega r^0 g^0 dx \leq \varepsilon^2$, take $p = p^{n+1}$; else, compute

$$\gamma_n = \int_\Omega r^{n+1}g^{n+1}dx / \int_\Omega r^n g^n dx, \tag{281}$$

and update w^n by

$$w^{n+1} = g^{n+1} + \gamma_n w^n. \tag{282}$$

Do $n = n + 1$ and return to (271). $\square$

Remark 23. Each iteration of algorithm (263)-(282) requires the solution of *one* elliptic system for the operator $\mathbf{v} \to \alpha\mathbf{v} - \nu\Delta\mathbf{v}$. As already mentioned, for flow at *large Reynolds number* where $\alpha \sim 1/\Delta t$ is large and ν is *small*, the discrete analogues to the above operator are fairly *well conditioned, symmetric and positive definite matrices*, making the *iterative* solution of the corresponding linear systems quite inexpensive. We also have to solve the *Poisson problems* (one among (266), (267), and (268), and another one among (277), (278), and (279)). We shall discuss this aspect of the practical implementation of algorithm (263)-(282) later. Actually, it follows from, e.g., Glowinski [4] (Chapter III, Sections 14.4 and 14.5), that the Poisson problems (266), (268) and (277), (279) are *well-posed* if Ω is bounded. Suppose now that $\Gamma_0 = \Gamma$; assuming that relation (215) holds (which is *necessary* for problem (208) to have a solution), it follows from, e.g., Glowinski [4] (Chapter III, Section 14.3), that the *Poisson-Neumann problem* (267) is well-posed, since (215) implies

$$\int_\Omega \nabla \cdot \mathbf{u}^0 dx = \int_\Gamma \mathbf{g}_0 \cdot \mathbf{n} d\Gamma = 0.$$

A similar result holds for the *Poisson-Neumann problem* (278), since $\bar{\mathbf{u}}^n \in V_0(= (H_0^1(\Omega))^d$, here) implies

$$\int_\Omega \nabla \cdot \bar{\mathbf{u}}^n dx = \int_\Gamma \bar{\mathbf{u}}^n \cdot \mathbf{n} d\Gamma = 0, \ \forall n \geq 0. \qquad \square$$

Remark 24. Algorithm (263)-(282) has proved to be quite effective for solving a large variety of Navier-Stokes problems, for a large range of Reynolds numbers. To be more precise, with ε of the order of 10^{-7} in the stopping criterion, it is very rare that more than *ten* iterations of algorithm (263)-(282) are needed to solve the generalized Stokes problem (208), even for complicated three dimensional flow problems, requiring several million of grid points for the space discretization. This high level of performances definitely justifies the choice of the operator S defined by (260), as *preconditioner*. From this facts, we feel obliged to quote Dennis and Schnabel [64] on the convergence of *conjugate gradient algorithms* (in this quotation p is the *number of iterations* necessary to achieve the convergence and n is the *dimension* of the optimization problem):

> "It is not unusual for strictly convex quadratic arising from discretized partial differential equations to be solved with $p \sim n/10^3$. Such spectacularly successful preconditioning nearly always comes from deep insight into the problem and not from matrix theoretic considerations. They often come from discretizing and solving a simplified problem."

There is nothing to add to the above quotation.   □

5 Finite Element Approximation of the Navier-Stokes Equations

We have discussed in Section 2 the *time discretization* by operator-splitting of the *Navier-Stokes equations* modeling *incompressible viscous flow*, these equations being completed by convenient *initial* and *boundary conditions*. In order to implement on computers the solution methods described in Sections 2, 3, and 4, we still have to address the *space discretization* issue; in this note we will focus on *finite element methods*. There exists a quite large literature concerning the finite element approximation of the Navier-Stokes equations; concentrating on books and review articles, let us mention Temam [19] (Chapter 3), Thomasset [65], Peyret and Taylor [66] (Chapter 7), Glowinski [8] (Chapter 7), Girault and Raviart [59], Cuvelier, Segal and Van Steenhoven [67], Gunzburger [69], Pironneau [10], Fletcher [70, 71], Gunzburger and Nicolaides [72], Fortin [68], Quartapelle [73], Hebeker, Rannacher and Wittum [74], Quarteroni and Valli [42] (Chapter 13), Brenner and Scott [26] (Chapter 11), Marion and Temam [23], Gresho and Sani [75]; the above list is far from complete. The basic reference on the mathematical analysis of finite element approximations for the *steady Navier-Stokes* equations is still Girault and Raviart [59], to be completed by Fortin [68], where finite element approximations not available in Girault and Raviart [59] are discussed. To our knowledge, there is no book form analogue of Girault and Raviart [59], concerning the finite element approximation of the *time dependent Navier-Stokes equations*.

What about the mathematical analysis of solution methods for the Navier-Stokes equations, combining finite element approximations and time discretization by operator-splitting?

There is clearly an abundance of such methods and, indeed, most modern Navier-Stokes solvers use some form of time discretization by operator-splitting in order to treat the *incompressibility condition*. These splitting methods can be roughly divided in two families:

The *first family* of splitting methods for the Navier-Stokes equations is related to those methods described in Section 2. The convergence and stability properties of these methods are discussed in Fernandez–Cara and Beltran [76] and Kloucek and Rys [77], the last article discussing mainly the θ-scheme introduced in Section 2.

The *second family* is related to the splitting methods of Marchuk and Yanenko - also known as *fractional step methods* - for which basic references are Yanenko [27] and Marchuk [28, 29]. These methods have been applied to the solution of the Navier-Stokes equations for incompressible viscous fluid flow by, e.g., Chorin [38, 39] and Temam [40, 41], the space discretization being by *finite differences* in the above references. A thorough discussion of fractional step methods for the Navier-Stokes equations can be found in Temam [19] (Chapter 3) and Marion and Temam [23] (Chapter 3) (see also the references therein).

In the present section, we shall focus on implementation issues when the θ-schemes of Section 2 is combined with *low-order finite element approximations* à la Bercovier-Pironneau (see Bercovier and Pironneau [79]) and *Hood-Taylor* (see Hood and Taylor [78]). In the later section, we will focus on implementation issues related to the splitting methods of Marchuk and Yanenko with the Bercovier-Pironneau finite element method. We are giving a special attention to the Hood and Taylor finite element methods for the following reasons:

(i) They are easy to implement, particularly in combination with the *time discretizations by operator-splitting* described in Section 2, the *least-squares/conjugate gradient algorithms* described in Section 3 and the *Stokes solvers* discussed in Section 4.

(ii) They are at the basis of some production codes for the simulation of incompressible viscous fluid flow, such as *N3S* developed by *Electricité de France* (EDF) and *FASTFLO* developed by the *CSIRO*, in Australia.

5.1 Finite Element Methods for the Stokes Problem

Some observations.

It is a fairly general opinion that the main difficulty related to the space approximation of the Navier-Stokes equations, in the *pressure-velocity* formulation, is the treatment of the *incompressibility condition*

$$\nabla \cdot \mathbf{u} = 0. \tag{283}$$

In order to show that the *boundary conditions* play also a role in these difficulties, let us consider first the *periodic Stokes problem*,

$$\begin{cases} \alpha\mathbf{u} - \nu\Delta\mathbf{u} + \nabla p = \mathbf{f} \text{ in } \Omega, \\ \nabla \cdot \mathbf{u} = 0 \text{ in } \Omega, \\ \mathbf{u}, \ \nabla\mathbf{u} \text{ and } p \text{ periodic at } \Gamma, \end{cases} \tag{284}$$

with $\alpha > 0$, $\nu > 0$, $\Omega = (0,1)^d$ and $\Gamma = \partial\Omega$; in the present context, we say that a function v is *periodic at Γ* if

$$\begin{cases} v(x_1, ...x_{i-1}, 0, x_{i+1}, ...x_d) = v(x_1, ...x_{i-1}, 1, x_{i+1}, ...x_d), \\ \forall i = 1, ...d, \ \forall x_j \in (0,1), \ \forall j = 1, ...d, \ j \neq i. \end{cases} \tag{285}$$

Solving problem (284) is quite easy; we compute first the *pressure p* from

$$\begin{cases} \Delta p = \boldsymbol{\nabla} \cdot \mathbf{f} \ in \ \Omega, \\ p, \ \boldsymbol{\nabla} p \ periodic \ at \ \Gamma, \end{cases} \tag{286}$$

and then the velocity $\mathbf{u}$ from

$$\begin{cases} \alpha\mathbf{u} - \nu\Delta\mathbf{u} = \mathbf{f} - \boldsymbol{\nabla}\mathbf{p} \ in \ \Omega, \\ \mathbf{u}, \ \boldsymbol{\nabla}\mathbf{u} \ \textbf{periodic at } \Gamma. \end{cases} \tag{287}$$

Suppose that $\mathbf{f}$ is *sufficiently smooth* and is also *periodic* at Γ; then, problems (286) and (287) are *well-posed* in $H^1(\Omega)/\mathbb{R}$ and $(H^1(\Omega))^d$, respectively. Now, denote $\boldsymbol{\nabla} \cdot \mathbf{u}$ by φ; it follows from (286), (287) that φ verifies

$$\begin{cases} \alpha\varphi - \nu\Delta\varphi = 0 \ in \ \Omega, \\ \varphi \ and \ \boldsymbol{\nabla}\varphi \ periodic \ at \ \Gamma, \end{cases} \tag{288}$$

whose unique solution is $\varphi = 0$, i.e., $\boldsymbol{\nabla} \cdot \mathbf{u} = \mathbf{0}$ on Ω. We have thus shown that problem (284) has a unique solution in $(H^1(\Omega))^d \times (H^1(\Omega)/\mathbb{R})$; this solution can be obtained via the solution of problems (286), (287) which are quite classical *elliptic* problems. Variational formulations for problems (286), (287) are given by

$$\begin{cases} p \in H_P^1(\Omega), \\ \displaystyle\int_\Omega \boldsymbol{\nabla} p \cdot \boldsymbol{\nabla} q dx = \int_\Omega \mathbf{f} \cdot \boldsymbol{\nabla} q dx, \ \forall q \in H_P^1(\Omega), \end{cases} \tag{289}$$

$$\begin{cases} \mathbf{u} \in (\mathbf{H}_\mathbf{P}^1(\Omega))^\mathbf{d}, \\ \alpha \displaystyle\int_\Omega \mathbf{u} \cdot \mathbf{v}dx + \nu \int_\Omega \boldsymbol{\nabla}\mathbf{u} : \boldsymbol{\nabla}\mathbf{v}dx = \int_\Omega \mathbf{f} \cdot \mathbf{v}dx + \int_\Omega \mathbf{p}\boldsymbol{\nabla} \cdot \mathbf{v}dx, \\ \qquad\qquad \forall\mathbf{v} \in (H_P^1(\Omega))^d, \end{cases} \tag{290}$$

respectively, with, in (289), (290), H_P^1 defined by

$$H_P^1(\Omega) = \{q | q \in H^1(\Omega), \ q \ periodic \ at \ \Gamma\}. \tag{291}$$

Solving problem (284), by *Galerkin type methods*, via the equivalent variational formulation (289), (290) is quite easy. We introduce first two families

$\{P_h\}_h$ and $\{V_h\}_h$ of *finite dimensional spaces*; we suppose that these families verify

$$P_h \subset H^1_P(\Omega), \quad \forall h, \; V_h \subset (H^1_P(\Omega))^d, \; \forall h, \tag{292}$$

$$\forall q \in H^1_P(\Omega), \quad \exists \{q_h\}_h : \; q_h \in P_h,$$
$$\forall h, \lim_{h \to 0} ||q_h - q||_{H^1(\Omega)} = 0, \tag{293}$$

$$\forall \mathbf{v} \in (H^1_P(\Omega))^d, \; \exists \{\mathbf{v}_h\}_h : \; \mathbf{v}_h \in V_h,$$
$$\forall h, \lim_{h \to 0} ||\mathbf{v}_h - \mathbf{v}||_{(H^1(\Omega))^d} = 0. \tag{294}$$

Starting from the variational formulation (289), (290), we approximate problem (284) by

$$\begin{cases} p_h \in P_h, \\ \displaystyle\int_\Omega \boldsymbol{\nabla} p_h \cdot \boldsymbol{\nabla} q_h \, dx = \int_\Omega \mathbf{f}_h \cdot \boldsymbol{\nabla} q_h \, dx, \; \forall q_h \in P_h, \end{cases} \tag{295}$$

$$\begin{cases} \mathbf{u_h} \in V_h, \\ \displaystyle\int_\Omega (\alpha \mathbf{u_h} \cdot \mathbf{v_h} + \nu \boldsymbol{\nabla} \mathbf{u_h} : \boldsymbol{\nabla} \mathbf{v_h}) d\mathbf{x} = \int_\Omega \mathbf{f_h} \cdot \mathbf{v_h} d\mathbf{x} + \int_\Omega p_\mathbf{h} \boldsymbol{\nabla} \cdot \mathbf{v_h} d\mathbf{x}, \\ \qquad \forall \mathbf{v}_h \in V_h, \end{cases} \tag{296}$$

where, in (295), (296), $\mathbf{f}_h$ is an approximation of $\mathbf{f}$ such that $\lim_{h \to 0} ||\mathbf{f}_h - \mathbf{f}||_{(L^2(\Omega))^d} = 0$.

It is a fairly easy exercise to prove that problems (295) and (296) are well-posed in $P_h/\mathbb{R}$ and V_h, respectively, and also that

$$\lim_{h \to 0} \{\mathbf{u_h}, \mathbf{p_h}\} = \{\mathbf{u}, \mathbf{p}\} \text{ in } (\mathbf{H}^1(\Omega))^{\mathbf{d+1}}, \tag{297}$$

where, in (297), $\{\mathbf{u}, \mathbf{p}\}$ is a solution of problem (284); to prove the convergence result (297) we can use the techniques discussed in, e.g., Strang and Fix [80], Ciarlet [24], Raviart and Thomas [37], Glowinski [8] (Appendix 1), Ciarlet [81] (Chapter 3) and Brenner and Scott [26] (Chapter 5).

From the above results, it appears that approximating the "periodic" Stokes problem (284) is a rather simple issue. Indeed, we can combine any *pressure* approximation to any *velocity* one, as long as properties (292)-(294) are verified. Thus, pressure and velocity approximations can be of different nature, use different meshes and/or basis functions, etc. On the other hand, as we shall see in the following section, approximating the *Stokes-Dirichlet* problem

$$\begin{cases} \alpha \mathbf{u} - \nu \Delta \mathbf{u} + \boldsymbol{\nabla} \mathbf{p} = \mathbf{f} \text{ in } \Omega, \\ \boldsymbol{\nabla} \cdot \mathbf{u} = \mathbf{0} \text{ in } \Omega, \\ \mathbf{u} = \mathbf{g} \text{ on } \Gamma \; (\text{with } \displaystyle\int_\Gamma \mathbf{g} \cdot \mathbf{n} d\Gamma = \mathbf{0}), \end{cases} \tag{298}$$

or the *Stokes-Neumann* problem

$$\begin{cases} \alpha\mathbf{u} - \nu\Delta\mathbf{u} + \boldsymbol{\nabla}\mathbf{p} = \mathbf{f} \text{ in } \Omega, \\ \boldsymbol{\nabla} \cdot \mathbf{u} = \mathbf{0} \text{ in } \Omega, \\ \nu\dfrac{\partial\mathbf{u}}{\partial n} - \mathbf{n}p = \mathbf{g} \text{ on } \Gamma, \end{cases} \qquad (299)$$

is a much more complicated matter, since *compatibility conditions* between the velocity and pressure approximations seem to be required if one wants to avoid spurious oscillations. In Glowinski [82] (Section 5.2), the mechanism producing numerical instabilities has been investigated on a particular case of the *Stokes-Dirichlet* problem (298) where $\Omega = (0,1) \times (0,1)$ via *Fourier Analysis*. To overcome these numerical instabilities we can use one of the following approaches

(a) *Use different type of approximations for pressure and velocity.*
(b) *Use the same type of approximation for pressure and velocity, combined with a regularization procedure.*

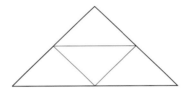

Fig. 5.1. Dividing $T \in \mathcal{T}_h$ to define $\mathcal{T}_{h/2}$

Approach (a) is well known and will be further discussed in this section. The main idea here is to construct pressure spaces which are "poor" in high frequency modes, compared to the velocity space. Figure 5.1 suggests an obvious remedy to spurious oscillations which is to use a pressure grid which is *twice coarser* than the velocity one, and then use approximations of the same type on both grids. This observation makes sense for finite difference, finite element, spectral, pseudo-spectral, and wavelet approximations of problem (298); the well-known (and converging) finite element method (introduced in Bercovier and Pironneau [79]) obtained by using a *continuous piecewise linear* approximation of the *pressure* (resp., of the *velocity*) on a triangulation $\mathcal{T}_h$ (resp., $\mathcal{T}_{h/2}$, obtained from $\mathcal{T}_h$ by joining as shown in Figure 5.1 the midpoints in any $T \in \mathcal{T}_h$) definitely follows the above rule. Beside the above reference, this method is discussed in, e.g., Glowinski [8] (Chapter 7), Glowinski [32, 33, 82], Bristeau, Glowinski, Mantel, Periaux and Perrier [9], Girault and Raviart [59], Bristeau, Glowinski and Periaux [83], Dean, Glowinski and Li [84], Pironneau [10], Gunzburger [69], Brezzi and Fortin [25], Glowinski and Pironneau [85], Fortin [68] (some of the above references show also numerical results obtained with it). Actually, the *Bercovier-Pironneau method* is a simple variation (easier to implement but less accurate) of the celebrated *Hood-Taylor method* (introduced in Hood and Taylor [78]) where *pressure* and *velocity* are approximated on the same triangulation by continuous approximations which are *piecewise linear* and *piecewise quadratic*, respectively.

Approach (b), introduced in Hughes, Franca and Balestra [86] (see also Douglas and Wang [87], Fortin [68], Cai and Douglas [88] and the references therein) leads essentially to *Tychonoff regularization procedures*, an obvious one being to "regularize" (one also says "stabilize") equation (219) by the following problem (written in *variational* form)

$$\begin{cases} p_\varepsilon \in H^1(\Omega), \\ \varepsilon \int_\Omega \boldsymbol{\nabla} p_\varepsilon \cdot \boldsymbol{\nabla} q dx + \int_\Omega (Ap_\varepsilon)q dx = -\int_\Omega \boldsymbol{\nabla} \cdot \mathbf{u_0 q} dx, \ \forall \mathbf{q} \in \mathbf{H^1}(\Omega), \end{cases} \quad (300)$$

where, in (300), ε is a *positive* parameter. Very good results have been obtained with approach (b) (see, e.g., Hughes, Franca and Balestra [86]), however, we prefer approach (a) for the following reasons:

(i) It is *parameter free*, unlike the second approach which requires the adjustment of the regularization parameter.

(ii) Quite often, the mesh size is adjusted, globally or locally, on the basis of the velocity behavior (boundary and shear layer thickness, for example). Therefore, compared to approach (a), approach (b) will be four times more costly (eight times for three-dimensional problems) from the pressure point of view, without further gains in accuracy.

(iii) Tychonoff regularization procedures are systematic methods for stabilizing ill-posed problems; in most cases, the adjustment of the regularization parameter is a delicate problem in itself, therefore, if there exist alternatives which are parameter free, we definitely think that the latter are preferable, particularly if they are based on an analysis of the mechanism producing the unwanted oscillations. Actually, the author of this article is a strong believer of Tychonoff regularization procedures when there is no alternative available to stabilize an ill-posed problem; indeed, we have been using such a procedure to solve *boundary control problems* for the *wave equation* (see Dean, Glowinski and Li [84], Glowinski, Li, and Lions [90]); however, as a consequence of our investigations concerning the Stokes problem, we have introduced, in Glowinski and Li [89], new solution methods for the above control problems which are more efficient than those discussed in Dean, Glowinski and Li [84], and Glowinski, Li, and Lions [90] (results obtained with the new method are also shown in Glowinski and Lions [91]).

Discrete spaces

We suppose that Ω is a bounded polygonal domain of $\mathbb{R}^2$ (cases where domain has a curved boundary have been discussed in Glowinski [4] (Chapter 5)). With $\mathcal{T}_h$ a standard finite element *triangulation* of Ω (see, e.g., Ciarlet [24, 81], Raviart and Thomas [37], Glowinski [8] (Appendix 1) for this notion) and h the maximal length of the edges of $\mathcal{T}_h$, we introduce the following discrete spaces (with P_k the space of the polynomials in two variables of degree $\leq k$):

$$P_h = \{q_h | q_h \in C^0(\overline{\Omega}), \ q_h|_T \in P_1, \ \forall T \in \mathcal{T}_h\}, \tag{301}$$

$$V_h = \{\mathbf{v}_h | \mathbf{v}_h \in (C^0(\overline{\Omega}))^2, \mathbf{v}_h|_T \in (P_2)^2, \ \forall T \in \mathcal{T}_h\}. \tag{302}$$

If the *boundary conditions* imply $\mathbf{u} = \mathbf{g_0}$ on Γ_0, we shall need the space V_{0h} defined by

$$V_{0h} = \{\mathbf{v}_h | \mathbf{v}_h \in V_h, \ \mathbf{v}_h = \mathbf{0} \text{ on } \Gamma\}, \ if \ \Gamma_0 = \Gamma, \tag{303}$$

and by

$$V_{0h} = \{\mathbf{v}_h | \mathbf{v}_h \in V_h, \ \mathbf{v}_h = \mathbf{0} \text{ on } \Gamma_0\}, \ if \ \int_{\Gamma_0} d\Gamma > 0, \Gamma_0 \neq \Gamma; \tag{304}$$

if we are in the situation associated with (303) it is of fundamental importance to have the points at the interface of Γ_0 and $\Gamma_1(= \Gamma \backslash \Gamma_0)$ as vertices of $\mathcal{T}_h$.

Another useful variant of V_h (and then V_{0h}), the *Bercovier-Pironneau* velocity space, is obtained as follows:

$$V_h = \{\mathbf{v}_h | \mathbf{v}_h \in (C^0(\overline{\Omega}))^2, \ \mathbf{v}_h|_T \in (P_1)^2, \ \forall T \in \mathcal{T}_{h/2}\}. \tag{305}$$

In (305), $\mathcal{T}_{h/2}$ is (as in previous subsection) the triangulation of Ω obtained from $\mathcal{T}_h$ by joining the mid-points of the edges of $T \in \mathcal{T}_h$ (see Figure 5.1); for the same triangulation $\mathcal{T}_h$, we have the same global number of degrees of freedom if we use V_h defined by either (302) or (305), space P_h being the same; however, the matrices encountered in the second case are more compact and sparser.

Remark 25. For obvious reasons, the finite element approximations of the Stokes problem based on the pair $\{P_h, V_h\}$ defined by (301), (302) (resp., (301), (305)) is called a P_1/P_2 *approximation* (resp., a P_1-*iso*-P_2/P_1 *approximation*). $\square$

Approximation of the Boundary Conditions

If the boundary conditions are defined by

$$\mathbf{u} = \mathbf{g} \text{ on } \Gamma, \text{ with } \int_\Gamma \mathbf{g} \cdot \mathbf{n} d\Gamma = 0, \tag{306}$$

it is of fundamental importance to approximate $\mathbf{g}$ by $\mathbf{g}_h$ so that

$$\int_\Gamma \mathbf{g}_h \cdot \mathbf{n} d\Gamma = 0. \tag{307}$$

We shall discuss the simple case where Ω is *polygonal* domain. For the curved boundary case, one can employ the methods discussed in Glowinski [3] (Chapter 5) (actually these methods have proved useful when solving *free surface* problems as shown in Glowinski and Juarez [102]).

We suppose that $\mathbf{g}$ is *continuous* on Γ. Then we have that $\mathbf{n}$ will be *piecewise constant* on Γ. Starting from V_h defined by either (302) or (305), we define the *boundary space* γV_h by

$$\gamma V_h = \{\boldsymbol{\mu}_h | \boldsymbol{\mu}_h = \mathbf{v}_h|_\Gamma, \ \mathbf{v}_h \in V_h\}, \tag{308}$$

i.e., $gammaV_h$ is the space of the traces on Γ of the functions $\mathbf{v}_h$ belonging to V_h. Actually, if V_h is defined by (302), γV_h is also the space of the functions *continuous* over Γ, taking their values in $\mathbb{R}^2$ and *quadratic* over the edges of $\mathcal{T}_h$ contained in Γ; similarly, if V_h is defined by (305) we have

$$\gamma V_h = \{\boldsymbol{\mu}_h | \boldsymbol{\mu}_h \in (C^0(\Gamma))^2, \ \boldsymbol{\mu}_h \ is \ \ affine \ over \ the \ edges \ of \tag{309}$$
$$\mathcal{T}_{h/2} \ contained \ in \ \Gamma\}. \tag{310}$$

Our problem is to construct an approximation $\mathbf{g}_h$ of $\mathbf{g}$ such that

$$\mathbf{g}_h \in gammaV_h, \ \int_\Gamma \mathbf{g}_h \cdot \mathbf{n} d\Gamma = 0. \tag{311}$$

If $\pi_h \mathbf{g}$ is the unique unique element of γV_h, obtained by piecewise linear or piecewise quadratic interpolation of $\mathbf{g}$ over Γ, i.e., Γ, i.e., obtained from the values taken by $\mathbf{g}$ at those vertices of $\mathcal{T}_h$ (or $\mathcal{T}_{h/2}$) belonging to Γ, we usually have $\int_\Gamma \pi_h \mathbf{g} \cdot \mathbf{n} d\Gamma \neq 0$. To overcome this difficulty we may proceed as follows:

(i) We define an approximation $\mathbf{n}_h$ of $\mathbf{n}$ as the solution of the following *linear variational problem* in $gammaV_h$

$$\begin{cases} \mathbf{n}_h \in gammaV_h, \\ \int_\Gamma \mathbf{n}_h \cdot \boldsymbol{\mu}_h d\Gamma = \int_\Gamma \mathbf{n} \cdot \boldsymbol{\mu}_h d\Gamma, \ \forall \boldsymbol{\mu}_h \in gammaV_h. \end{cases} \tag{312}$$

Problem (312) is *equivalent* to a linear system whose matrix is *sparse, symmetric positive definite*, very *well-conditioned* and *easy to compute* (also, problem (312) needs to be solved *only once* if the finite element mesh is *fixed*). Observe also that the fact that $\mathbf{n}$ is constant, on each edge of $\mathcal{T}_h$ contained in Γ, makes the calculation of the right hand side of the above equivalent linear system very easy (the details can be found in Glowinski [4] (Chapter 5)).

(ii) Define $\mathbf{g}_h$ by

$$\mathbf{g}_h = \pi_h \mathbf{g} - \left(\int_\Gamma \pi_h \mathbf{g} \cdot \mathbf{n} d\Gamma / \int_\Gamma \mathbf{n} \cdot \mathbf{n}_h d\Gamma \right) \mathbf{n}_h. \tag{313}$$

It is easy to check that (312), (313) imply that $\mathbf{g}_h$ verifies the *flux condition* (311).

Formulation of the Discrete Stokes Problem

In the following we shall denote by Ω_h the *computational domain* and by Γ_h its boundary even though we have considered here that Ω is *polygonal* and hence $\Omega_h = \Omega$ and $\Gamma_h = \Gamma$.

The Dirichlet case.

The *Stokes problem*, considered here, has the following formulation:

$$\begin{cases} \alpha \mathbf{u} - \nu \Delta \mathbf{u} + \nabla p = \mathbf{f} \text{ in } \Omega, \\ \nabla \cdot \mathbf{u} = 0 \text{ in } \Omega, \\ \mathbf{u} = \mathbf{g} \text{ on } \Gamma, \end{cases} \tag{314}$$

with $\mathbf{f} \in (H^{-1}(\Omega))^d$ and $\mathbf{g} \in (H^{1/2}(\Gamma))^d$, with $\int_\Gamma \mathbf{g} \cdot \mathbf{n} d\Gamma = 0$. It follows from Section 4, that problem (314) has a *unique solution* in $V_g \times (L^2(\Omega)/\mathbb{R})$, with

$$V_g = \{\mathbf{v} | \mathbf{v} \in (H^1(\Omega))^d, \ \mathbf{v} = \mathbf{g} \text{ on } \Gamma\}. \tag{315}$$

Problem (314) can also be formulated as

$$\begin{cases} \mathbf{u} \in \mathbf{V_g}, \ p \in \mathbf{L^2}(\Omega), \\ \alpha \int_\Omega \mathbf{u} \cdot \mathbf{v} dx + \nu \int_\Omega \nabla \mathbf{u} :: \nabla \mathbf{v} dx - \int_\Omega p \nabla \cdot \mathbf{v} dx = <\mathbf{f}, \mathbf{v}>, \ \forall \mathbf{v} \in \mathbf{V_0}, \\ \int_\Omega q \nabla \cdot \mathbf{u} dx = 0, \ \forall q \in \mathbf{L^2}(\Omega), \end{cases} \tag{316}$$

where, in (316), the *test function space* V_0 is defined by

$$V_0 = (H_0^1(\Omega))^d, \tag{317}$$

and where $< \cdot, \cdot >$ denotes the duality pairing between $(H^{-1}(\Omega))^d$ and V_0.
Next, let us define V_{0h} and V_{g_h} by

$$V_{0h} = \{\mathbf{v}_h | \mathbf{v}_h \in V_h, \ \mathbf{v}_h = \mathbf{0} \text{ on } \Gamma_h\} \tag{318}$$
$$V_{g_h} = \{\mathbf{v}_h | \mathbf{v}_h \in V_h, \ \mathbf{v}_h = \mathbf{g}_h \text{ on } \Gamma_h\} \tag{319}$$

with, in (318) and (319), V_h and $\mathbf{g}_h$ defined as in Sections 5.1 and 5.1, respectively; we have in particular $\int_{\Gamma_h} \mathbf{g}_h \cdot \mathbf{n}\, d\Gamma_h = 0$. We approximate the *Stokes-Dirichlet problem* (314) by

$$
\begin{cases}
\mathbf{u_h} \in V_{\mathbf{g_h}},\ \mathbf{p_h} \in P_h, \\
\alpha \displaystyle\int_{\Omega_h} \mathbf{u_h} \cdot \mathbf{v_h dx} + \nu \int_{\Omega_h} \boldsymbol{\nabla}\mathbf{u_h} : \boldsymbol{\nabla}\mathbf{v_h dx} - \int_{\Omega_h} \mathbf{p_h}\boldsymbol{\nabla} \cdot \mathbf{v_h dx} \\
\qquad =< \mathbf{f}_h, \mathbf{v}_h >_h,\ \forall \mathbf{v}_h \in V_{0h}, \\
\displaystyle\int_{\Omega_h} q_h \boldsymbol{\nabla} \cdot \mathbf{u_h dx} = 0,\ \forall \mathbf{q_h} \in P_h
\end{cases}
\tag{320}
$$

with the space P_h as in Sections 5.1; in (320) $\mathbf{f}_h$ is an approximation of $\mathbf{f}$ and $< \cdot, \cdot >_h$ denotes the duality pairing between $(H^{-1}(\Omega_h))^d$ and $(H_0^1(\Omega_h))^d$. The *well-posedness* of problem (320) will be addressed in next subsection, which will contain also some comments on the *convergence* of the pair $\{\mathbf{u_h}, \mathbf{p_h}\}$ as $h \to 0$.

The case of the mixed boundary conditions.

The *Stokes problem*, considered now, has the following formulation

$$
\begin{cases}
\alpha\mathbf{u} - \nu\Delta\mathbf{u} + \boldsymbol{\nabla}\mathbf{p} = \mathbf{f} \text{ in } \Omega, \\
\boldsymbol{\nabla} \cdot \mathbf{u} = \mathbf{0} \text{ in } \Omega, \\
\mathbf{u} = \mathbf{g}_0 \text{ on } \Gamma_0,\ \nu\dfrac{\partial\mathbf{u}}{\partial\mathbf{n}} - \mathbf{n}p = \mathbf{g}_1 \text{ on } \Gamma_1.
\end{cases}
\tag{321}
$$

In order to avoid too many technicalities we shall assume that, in (321), we have $\mathbf{g}_0 = \tilde{\mathbf{g}}_0|_{\Gamma_0}$ with $\tilde{\mathbf{g}}_0 \in (H^1(\Omega))^d$, $\mathbf{g}_1 \in (L^2(\Gamma_1))^d$, and $\mathbf{f} \in (L^2(\Omega))^d$. A *variational formulation* of problem (321) is given by

$$
\begin{cases}
\mathbf{u} \in V_{\mathbf{g_0}},\ \mathbf{p} \in \mathbf{L}^2(\Omega), \\
\alpha \displaystyle\int_\Omega \mathbf{u} \cdot \mathbf{v dx} + \nu \int_\Omega \boldsymbol{\nabla}\mathbf{u} : \boldsymbol{\nabla}\mathbf{v dx} - \int_\Omega \mathbf{p}\boldsymbol{\nabla} \cdot \mathbf{v dx} \\
\qquad = \displaystyle\int_\Omega \mathbf{f} \cdot \mathbf{v} dx + \int_{\Gamma_1} \mathbf{g}_1 \cdot \mathbf{v} d\Gamma, \forall \mathbf{v} \in V_0, \\
\displaystyle\int_\Omega q\boldsymbol{\nabla} \cdot \mathbf{u dx} = 0,\ \forall \mathbf{q} \in \mathbf{L}^2(\Omega),
\end{cases}
\tag{322}
$$

where, in (322), V_{g_0} and V_0 are defined by

$$
V_{g_0} = \{\mathbf{v}|\mathbf{v} \in (H^1(\Omega))^d,\ \mathbf{v} = \mathbf{g}_0 \text{ on } \Gamma_0\},
\tag{323}
$$
$$
V_0 = \{\mathbf{v}|\mathbf{v} \in (H^1(\Omega))^d,\ \mathbf{v} = \mathbf{0} \text{ on } \Gamma_0\},
\tag{324}
$$

respectively; if $\Gamma_0 = \emptyset$, then $V_0 = V_{g_0} = (H^1(\Omega))^d$ and $\Gamma_1 = \Gamma$.

Following (322) we approximate the *Stokes problem* (321) by

$$\begin{cases} \mathbf{u_h} \in \mathbf{V_{g_{0h}}}, \ \mathbf{p_h} \in \mathbf{P_h}; \ \forall \mathbf{v_h} \in \mathbf{V_{0h}} \text{ and } \mathbf{q_h} \in \mathbf{P_h}, \text{ we have} \\[2mm] \alpha \int_{\Omega_h} \mathbf{u_h} \cdot \mathbf{v_h} dx + \nu \int_{\Omega_h} \nabla \mathbf{u_h} : \nabla \mathbf{v_h} dx - \int_{\Omega_h} \mathbf{p_h} \nabla \cdot \mathbf{v_h} dx \\[2mm] \qquad = \int_{\Omega_h} \mathbf{f_h} \cdot \mathbf{v_h} dx + \int_{\Gamma_{1h}} \mathbf{g}_{1h} \cdot \mathbf{v_h} d\Gamma_h, \\[2mm] \int_{\Omega_h} q_h \nabla \cdot \mathbf{u_h} dx = \mathbf{0}. \end{cases} \tag{325}$$

In (325) the space P_h is defined as in Section 5.1, while

$$V_{g_{0h}} = \{\mathbf{v}_h | \mathbf{v}_h \in V_h, \ \mathbf{v}_h = \mathbf{g}_{0h} \ on \ \Gamma_{0h}\}, \tag{326}$$
$$V_{0h} = \{\mathbf{v}_h | \mathbf{v}_h \in V_h, \ \mathbf{v}_h = \mathbf{0} \ on \ \Gamma_{0h}\} \tag{327}$$

with V_h defined as in Section 5.1. The functions $\mathbf{f}_h$, $\mathbf{g}_{0h}$ and $\mathbf{g}_{1h}$ are approximations of $\mathbf{f}$, $\mathbf{g}_0$ and $\mathbf{g}_1$, respectively; Γ_{ih} approximates $\Gamma_i, \forall \ i = 0, 1$.

On the Convergence of the Finite Element Approximations of the Stokes Problem

In this subsection, we are going to discuss the *convergence* - as $h \to 0$ - of the *finite element approximations* of the Stokes problems, introduced in the preceding paragraphs. *Convergence* is, at the same time, a delicate and well-documented issue. It is our opinion that the celebrated article by Crouzeix and Raviart [92] was really the first one to address the convergence issues in a systematic, rigorous and general way; also, this article introduced novel (at the time) approximations of the Stokes problem which are still used nowadays by some practitioners. A very complete discussion of the convergence properties of various finite element approximations to the Stokes and steady Navier-Stokes equations can be found in the book by Girault and Raviart [59], which is still a basic (if not *the* basic) reference on the subject. However, the reader should also consult Brezzi and Fortin [25] (Chapter 6) and the review article by Fortin [68] which discusses - among other things - finite element approximations of the Stokes and Navier-Stokes equations not available in the mid-eighties (the following references are also worth consulting: Temam [19] (Chapter 1), Glowinski [8] (Chapter 7 and Appendix 3), Gunzburger [69] (Part 1), Pironneau [10] (Chapter 4), Brenner and Scott [26] (Chapter 10)).

For simplicity, in the following we shall consider only the *Stokes-Dirichlet problem* with $\mathbf{g} = \mathbf{0}$ *on* Γ; we have then, from (314),

$$\begin{cases} \alpha \mathbf{u} - \nu \Delta \mathbf{u} + \nabla \mathbf{p} = \mathbf{f} \text{ in } \Omega, \\ \nabla \cdot \mathbf{u} = \mathbf{0} \text{ on } \Omega, \\ \mathbf{u} = \mathbf{0} \text{ on } \Gamma. \end{cases} \tag{328}$$

A *variational formulation* of problem (328) is given by

$$\begin{cases} \mathbf{u} \in \mathbf{V_0}, \ \mathbf{p} \in \mathbf{L^2}(\Omega), \\ \alpha \int_{\Omega} \mathbf{u} \cdot \mathbf{v} d\mathbf{x} + \nu \int_{\Omega} \boldsymbol{\nabla} \mathbf{u} :: \boldsymbol{\nabla} \mathbf{v} d\mathbf{x} - \int_{\Omega} \mathbf{p} \boldsymbol{\nabla} \cdot \mathbf{v} d\mathbf{x} =< \mathbf{f}, \mathbf{v} >, \ \forall \mathbf{v} \in \mathbf{V_0}, \\ \int_{\Omega} q \boldsymbol{\nabla} \cdot \mathbf{u} d\mathbf{x} = \mathbf{0}, \ \forall \mathbf{q} \in \mathbf{L^2}(\Omega), \end{cases}$$
(329)

with $V_0 = (H_0^1(\Omega))^d$ and $< \cdot, \cdot >$ the duality pairing between $(H^{-1}(\Omega))^d$ and $(H_0^1(\Omega))^d$. We know from Section 4, that problem (328), (329) is equivalent to the following saddle-point problem in $V_0 \times L^2(\Omega)$:

$$\begin{cases} Find \ \{\mathbf{u}, \mathbf{p}\} \in \mathbf{V_0} \times \mathbf{L^2}(\Omega), \ \textbf{so that} \\ \mathcal{L}(\mathbf{u}, \mathbf{q}) \leq \mathcal{L}(\mathbf{u}, \mathbf{p}) \leq \mathcal{L}(\mathbf{v}, \mathbf{p}), \ \forall \{\mathbf{v}, \mathbf{q}\} \in \mathbf{V_0} \times \mathbf{L^2}(\Omega), \end{cases}$$
(330)

with the *Lagrangian functional* $\mathcal{L}$ defined, $\forall \{\mathbf{v}, q\} \in (H^1(\Omega))^d \times L^2(\Omega)$, by

$$\mathcal{L}(\mathbf{v}, q) = \frac{1}{2} \int_{\Omega} (\alpha |\mathbf{v}|^2 + \nu |\boldsymbol{\nabla} \mathbf{v}|^2) dx - \int_{\Omega} q \boldsymbol{\nabla} \cdot \mathbf{v} dx - < \mathbf{f}, \mathbf{v} > .$$
(331)

The saddle-point problem (329), (330) is a member of the following family of *generalized linear saddle-point problems*

$$\begin{cases} Find \ \{u, \lambda\} \in X \times \Lambda \ so \ that \\ a(u, v) + b(v, \lambda) =< l, v >, \ \forall v \in X, \\ b(u, \mu) =< \mathcal{X}, \mu >, \ \forall \mu \in \Lambda, \end{cases}$$
(332)

where, in (332):

- X and Λ are two *real Hilbert spaces*, with X' and Λ' their respective *dual spaces*;
- $a : X \times X \to \mathbb{R}$ is *bilinear* and *continuous* (possibly non-symmetric);
- $b : X \times \Lambda \to \mathbb{R}$ is *bilinear* and *continuous*,
- $< \cdot, \cdot >$ denotes the *duality pairing* between either X' and X or Λ' and Λ,
- $l \in X'$ and $\mathcal{X} \in \Lambda'$.

Using the *Riesz Theorem* we can associate to the bilinear functionals $a(\cdot, \cdot)$ and $b(\cdot, \cdot)$ two operators A and B so that

$$\begin{cases} A \in \mathcal{L}(X, X'), \\ < Av, w >= a(v, w), \ \forall v, w \in X, \end{cases}$$

$$\begin{cases} B \in \mathcal{L}(\Lambda, \Lambda'), \\ < Bv, \mu >= b(v, \mu), \ \forall v \in X, \ \forall \mu \in \Lambda. \end{cases}$$

The generalized saddle-point problem (332) takes then the equivalent operator formulation

$$\begin{cases} Au + B'\lambda = l, \\ Bu = \mathcal{X}, \end{cases} \tag{333}$$

where, in (333), $B'(\in \mathcal{L}(\Lambda, X'))$ is the dual (transpose) operator of B, i.e.

$$< Bv, \mu >=< B'\mu, v >, \ \forall\{v, \mu\} \in X \times \Lambda.$$

Remark 26. If the bilinear functional $a(\cdot, \cdot)$ is *symmetric* problem (332), (333) is equivalent to the *genuine* saddle-point problem

$$\begin{cases} \{u, \lambda\} \in X \times \Lambda, \\ L(u, \mu) \le L(u, \lambda) \le L(v, \lambda), \ \forall\{v, \mu\} \in X \times \Lambda, \end{cases} \tag{334}$$

with, in (334), the Lagrangian L defined by

$$L(v, \mu) = \frac{1}{2}a(v, v) + b(v, \mu)- < \mathcal{X}, \mu > - < l, v >, \ \forall\{v, \mu\} \in X \times \Lambda.$$

$\square$

We can also easily show (using the techniques employed in Section 4) that the vector u in (334) is also the solution of the following *constrained minimization* problem

$$\begin{cases} u \in V(\mathcal{X}), \\ j(u) \le j(v), \ \forall v \in V(\mathcal{X}), \end{cases} \tag{335}$$

with, in (335), the functional $j(\cdot)$ and the space $V(\mathcal{X})$ defined by

$$j(v) = \frac{1}{2}a(v, v)- < l, v >, \ \forall v \in X,$$

$$V(\mathcal{X}) = \{v|v \in X, \ b(v, \mu) \ = < \mathcal{X}, \mu >, \ \forall \mu \in \Lambda\},$$

respectively (we clearly have, for $V(\mathcal{X})$, the alternative definition

$$V(\mathcal{X}) = \{v|v \in X, Bv = \mathcal{X}\}).$$

Vector λ can be seen as a *Lagrange multiplier* associated with the linear relation $Bv = \mathcal{X}$.

Remark 27. We can easily show that the component u of the solution of problem (332), (333) is also a solution of the following linear variational problem in $V(\mathcal{X})$ ($V(\mathcal{X})$ has been defined in the above remark):

$$\begin{cases} u \in V(\mathcal{X}), \\ a(u, v) =< l, v >, \ \forall v \in V_0, \end{cases} \tag{336}$$

where $V_0 = \ker(B)$, i.e.

$$V_0 = \{v|v \in X, \ b(v, \mu) = 0, \ \forall \mu \in \Lambda\}.$$

$\square$

With space V_0 still being the *kernel* of operator B, let us define $\pi \in \mathcal{L}(X', V_0')$ by

$$< \pi f, v > = < f, v >, \ \forall f \in X', \ \forall v \in V_0.$$

Concerning the *uniqueness* and the *existence* of a solution to problem (332), (333) we have the following

Theorem 8. *Problem (332), (333) is well-posed (i.e., operator* $\begin{pmatrix} A & B' \\ B & 0 \end{pmatrix}$ *is an isomorphism from* $X \times \Lambda$ *onto* $X' \times \Lambda'$*) if and only if the following conditions hold:*

(i) operator πA *is an isomorphism from* V_0 *onto* V_0'*;*
(ii) there exists a constant $\beta > 0$ *such that*

$$\inf_{\mu \in \Lambda \setminus \{0\}} \ \sup_{v \in X \setminus \{0\}} \ \frac{b(v, \mu)}{||v||_X ||\mu||_\Lambda} \ge \beta. \tag{337}$$

(Condition (337) is known as an inf-sup condition).

For a proof of Theorem 8 see, e.g., Girault and Raviart [59] (Chapter 1, Section 4); actually in the above reference one can also find a proof of the following

Corollary 1. *Suppose that the bilinear functional* $a(\cdot, \cdot)$ *is* V*-elliptic, i.e., there exists a constant* $\alpha > 0$ *such that*

$$a(v, v) \ge \alpha ||v||_X^2, \ \forall v \in X.$$

Then, problem (332), (333) is well-posed if and only if the bilinear functional $b(\cdot, \cdot)$ *satisfies the inf-sup condition (337).*

Before going further, we think that it may be worthwhile to check if either Theorem 8 or Corollary 1 apply to the solution of the Stokes-Dirichlet problem (328), (329); it is indeed the case as shown by the following

Corollary 2. *It follows from Corollary 1 that problem (328), (329) has a unique solution in* $(H_0^1(\Omega))^d \times L_0^2(\Omega)$*, where*

$$L_0^2(\Omega) = \{q | q \in L^2(\Omega), \int_\Omega q \, dx = 0\}.$$

PROOF: The above result has been shown already in Section 4. The other proof can be obtained as a direct consequence of Corollary 1 (see Glowinski [4] (Chapter 5)).

Let us discuss now the *approximation* of the *generalized saddle-point problem* (332). With h a *discretization parameter*, we introduce two finite-dimensional spaces X_h and Λ_h, so that

$$X_h \subset X \text{ and } \Lambda_h \subset \Lambda. \tag{338}$$

Next, to each $\mathcal{X} \in \Lambda'$ we associate $V_h(\mathcal{X})$ - a discrete analogue of $V(\mathcal{X})$ - defined by

$$V_h(\mathcal{X}) = \{v_h | v_h \in X_h, \ b(v_h, \mu_h) = <\mathcal{X}, \mu_h>, \ \forall \mu_h \in \Lambda_h\}, \tag{339}$$

and we denote $V_h(0)$ by V_{0h}, i.e.

$$V_{0h} = \{v_h | v_h \in X_h, b(v_h, \mu_h) = 0, \ \forall \mu_h \in \Lambda_h\}. \tag{340}$$

We observe that, in general, $V_h(\mathcal{X}) \not\subset V(\mathcal{X})$ and $V_{0h} \not\subset V_0$ (with V_0 as in Remark 27).

We approximate, then, problem (332) by

$$\begin{cases} Find \ \{u_h, \lambda_h\} \in X_h \times \Lambda_h \ so \ that \\ a(u_h, v_h) + b(v_h, \lambda_h) = <l, v_h>, \ \forall v_h \in X_h, \\ b(u_h, \mu_h) = <\mathcal{X}, \mu_h>, \ \forall \mu_h \in \Lambda_h. \end{cases} \tag{341}$$

If $\{u_h, \lambda_h\}$ is a solution of problem (341), we can easily show that u_h is also a solution of the following finite dimensional linear variational problem

$$\begin{cases} u_h \in V_h(\mathcal{X}), \\ a(u_h, v_h) = <l, v_h>, \ \forall v_h \in V_{0h}; \end{cases} \tag{342}$$

problem (342) is clearly a discrete analogue of problem (336). Define now the norms $||a||$ and $||b||$ of the bilinear functionals $a(\cdot, \cdot)$ and $b(\cdot, \cdot)$ by

$$||a|| = \sup \frac{|a(v, w)|}{||v||_X ||w||_X}, \ \{v, w\} \in (X \backslash \{0\})^2 \tag{343}$$

and

$$||b|| = \sup \frac{|b(v, \mu)|}{||v||_X ||\mu||_\Lambda}, \{v, \mu\} \in (X \backslash \{0\}) \times (\Lambda \backslash \{0\}), \tag{344}$$

respectively; concerning the approximation of the solutions $\{u, \lambda\}$ of problem (332) by the solutions $\{u_h, \lambda_h\}$ of problem (341), we have then the following

Theorem 9. *1. Assume that the following conditions are verified*

(i) space $V_h(\mathcal{X})$ is not empty;
(ii)there exists a positive constant α^* such that

$$a(v_h, v_h) \mathbf{ge} \alpha^* ||v_h||_X^2, \ \forall v_h \in V_{0h}. \tag{345}$$

Then, problem (342) has a unique solution $u_h \in V_h(\mathcal{X})$ and there exists a constant C_1 depending only of $\alpha^*, ||a||$ and $||b||$ such that the following error estimate holds:

$$||u - u_h||_X \le C_1 \left(\inf_{v_h \in V_h(\mathcal{X})} ||u - v_h||_X + \inf_{\mu_h \in \Lambda_h} ||\lambda - \mu_h||_\Lambda \right). \qquad (346)$$

2. *Assume that hypothesis (ii) holds and, in addition, that:*
   (iii)there exists a positive constant β^* such that

$$\inf_{\mu_h \in \Lambda_h \setminus \{0\}} \sup_{v_h \in X_h \setminus \{0\}} \frac{b(v_h, \mu_h)}{||v_h||_X ||\mu_h||_\Lambda} \mathbf{ge} \beta^*. \qquad (347)$$

Then, $V_h(\mathcal{X}) \ne \emptyset$ and there exists a unique λ_h in Λ_h such that $\{u_h, \lambda_h\}$ is the unique solution of problem (341). Furthermore, there exists a constant C_2, depending only of $\alpha^*, \beta^*, ||a||$ and $||b||$, such that

$$||u - u_h||_X + ||\lambda - \lambda_h||_\Lambda \le C_2 (\inf_{v_h \in X_h} ||u - v_h||_X + \inf_{\mu_h \in \Lambda_h} ||\lambda - \mu_h||_\Lambda). \qquad (348)$$

For a proof of the above theorem, see Girault and Raviart [59] (pp. 114-116) (see also Roberts and Thomas [93] (Chapter 3) and Brezzi and Fortin [25] (Chapter 2); actually, the two above references contain a discussion of the effects of *numerical integration* on the error estimates, a most important practical issue).

Before discussing the convergent results, we have to introduce some (fairly classical) definitions, namely:

Definition 5.1: A family $\{\mathcal{T}_h\}_h$ of *triangulations* of Ω is said to be *regular* if there exists $\theta_0, 0 < \theta_0 \le \pi/3$, such that

$$\theta_T \mathbf{ge} \theta_0, \quad \forall T \in \mathcal{T}_h, \ \forall h, \qquad (349)$$

where, in (349), θ_T is the *smallest angle* of triangle T.

Definition 5.2: A family $\{\mathcal{T}_h\}_h$ of *triangulations* of Ω is said to be *uniformly regular* if it is *regular* and if there exists σ, $\sigma \mathbf{ge} 1$, such that

$$\max_{T \in \mathcal{T}_h} h_T / \min_{T \in \mathcal{T}_h} h_T \le \sigma, \quad \forall h, \qquad (350)$$

where, in (350), h_T is the length of the largest edge(s) of triangle T.

Remark 28. In Definitions 5.1 and 5.2, we have been assuming that Ω is a polygonal domain of $\mathbb{R}^2$ such that $\overline{\Omega} = \bigcup_{T \in \mathcal{T}_h} T, \forall h$. Actually, the above two definitions can be generalized to two-dimensional domains with *curved boundaries* and also to *three-dimensional* domains with *curved* or *polyhedral* boundaries, as shown in, e.g., Ciarlet [81] (Chapter 6, Section 37). If Ω is a *polyhedral* domain of $\mathbb{R}^3$ and $\mathcal{T}_h$ a "triangulation" of Ω (i.e., $T \in \mathcal{T}_h \mathbb{R}ightarrow T$ is a tetrahedron) so that $\overline{\Omega} = \bigcup_{T \in \mathcal{T}_h} T$, we say that the family $\{\mathcal{T}_h\}_h$ is *regular* if there exists $\sigma_1 > 0$ such that

$$h_T / \rho_T \le \sigma_1, \quad \forall T \in \mathcal{T}_h, \ \forall h, \qquad (351)$$

with h_T the length of the *largest edge(s)* of tetrahedron T, and ρ_T the *diameter of the sphere inscribed* in T. Similarly, we say that the family $\{\mathcal{T}_h\}_h$ is *uniformly regular* if it is *regular* and if there exists $\sigma_2, \sigma_2 \mathbf{ge}1$, so that

$$\max_{T \in \mathcal{T}_h} h_T / \min_{T \in \mathcal{T}_h} h_T \le \sigma_2, \ \forall T \in \mathcal{T}_h, \ \forall h, \tag{352}$$

with h_T as just above. $\quad \square$

Following Girault and Raviart [59] (Chapter 2, Section 4), and Brezzi and Fortin [25] (Chapter 6), we are going to provide convergence results for *finite element approximations* of the *Stokes-Dirichlet problem* (328). We shall focus our attention on the *Hood-Taylor* and *Bercovier-Pironneau* approximations described in Section 5.1; convergence results concerning other finite element approximations of the Stokes problem can be found in, e.g., the two above references and in Fortin [68] (see also the references therein).

Since Ω is polygonal it follows from Section 5.1 that the Stokes problem (328) is approximated by

$$\begin{cases} \{\mathbf{u_h}, \mathbf{p_h}\} \in \mathbf{V_{0h}} \times \mathbf{P_h}; \ \forall \{\mathbf{v}, \mathbf{q_h}\} \in \mathbf{V_{0h}} \times \mathbf{P_h} \text{ we have} \\ \alpha \int_\Omega \mathbf{u_h} \cdot \mathbf{v_h} dx + \nu \int_\Omega \nabla \mathbf{u_h} :: \nabla \mathbf{v_h} dx - \int_\Omega \mathbf{p_h} \nabla \cdot \mathbf{v_h} dx = <\mathbf{f}, \mathbf{v_h}>, \\ \int_\Omega \nabla \cdot \mathbf{u_h} \mathbf{q_h} dx = 0, \end{cases} \tag{353}$$

with

$$P_h = \{q_h | q_h \in C^0(\overline{\Omega}), q_h|_T \in P_1, \ \forall T \in \mathcal{T}_h\} \tag{354}$$

and

$$V_{0h} = \{v_h | v_h \in (C^0(\overline{\Omega}))^2, \ v_h|_T \in (P_2)^2, \ \forall T \in \mathcal{T}_h, \ v_h = 0 \text{ on } \Gamma\}. \tag{355}$$

In Girault and Raviart [59] (Chapter 2, Section 4.2), it is shown that the approximate Stokes-Dirichlet problem (353) has a *unique* solution in $V_{0h} \times P_{0h}$ if *no triangle of $\mathcal{T}_h$ has more than one edge contained* in Γ and that the following convergence theorem holds:

Theorem 10. *Let Ω be a bounded polygonal domain of $\mathbb{R}^2$ and suppose that the solution $\{\mathbf{u}, \mathbf{p}\}$ of the Stokes problem (328) verifies*

$$\mathbf{u} \in (\mathbf{H^{k+1}}(\Omega) \cap \mathbf{H_0^1}(\Omega))^2, \ \mathbf{p} \in \mathbf{H^k}(\Omega) \cap \mathbf{L_0^2}(\Omega), \ \mathbf{k = 1 \text{ or } 2}.$$

If the family $\{\mathcal{T}_h\}_h$ is regular and if, $\forall h$, no triangle of $\mathcal{T}_h$ has more than one edge contained in Γ, the solution $\{\mathbf{u_h}, \mathbf{p_h}\}$ of problem (353), with P_h and V_{0h} defined by (354) and (355), respectively, verifies,

$$\|\mathbf{u_h} - \mathbf{u}\|_{(\mathbf{H_0^1}(\Omega))^2} + \|\mathbf{p} - \mathbf{p_h}\|_{\mathbf{L^2}(\Omega)} \leq \mathbf{C_1 h^k}(\|\mathbf{u}\|_{(\mathbf{H^{k+1}}(\Omega))^2} + \|\mathbf{p}\|_{\mathbf{H^k}(\Omega)}).$$
(356)

If Ω is convex, we also have

$$\|\mathbf{u_h} - \mathbf{u}\|_{(\mathbf{L^2}(\Omega))^2} \leq \mathbf{C_2 h^{k+1}}(\|\mathbf{u}\|_{(\mathbf{H^{k+1}}(\Omega))^2} + \|\mathbf{p}\|_{\mathbf{H^k}(\Omega)}).$$
(357)

Finally, if the family $\{T_h\}_h$ is uniformly regular (but Ω not necessarily convex) we also have

$$\|p_h - p\|_{H^1(\Omega)} \leq C_3 h^{k-1}(\|\mathbf{u}\|_{(\mathbf{H^{k+1}}(\Omega))^2} + \|\mathbf{p}\|_{\mathbf{H^k}(\Omega)}).$$
(358)

In (356)-(358), C_1, C_2 and C_3 are positive constants.

A first proof of the above theorem was given by Bercovier and Pironneau [79]; this proof was improved by Verfurth [94] and further improved by Girault and Raviart [59] (Chapter 2, Section 4.2) (see also Brezzi and Fortin [25] (Chapter 6, Section 6)). We shall conclude this paragraph with the *Bercovier-Pironneau* approximation of the Stokes problem (328); from Section 5.1, this approximation is defined by (353), with P_h given by (354) and V_{0h} by

$$V_{0h} = \{\mathbf{v}_h | \mathbf{v}_h \in (C^0(\overline{\Omega}))^2, \mathbf{v}_h|_T \in (P_1)^2, \forall T \in T_{h/2}, \mathbf{v}_h = \mathbf{0} \text{ on } \Gamma\}, \quad (359)$$

with, in (359), $T_{h/2}$ obtained from T_h by dividing each triangle T of T_h in four similar triangles, by joining the mid-points of the edges of T (as shown in Figure 5.1). It follows from Girault and Raviart [59] (Chapter 2, Section 4.2) that if no triangle of T_h has more than one edge contained in Γ, then problem (353) has a unique solution in $V_{0h} \times P_{0h}$ and the following convergence theorem holds:

Theorem 11. *Let Ω and $\{T_h\}_h$ be as in Theorem 10 and suppose that the solution $\{\mathbf{u}, \mathbf{p}\}$ of problem (328) verifies*

$$\mathbf{u} \in (\mathbf{H^2}(\Omega) \cap \mathbf{H_0^1}(\Omega))^2, \ \mathbf{p} \in \mathbf{H^1}(\Omega) \times \mathbf{L_0^2}(\Omega).$$

Then, the solution $\{\mathbf{u_h}, \mathbf{p_h}\}$ of problem (353), with P_h and V_{0h} defined by (354) and (359), respectively, verifies

$$\|\mathbf{u_h} - \mathbf{u}\|_{(\mathbf{H_0^1}(\Omega))^2} + \|\mathbf{p_h} - \mathbf{p}\|_{\mathbf{L^2}(\Omega)} \leq \mathbf{C_1 h}(\|\mathbf{u}\|_{(\mathbf{H^2}(\Omega))^2} + \|\mathbf{p}\|_{\mathbf{H^1}(\Omega)}).$$
(360)

Moreover, if Ω is convex, we have the following L^2-error estimate

$$\|\mathbf{u_h} - \mathbf{u}\|_{(\mathbf{L^2}(\Omega))^d} \leq \mathbf{C_2 h^2}(\|\mathbf{u}\|_{(\mathbf{H^2}(\Omega))^2} + \|\mathbf{p}\|_{\mathbf{H^1}(\Omega)}).$$
(361)

In (360), (361), C_1 and C_2 are two positive constants.

Remark 29. We have discussed several *finite element approximations* of the *Stokes problems* (314) and (321) . Once a formulation such as (320) (or (325)) has been obtained, several practical issues still have to be addressed, among them the derivation of the linear systems equivalent to the discrete Stokes problems, and then the numerical solution of these systems. Those issues have been discussed in details in Glowinski [4] (Chapter 5), especially when deriving the equivalent linear systems how to obtain the accurate evaluation of *multiple integrals* over the elements of $\mathcal{T}_h$ (or $\mathcal{T}_{h/2}$), or over the element of reference $\hat{T}$. The discussion starts from the finite element approximation of the *Stokes-Dirichlet* problem (298) by the *Hood-Taylor* and *Bercovier-Pironneau* methods, assuming that Ω is a bounded polygonal domain of $\mathbb{R}^2$, and then the mini-element of *Arnold-Brezzi-Fortin*, the case of *curved boundaries* and finally the Stokes problem with *other boundary conditions* than Dirichlet.   □

5.2 Finite Element Implementation of the θ-Scheme

We are going to discuss in this section the *full discretization* of the *Navier-Stokes equations*

$$\frac{\partial \mathbf{u}}{\partial t} - \nu \triangle \mathbf{u} + (\mathbf{u} \cdot \boldsymbol{\nabla})\mathbf{u} + \boldsymbol{\nabla} p = \mathbf{f} \; in \; \Omega \times (0, T), \qquad (362)$$

$$\boldsymbol{\nabla} \cdot \mathbf{u} = 0 \; in \; \Omega \times (0, T), \qquad (363)$$

$$\mathbf{u}(0) = \mathbf{u}_0 (with \; \boldsymbol{\nabla} \cdot \mathbf{u}_0 = 0), \qquad (364)$$

$$\mathbf{u} = \mathbf{g}_0 \; on \; \Gamma_0 \times (0, T), \nu \frac{\partial \mathbf{u}}{\partial n} - \mathbf{n}p = \mathbf{g}_1 \; on \; \Gamma_1 \times (0, T). \qquad (365)$$

To approximate problem (362) – (365) we shall combine the finite element methods discussed in the previous subsection with the θ-scheme described by relations (80) - (88). We have seen in Section 2 that a "good" choice for θ, α, β is given by

$$\theta = 1 - 1/\sqrt{2}, \alpha = (1 - 2\theta)/(1 - \theta), \beta = \theta/(1 - \theta). \qquad (366)$$

A safe way to achieve the *full discretization* of the time dependent Navier-Stokes equations (362) – (365) is to proceed as follows (this approach applies, obviously, to other problems):

(i) Keeping time continuous we shall use the *finite element spaces* introduced in the previous subsection to space discretize the Navier-Stokes equations. We obtain then a system coupling *ordinary differential equations* and *algebraic equations.*

(ii) We shall apply the *operator splitting-methods* of Section 2, to the time-discretization of the above system of algebraic and ordinary differential equations.

At first we shall consider the *pure Dirichlet* case (i.e., the particular case of (365) where $\Gamma_1 = \emptyset$) and, then, mixed boundary conditions such as (365).

Space Approximation of the Time Dependent Navier-Stokes Equations

The Dirichlet case.

The problem that we consider is defined by (362) – (364), completed by

$$\mathbf{u} = \mathbf{g} \; on \; \partial\Omega \times (0,T). \tag{367}$$

To have a *well-posed* problem we assume that

$$\int_{\partial\Omega} \mathbf{g}(t) \cdot \mathbf{n} d\Gamma = 0 \; on \; (0,T), \tag{368}$$

and also, in principle,

$$\mathbf{u}_0 \cdot \mathbf{n} = \mathbf{g}(0) \cdot \mathbf{n} \; on \; \partial\Omega \tag{369}$$

(we say "in principle" since some of the test problems does not verify (369), without too much damage on the computational procedure and on the computed solution).

Assuming that $\Omega \subset \mathbb{R}^2$, we *space-approximate* problem (362) – (364), (367) by

$$Find \; \{\mathbf{u}_h(t), p_h(t)\} \in V_h \times P_h, \forall \; t \in (0,T), \; such \; that$$

$$\begin{cases} \int_{\Omega_h} \dot{\mathbf{u}}_h \cdot \mathbf{v}_h dx + \nu \int_{\Omega_h} \boldsymbol{\nabla}\mathbf{u}_h : \boldsymbol{\nabla}\mathbf{v}_h dx + \int_{\Omega_h} (\mathbf{u}_h \cdot \boldsymbol{\nabla})\mathbf{u}_h \cdot \mathbf{v}_h dx \\ - \int_{\Omega_h} p_h \boldsymbol{\nabla} \cdot \mathbf{v}_h dx = \int_{\Omega_h} \mathbf{f}_h \cdot \mathbf{v}_h dx, \forall \; \mathbf{v}_h \in V_{0h}, \end{cases} \tag{370}$$

$$\int_{\Omega_h} \boldsymbol{\nabla} \cdot \mathbf{u}_h q_h dx = 0, \forall \; q_h \in P_h, \tag{371}$$

$$\mathbf{u}_h(t) = \mathbf{g}_h(t) \; on \; \partial\Omega_h (with \; \mathbf{g}_h(t) \in gammaV_h), \tag{372}$$

$$\mathbf{u}_h(0) = \mathbf{u}_{0h} (with \; \mathbf{u}_{0h} \in V_h). \tag{373}$$

In (370) - (373):

- We have $\Omega_h = \Omega$ and $\partial\Omega_h = \partial\Omega$ if Ω is polygonal. For the cases where Ω is not polygonal, see the discussion in Glowinski [4] for the *isoparametric* generalization of the *Hood-Taylor* (resp., *Bercovier-Pironneau*) approximation.
- The finite element *velocity* and *pressure* spaces V_h and P_h are as in Section 5.1 and, here,

$$V_{0h} = V_h \cap (H_0^1(\Omega_h))^2 = \{\mathbf{v}_h | \mathbf{v}_h \in V_h, \mathbf{v}_h = \mathbf{0} \; on \; \partial\Omega_h\}.$$

- We have used the notation $\dot{\mathbf{u}}_h$ for $\dfrac{\partial \mathbf{u}_h}{\partial t}$.
- The functions $\mathbf{f}_h, \mathbf{u}_{0h}$ and $\mathbf{g}_h$ are convenient approximations of $\mathbf{f}, \mathbf{u}_0$ and $\mathbf{g}$, respectively. Function $\mathbf{g}_h$ has to verify

$$\int_{\partial \Omega_h} \mathbf{g}_h(t) \cdot \mathbf{n} d\Gamma_h = 0 \ on \ (0, T); \tag{374}$$

  to construct, from $\mathbf{g}$, an approximation $\mathbf{g}_h$ verifying (374) we shall use the methods discussed in Section 5.1.
- The boundary space $gammaV_h$ is defined as in Section 5.1.

The case of the mixed boundary conditions (365).

In this case the boundary conditions are given by

$$\mathbf{u} = \mathbf{g}_0 \ on \ \Gamma_0 \times (0, T), \nu \frac{\partial \mathbf{u}}{\partial n} - \mathbf{n} p = \mathbf{g}_1 \ on \ \Gamma_1 \times (0, T),$$

leading to the following approximate problem:

$$Find \ \{\mathbf{u}_h(t), p_h(t)\} \in V_h \times P_h, \forall \ t \in (0, T), \ such \ that$$

$$\begin{cases} \displaystyle \iint_{\Omega_h} \dot{\mathbf{u}}_h \cdot \mathbf{v}_h dx + \nu \int_{\Omega_h} \boldsymbol{\nabla} \mathbf{u}_h : \boldsymbol{\nabla} \mathbf{v}_h dx + \int_{\Omega_h} (\mathbf{u}_h \cdot \boldsymbol{\nabla}) \mathbf{u}_h \cdot \mathbf{v}_h dx \\ \displaystyle - \int_{\Omega_h} p_h \boldsymbol{\nabla} \cdot \mathbf{v}_h dx = \int_{\Omega_h} \mathbf{f}_h \cdot \mathbf{v}_h dx + \int_{\Gamma_{1h}} \mathbf{g}_{1h} \cdot \mathbf{v}_h d\Gamma_h, \forall \ \mathbf{v}_h \in V_{0h}, \end{cases} \tag{375}$$

$$\int_{\Omega_h} \boldsymbol{\nabla} \cdot \mathbf{u}_h q_h dx = 0, \forall \ q_h \in P_h, \tag{376}$$

$$\mathbf{u}_h(t) = \mathbf{g}_{0h}(t) \ on \ \Gamma_{0h}, \tag{377}$$

$$\mathbf{u}_h(0) = \mathbf{u}_{0h} (with \ \mathbf{u}_{0h} \in V_h); \tag{378}$$

in (375) the space V_{0h} is defined as in Section 5.1, the other notation being self-explanatory.

Expanding $\mathbf{u}_h$ and p_h on vector bases of V_h and P_h, respectively, and taking for the test functions $\mathbf{v}_h$ and q_h *all* the elements of the vector bases of V_{0h} and P_h, formulations (370) – (373) and (375) – (378) will produce a system of *ordinary differential equations* with respect to t coupled to the linear relations associated to the discrete incompressibility condition. Applying to these *algebraic-differential* problems the time discretization methods by operator splitting of Section 2 is straightforward, as we shall see hereafter, where we shall focus on the θ-scheme of Section 2 in order to derive the fully discrete analogs of schemes (80) - (88) including the particular case where $\Gamma_1 = \emptyset$ (pure Dirichlet boundary conditions).

Full discretization by the θ-scheme: Case of the Dirichlet boundary conditions.

The *algebraic-differential system* to time-discretize is (370) – (373). We obtain then

$$\mathbf{u}_h^0 = \mathbf{u}_{0h}; \tag{379}$$

then, for $n \geq 0, \mathbf{u}_h^n$ *being known, we compute* $\{\mathbf{u}_h^{n+\theta}, p_h^{n+\theta}\} \in V_h \times P_h$, *then* $\mathbf{u}_h^{n+1-\theta} \in V_h$, *and finally* $\{\mathbf{u}_h^{n+1}, p_h^{n+1}\} \in V_h \times P_h$ *by solving the following discrete elliptic systems*

$$
\begin{cases}
\displaystyle\int_{\Omega_h} \frac{\mathbf{u}_h^{n+\theta} - \mathbf{u}_h^n}{\theta \Delta t} \cdot \mathbf{v}_h dx + \alpha\nu \int_{\Omega_h} \boldsymbol{\nabla}\mathbf{u}_h^{n+\theta} : \boldsymbol{\nabla}\mathbf{v}_h dx - \int_{\Omega_h} p_h^{n+\theta} \boldsymbol{\nabla} \cdot \mathbf{v}_h dx = \\
\displaystyle\int_{\Omega_h} \mathbf{f}_h^{n+\theta} \cdot \mathbf{v}_h dx - \beta\nu \int_{\Omega_h} \boldsymbol{\nabla}\mathbf{u}_h^n : \boldsymbol{\nabla}\mathbf{v}_h dx - \int_{\Omega_h} (\mathbf{u}_h^n \cdot \boldsymbol{\nabla})\mathbf{u}_h^n \cdot \mathbf{v}_h dx, \forall \mathbf{v}_h \in V_{0h}, \\
\displaystyle\int_{\Omega_h} \boldsymbol{\nabla} \cdot \mathbf{u}_h^{n+\theta} q_h dx = 0, \forall q_h \in P_h, \\
\mathbf{u}_h^{n+\theta} = \mathbf{g}_h^{n+\theta} \text{ on } \partial\Omega_h,
\end{cases}
\tag{380}
$$

then

$$
\begin{cases}
\displaystyle\int_{\Omega_h} \frac{\mathbf{u}_h^{n+1-\theta} - \mathbf{u}_h^{n+\theta}}{(1 - 2\theta) \Delta t} \cdot \mathbf{v}_h dx + \beta\nu \int_{\Omega_h} \boldsymbol{\nabla}\mathbf{u}_h^{n+1-\theta} : \boldsymbol{\nabla}\mathbf{v}_h dx \\
\displaystyle + \int_{\Omega_h} (\mathbf{u}_h^{n+1-\theta} \cdot \boldsymbol{\nabla})\mathbf{u}_h^{n+1-\theta} \cdot \mathbf{v}_h dx = \int_{\Omega_h} \mathbf{f}_h^{n+\theta} \cdot \mathbf{v}_h dx \\
\displaystyle - \alpha\nu \int_{\Omega_h} \boldsymbol{\nabla}\mathbf{u}_h^{n+\theta} : \boldsymbol{\nabla}\mathbf{v}_h dx - \int_{\Omega_h} p_h^{n+\theta} \cdot \boldsymbol{\nabla}\mathbf{v}_h dx, \forall \mathbf{v}_h \in V_{0h}, \\
\mathbf{u}_h^{n+1-\theta} = \mathbf{g}_h^{n+1-\theta} \text{ on } \partial\Omega_h,
\end{cases}
\tag{381}
$$

and finally

$$
\begin{cases}
\displaystyle\int_{\Omega_h} \frac{\mathbf{u}_h^{n+1} - \mathbf{u}_h^{n+1-\theta}}{\theta \Delta t} \cdot \mathbf{v}_h dx + \alpha\nu \int_{\Omega_h} \boldsymbol{\nabla}\mathbf{u}_h^{n+1} : \boldsymbol{\nabla}\mathbf{v}_h dx \\
\displaystyle - \int_{\Omega_h} p_h^{n+1} \boldsymbol{\nabla} \cdot \mathbf{v}_h dx = \int_{\Omega_h} \mathbf{f}_h^{n+1} \cdot \mathbf{v}_h dx - \beta\nu \int_{\Omega_h} \boldsymbol{\nabla}\mathbf{u}_h^{n+1-\theta} : \boldsymbol{\nabla}\mathbf{v}_h dx \\
\displaystyle - \int_{\Omega_h} (\mathbf{u}_h^{n+1-\theta} \cdot \boldsymbol{\nabla})\mathbf{u}_h^{n+1-\theta} \cdot \mathbf{v}_h dx, \forall \mathbf{v}_h \in V_{0h}, \\
\displaystyle\int_{\Omega_h} \boldsymbol{\nabla} \cdot \mathbf{u}_h^{n+1} q_h dx = 0, \forall q_h \in P_h, \\
\mathbf{u}_h^{n+1} = \mathbf{g}_h^{n+1} \text{ on } \partial\Omega_h,
\end{cases}
\tag{382}
$$

respectively. In (379) – (382), the finite element spaces V_h, V_{0h}, and P_h are as in Section 5.1 for the Dirichlet case. For θ, α, β we advocate the values given by (366).

Full discretization by the θ-scheme : Case of the mixed boundary conditions.

The time discretization of problem $(375) - (378)$ leads to the following scheme:

$$\mathbf{u}_h^0 = \mathbf{u}_{0h}; \tag{383}$$

then, for $n \geq 0, \mathbf{u}_h^n$ being known, we compute $\{\mathbf{u}_h^{n+\theta}, p_h^{n+\theta}\} \in V_h \times P_h$, then $\mathbf{u}_h^{n+1-\theta} \in V_h$, and finally $\{\mathbf{u}_h^{n+1}, p_h^{n+1}\} \in V_h \times P_h$ by solving the following discrete elliptic systems

$$
\begin{cases}
\displaystyle\int_{\Omega_h} \frac{\mathbf{u}_h^{n+\theta} - \mathbf{u}_h^n}{\theta \Delta t} \cdot \mathbf{v}_h dx + \alpha \nu \int_{\Omega_h} \boldsymbol{\nabla}\mathbf{u}_h^{n+\theta} : \boldsymbol{\nabla}\mathbf{v}_h dx - \int_{\Omega_h} p_h^{n+\theta} \boldsymbol{\nabla} \cdot \mathbf{v}_h dx = \\
\displaystyle\int_{\Omega_h} \mathbf{f}_h^{n+\theta} \cdot \mathbf{v}_h dx + \int_{\Gamma_{1h}} \mathbf{g}_{1h}^{n+\theta} \cdot \mathbf{v}_h d\Gamma_h - \beta \nu \int_{\Omega_h} \boldsymbol{\nabla}\mathbf{u}_h^n : \boldsymbol{\nabla}\mathbf{v}_h dx \\
\displaystyle - \int_{\Omega_h} (\mathbf{u}_h^n \cdot \boldsymbol{\nabla})\mathbf{u}_h^n \cdot \mathbf{v}_h dx, \forall \mathbf{v}_h \in V_{0h}, \\
\displaystyle\int_{\Omega_h} \boldsymbol{\nabla} \cdot \mathbf{u}_h^{n+\theta} q_h dx = 0, \forall q_h \in P_h, \\
\mathbf{u}_h^{n+\theta} = \mathbf{g}_{0h}^{n+\theta} \text{ on } \Gamma_{0h},
\end{cases}
\tag{384}
$$

then

$$
\begin{cases}
\displaystyle\int_{\Omega_h} \frac{\mathbf{u}_h^{n+1-\theta} - \mathbf{u}_h^{n+\theta}}{(1-2\theta)\Delta t} \cdot \mathbf{v}_h dx + \beta \nu \int_{\Omega_h} \boldsymbol{\nabla}\mathbf{u}_h^{n+1-\theta} : \boldsymbol{\nabla}\mathbf{v}_h dx \\
\displaystyle + \int_{\Omega_h} (\mathbf{u}_h^{n+1-\theta} \cdot \boldsymbol{\nabla})\mathbf{u}_h^{n+1-\theta} \cdot \mathbf{v}_h dx = \int_{\Omega_h} \mathbf{f}_h^{n+\theta} \cdot \mathbf{v}_h dx \\
\displaystyle + \int_{\Gamma_{1h}} \mathbf{g}_{1h}^{n+\theta} \cdot \mathbf{v}_h d\Gamma_h - \alpha \nu \int_{\Omega_h} \boldsymbol{\nabla}\mathbf{u}_h^{n+\theta} : \boldsymbol{\nabla}\mathbf{v}_h dx - \int_{\Omega_h} p_h^{n+\theta} \boldsymbol{\nabla} \cdot \mathbf{v}_h dx, \forall \mathbf{v}_h \in V_{0h}, \\
\mathbf{u}_h^{n+1-\theta} = \mathbf{g}_{0h}^{n+1-\theta} \text{ on } \Gamma_{0h},
\end{cases}
\tag{385}
$$

and finally

$$
\begin{cases}
\displaystyle\int_{\Omega_h} \frac{\mathbf{u}_h^{n+1} - \mathbf{u}_h^{n+1-\theta}}{\theta \Delta t} \cdot \mathbf{v}_h dx + \alpha \nu \int_{\Omega_h} \boldsymbol{\nabla}\mathbf{u}_h^{n+1} : \boldsymbol{\nabla}\mathbf{v}_h dx \\
\displaystyle - \int_{\Omega_h} p_h^{n+1} \boldsymbol{\nabla} \cdot \mathbf{v}_h dx = \int_{\Omega_h} \mathbf{f}_h^{n+1} \cdot \mathbf{v}_h dx + \int_{\Gamma_{1h}} \mathbf{g}_{1h}^{n+1} \cdot \mathbf{v}_h d\Gamma_h \\
\displaystyle - \beta \nu \int_{\Omega_h} \boldsymbol{\nabla}\mathbf{u}_h^{n+1-\theta} : \boldsymbol{\nabla}\mathbf{v}_h dx - \int_{\Omega_h} (\mathbf{u}_h^{n+1-\theta} \cdot \boldsymbol{\nabla})\mathbf{u}_h^{n+1-\theta} \cdot \mathbf{v}_h dx, \forall \mathbf{v}_h \in V_{0h}, \\
\displaystyle\int_{\Omega_h} \boldsymbol{\nabla} \cdot \mathbf{u}_h^{n+1} q_h dx = 0, \forall q_h \in P_h, \\
\mathbf{u}_h^{n+1} = \mathbf{g}_{0h}^{n+1} \text{ on } \Gamma_{0h},
\end{cases}
\tag{386}
$$

respectively. In (383)–(386), the finite element spaces V_h, V_{0h}, and P_h are as in Section 5.1 for the case of mixed boundary conditions and for θ, α, β we advocate the values given by (366).

Remark 30. In order to solve the *discrete Stokes problems* (380), (382), (384), (386) and the *discrete-advection diffusion problems* (381), (385), one can use discrete variants of the conjugate gradient algorithms discussed in Sections 3 and 4. The implementation of these algorithms, which boils down to the solution of sequences of linear systems for symmetric and positive definite matrices, will be further discussed later. $\square$

Remark 31. If we replace the *nonlinear problem* (381) by the following (*linearized*) one

$$
\begin{cases}
\displaystyle\int_{\Omega_h} \frac{\mathbf{u}_h^{n+1-\theta} - \mathbf{u}_h^{n+\theta}}{(1-2\theta)\Delta t} \cdot \mathbf{v}_h dx + \beta\nu \int_{\Omega_h} \boldsymbol{\nabla}\mathbf{u}_h^{n+1-\theta} : \boldsymbol{\nabla}\mathbf{v}_h dx \\[2ex]
\displaystyle + \int_{\Omega_h} (\mathbf{u}_h^{n+\theta} \cdot \boldsymbol{\nabla})\mathbf{u}_h^{n+1-\theta} \cdot \mathbf{v}_h dx = \int_{\Omega_h} \mathbf{f}_h^{n+\theta} \cdot \mathbf{v}_h dx \\[2ex]
\displaystyle -\alpha\nu \int_{\Omega_h} \boldsymbol{\nabla}\mathbf{u}_h^{n+\theta} : \boldsymbol{\nabla}\mathbf{v}_h dx - \int_{\Omega_h} p_h^{n+\theta} \cdot \boldsymbol{\nabla}\mathbf{v}_h dx, \forall \mathbf{v}_h \in V_{0h}, \\[2ex]
\mathbf{u}_h^{n+1-\theta} = \mathbf{g}_h^{n+1-\theta} \ on \ \partial\Omega_h,
\end{cases}
$$

the new scheme is essentially as stable and accurate as the original scheme (379) – (382); on the other hand, it is less costly to solve the linearized one than problem (381) (for the same value of Δt, at least). Similar replacement can be done in (385).

The numerical integration of the advection term in (379)–(382) and (383)–(386) (also in the linearized ones) for the *Hood-Taylor*, *Bercovier-Pironneau* and *Arnold-Brezzi-Fortin* approximations of the Navier-Stokes equations have been discussed in Glowinski [4] (Chapter 5, Section 27). $\square$

5.3 Finite Element Implementation of the L^2-Projection/wave-like Equation Method

This section is dedicated to the numerical solution of the Navier-Stokes equations modeling incompressible viscous fluid flow by a methodology combining time discretization by a first order accurate operator-splitting, Stokes solvers à la Uzawa and a wave-like equation treatment of the advection. The goal is to apply this approach to simulate more complicated flow problems, such as rigid bodies moving freely in the fluid (see, e.g., Glowinski et al. [95, 96, 97, 98].

Following Chorin [38, 39], most "modern" Navier-Stokes solvers are based on operator splitting algorithms (see, e.g., in Marchuk [29] and Turek [99]) in order to force the incompressibility condition via a Stokes solver or a L^2-projection method.

Applying scheme à la Marchuk–Yanenko discussed in Section 2.3, we have the following scheme for the Dirichlet case (370) - (373) (after dropping some of the subscripts h and applying the backward Euler's method for time discretization):

$$\mathbf{u}^0 = \mathbf{u}_{0h} \text{ is given;} \tag{387}$$

for $n \geq 0$, $\mathbf{u}^n$ being known,

$$\begin{cases} \int_\Omega \dfrac{\mathbf{u}^{n+1/3} - \mathbf{u}^n}{\triangle t} \cdot \mathbf{v}\, dx - \int_\Omega p^{n+1/3} \boldsymbol{\nabla} \cdot \mathbf{v}\, dx = 0, \;\; \forall \mathbf{v} \in V_{0h}, \\[2mm] \int_\Omega q \boldsymbol{\nabla} \cdot \mathbf{u}^{n+1/3}\, dx = 0, \;\; \forall q \in L_h^2; \\[2mm] \mathbf{u}^{n+1/3} \in \mathbf{V}_{\mathbf{g}_h}^{n+1}, \;\; \mathbf{p}^{n+1/3} \in L_{0h}^2, \end{cases} \tag{388}$$

$$\begin{cases} \int_\Omega \dfrac{\partial \mathbf{u}(t)}{\partial t} \cdot \mathbf{v}\, dx + \int_\Omega (\mathbf{u}^{n+1/3} \cdot \boldsymbol{\nabla})\mathbf{u}(t) \cdot \mathbf{v}\, dx = 0 \text{ on } (\mathbf{t}^n, \mathbf{t}^{n+1}), \\[2mm] \qquad\qquad \forall \mathbf{v} \in V_{0h}^{n+1,-}, \\[2mm] \mathbf{u}(\mathbf{t}^n) = \mathbf{u}^{n+1/3}, \\[2mm] \mathbf{u}(t) \in \mathbf{V}_h, \;\; \mathbf{u}(t) = \mathbf{g}_h(\mathbf{t}^{n+1}) \text{ on } \Gamma_-^{n+1} \times (\mathbf{t}^n, \mathbf{t}^{n+1}), \end{cases} \tag{389}$$

$$\mathbf{u}^{n+2/3} = \mathbf{u}(\mathbf{t}^{n+1}), \tag{390}$$

$$\begin{cases} \int_\Omega \dfrac{\mathbf{u}^{n+1} - \mathbf{u}^{n+2/3}}{\triangle t} \cdot \mathbf{v}\, dx + \nu \int_\Omega \boldsymbol{\nabla}\mathbf{u}^{n+1} : \boldsymbol{\nabla}\mathbf{v}\, dx = 0, \\[2mm] \forall \mathbf{v} \in V_{0h}; \;\; \mathbf{u}^{n+1} \in \mathbf{V}_{\mathbf{g}_h}^{n+1}, \end{cases} \tag{391}$$

with:

(a) $V_{\mathbf{g}_h}^{n+1} = V_{\mathbf{g}_h(t^{n+1})}$,
(b) $\Gamma_-^{n+1} = \{\mathbf{x} \mid \mathbf{x} \in \Gamma, \; \mathbf{g}_h(\mathbf{x}, t^{n+1}) \cdot \mathbf{n}(\mathbf{x}) < 0\}$,
(c) $V_h = \{\mathbf{v}_h \mid \mathbf{v}_h \in (C^0(\overline{\Omega}))^2, \; \mathbf{v}_h|_T \in P_1 \times P_1, \; \forall T \in \mathcal{T}_h\}$,
(d) $V_{0h}^{n+1,-} = \{\mathbf{v} \mid \mathbf{v} \in V_h, \mathbf{v} = 0 \text{ on } \Gamma_-^{n+1}\}$.

   Problem (388) can be viewed as a degenerated (zero viscosity) discrete Stokes problem for which efficient solution methods already exist (e.g., the discrete analogue of the preconditioned conjugate gradient algorithm for the generalized Stokes problems discussed in Section 4.6). Problem (389) can be solved by a wave-like equation method discussed in Section 3.3. Similarly problem (391) is a discrete elliptic system whose iterative or direct solution is quite a classical problem.

5.4 On the Numerical Solution of the Discrete Subproblems

The solution of the subproblems, encountered at each time step of the operator splitting schemes described in Sections 5.2 and 5.3, can be computed by iterative methods which are the discrete analogues of the conjugate gradient methods discussed in Sections 3 and 4. In particular, we shall have to solve quite systematically the linear systems approximating the elliptic systems associated to the *Helmholtz operator* $\alpha I - \nu\triangle$. Also, some of the *Stokes solvers* discussed in Section 4 require the solution of *Poisson problems* for preconditioning purposes. From the above observations, it makes sense to discuss with some detail the numerical solution of the discrete Helmholtz and Poisson problems encountered at each step of the operator splitting schemes.

On the solution of the discrete Helmholtz equations.

If the boundary conditions are of the *Dirichlet type* only (i.e., if $\Gamma_0 = \Gamma(= \partial\Omega)$), we shall have to solve problems like

$$\alpha\mathbf{u} - \nu\triangle\mathbf{u} = \mathbf{f} \ in \ \Omega, \mathbf{u} = \mathbf{g} \ on \ \Gamma. \tag{392}$$

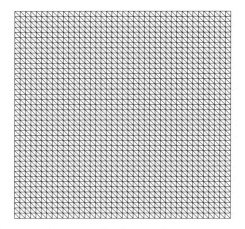

Fig. 5.2. An example of a regular triangulation.

Paradoxically, solving problem (392) is not very expensive for flow at *high Reynolds numbers*. Why? Because for such flow, the viscosity ν is small, and their fast dynamics requires small $\triangle t$, i.e., large values of α. Suppose for simplicity that $\Omega = (0,1)^2$, and also that one uses over Ω a regular triangulation like the one in Figure 5.2 where $h = 1/(I + 1)$ (I a positive integer). Suppose also that one uses continuous and piecewise linear approximations of the velocity over the above triangulation, and that integrals like $\int_\Omega \mathbf{v} \cdot \mathbf{w} dx$ are approximated using the trapezoidal rule. One obtains then the approximation of problem (392) associated to the traditional *five point finite difference scheme*, namely (with obvious notation)

$$\begin{cases} \alpha \mathbf{u}_{ij} + \dfrac{\nu}{h^2}(4\mathbf{u}_{ij} - \mathbf{u}_{i+1j} - \mathbf{u}_{i-1j} - \mathbf{u}_{ij+1} - \mathbf{u}_{ij-1}) = \mathbf{f}_{ij}, \\ 1 \le i, j \le I, \\ \mathbf{u}_{kl} = \mathbf{g}_{kl} \ if \ \{kh, lh\} \in \Gamma. \end{cases} \qquad (393)$$

It is well known that the matrix in (393) has for smallest and largest eigen-values

$$\lambda_{min} = \alpha + \frac{8\nu}{h^2} \, sin^2 \frac{\pi h}{2}, \quad \lambda_{max} = \alpha + \frac{8\nu}{h^2} sin^2 \frac{I\pi h}{2},$$

respectively. For *small values* of h, we clearly have

$$\lambda_{min} \approx \alpha + 2\pi^2 \nu, \lambda_{max} \approx \alpha + 8\nu/h^2,$$

implying that the *condition number* $\mathcal{N}$ of the above matrix verifies

$$\mathcal{N} = \lambda_{max}/\lambda_{min} \approx (\alpha + 8\nu/h^2)/(\alpha + 2\pi^2 \nu).$$

Suppose now that $\nu = 10^{-3}, h = 10^{-2}, \triangle t = 10^{-2}(\mathbb{R}ightarrow \alpha = 10^2)$; we have then

$$\mathcal{N} \approx 1.8. \qquad (394)$$

Suppose now that we solve the linear system (393) by a *nonpreconditioned conjugate gradient algorithm*. It follows then from (394) and from (126) that the distance, between the solution of problem (393) and the nth iterate, converges to zero at least as fast as

$$\left(\frac{\sqrt{1.8} - 1}{\sqrt{1.8} + 1} \right)^n = (.145898 \ldots)^n,$$

which corresponds to a high speed of convergence. A similar conclusion would hold for the *successive over-relaxation* method with optimal parameter. Actually, the convergence of the above methods is sufficiently fast (in that particular case, at least) that it makes useless further speeding up (by a multigrid method for example).

Remark 32. Suppose now that the finite element mesh used to solve problem (392) is *unstructured* (or at least less structured than the mesh shown on Figure 5.2). We advocate, then, to solve the discrete analogue of (392), namely (with obvious notation)

$$\mathbf{A}_h \mathbf{U}_h = \mathbf{F}_h, \qquad (395)$$

by a *conjugate gradient algorithm, preconditioned* by the *diagonal* $\mathbf{D}_h$ of matrix $\mathbf{A}_h$. $\square$

On the solution of the pressure related discrete Poisson problems.

The solution of the *discrete Stokes problems* (380), (382), (384), (386), by the discrete analogues of the *preconditioned conjugate gradient algorithms* discussed in Section 4, requires – at each iteration – the solution of a linear system approximating *Poisson problems* of the following types

$$
\begin{cases}
-\triangle\varphi = f \ in \ \Omega, \dfrac{\partial\varphi}{\partial n} = 0 \ on \ \Gamma, \displaystyle\int_\Omega \varphi dx = 0, \\
if \ \Gamma_0 = \Gamma (Stokes - Dirichlet \ case),
\end{cases}
\tag{396}
$$

and

$$
\begin{cases}
-\triangle\varphi = f \ in \ \Omega, \dfrac{\partial\varphi}{\partial n} = 0 \ on \ \ \Gamma_0, \varphi = 0 \ on \ \Gamma_1, \\
if \ \displaystyle\int_{\Gamma_i} d\Gamma > 0, \forall i = 0, 1 (Stokes \ problem \ with \ mixed \ boundary \ conditions).
\end{cases}
\tag{397}
$$

The matrices approximating the Laplace operators occurring in (396) and (397) do not enjoy the nice properties of the elliptic operator $\alpha I - \nu\triangle$ discussed above, concerning their condition number, and therefore the approximate solution of problems (396) and (397) may be costly (for three-dimensional problems, particularly). For *two − dimensional problems*, we advocate *direct methods* (à la *Cholesky*, for example) for solving these *discrete Poisson problems*. For *three−dimensional* flow problems, *multigrid methods* seem to be well-suited to solve problems (396) and (397); the multigrid solution of problems such as (396) and (397) has been discussed in, e.g., Glowinski [4] (Chapter 5).

Remark 33. The condition number of the finite element matrices approximating the Laplace operator in (396) and (397) behaves like h^{-2}.   □

Remark 34. To solve the linear system approximating (396), by the *method of Cholesky*, we shall proceed as follows:

(i) We delete one equation and set to zero the corresponding unknown.
(ii) We solve the remaining system by the method of Cholesky.
(iii)Let φ_h^* be the element of the pressure space P_h associated to the solution of the above linear system. Compute (via the trapezoidal rule) $m_h = \displaystyle\int_{\Omega_h} \varphi_h^* dx / meas.(\Omega_h)$ and denote by φ_h the function defined by

$$
\varphi_h = \varphi_h^* - m_h;
$$

we clearly have $\displaystyle\int_{\Omega_h} \varphi_h dx = 0.$ □

Remark 35. The discrete Poisson problems, approximating problems (396) and (397), have to be solved in the discrete pressure space P_h; if one uses the approximations defined by (301), (302) (*Hood − Taylor*), (301), (305) (*Bercovier − Pironneau*), we have 8 times more unknowns for velocity than for pressure (16 times more for three-dimensional flow).   □

6 Numerical Experiments

In this section we have considered two– and three–dimensional wall-driven cavity problems. The finite element spaces for the velocity field and pressure are P_1-*iso*-P_2 and P_1, respectively. The operator splitting scheme used in the simulation is the Marchuk–Yanenko scheme, i.e., the numerical results have been obtained via the algorithm (387)–(391). Numerical results obtained via the θ–scheme have been presented in Glowinski [82].

6.1 The Wall-driven Cavity Problem

The first test problem that we consider is the celebrated wall-driven cavity flow problem. We consider this specific test problem since it is very well documented, a basic (if not *the* basic) reference being Ghia, Ghia and Shin [105] (see also Schreiber and Keller [106]). We have then $\Omega = (0, 1) \times (0, 1)$ and $\mathbf{g}(\mathbf{x}, t)$ defined by

$$\mathbf{g}(\mathbf{x}, t) = \begin{cases} (f(x_1), 0)^T \ on \ \{\mathbf{x} \mid \mathbf{x} = (x_1, 1)^T, 0 < x_1 < 1\}, \\ \mathbf{0} \ elsewhere \ on \ \partial\Omega \end{cases} \tag{398}$$

where

$$f(x) = \begin{cases} \sin(x\pi/2a), & if \ 0 < x \leq a, \\ 1, & if \ a \leq x \leq 1 - a, \\ \sin((1 - x)\pi/2a), & if \ 1 - a \leq x < 1. \end{cases} \tag{399}$$

The above Dirichlet data has been smoothed very locally in the two upper corners (the parameter a is $1/32$ in all simulations reported in this section). In order to avoid possible difficulties associated to a genuine impulsive start we have multiplied $\mathbf{g}(\mathbf{x}, t)$ in (398) by $\theta(t)$ defined by $\theta(t) = 1 - e^{-50t}$ if $t \in (0, .15)$ and $\theta(t) = 1$ for $t \geq .15$.

In Ghia, Ghia and Shin [105] uniform grids were also chosen. Another reason we have chosen regular triangulations is that we wanted to use direct solvers like fast elliptic solvers based on cyclic reduction to solve elliptic problems from (388) and (391). When computing steady state solutions, we

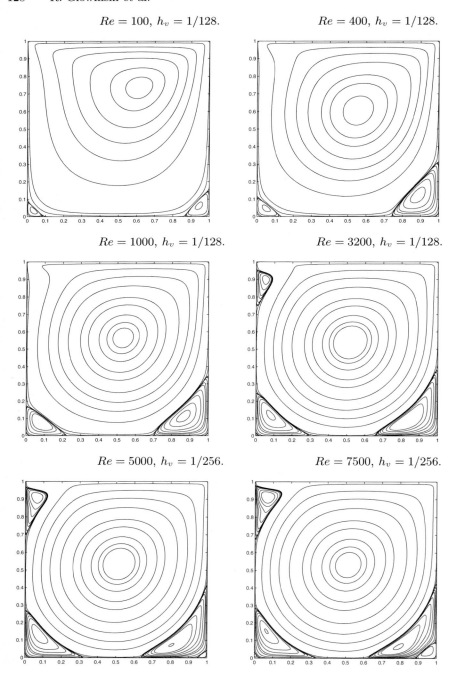

Fig. 6.1. Streamlines.

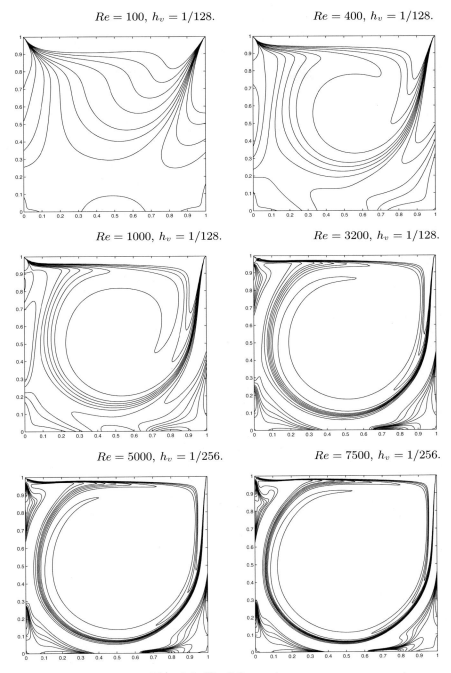

$Re = 100, \ h_v = 1/128.$

$Re = 400, \ h_v = 1/128.$

$Re = 1000, \ h_v = 1/128.$

$Re = 3200, \ h_v = 1/128.$

$Re = 5000, \ h_v = 1/256.$

$Re = 7500, \ h_v = 1/256.$

Fig. 6.2. Vorticity contours.

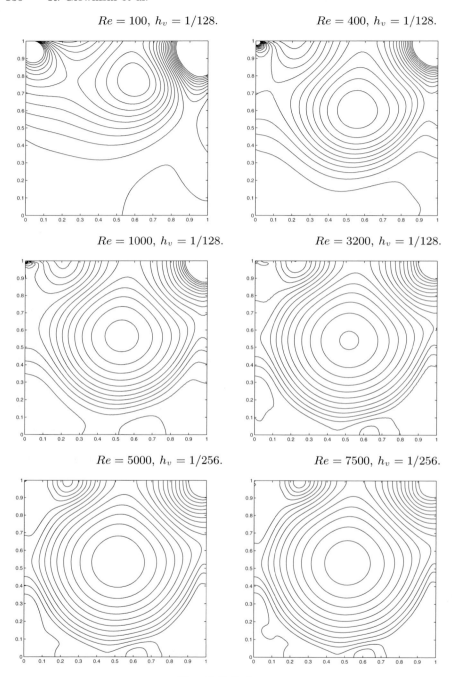

Fig. 6.3. Isobars.

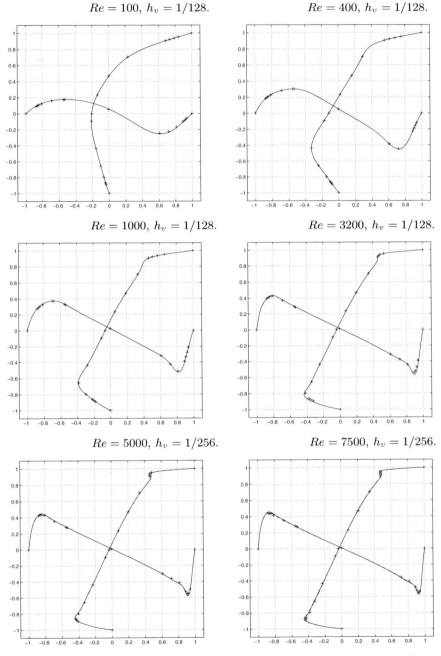

Fig. 6.4. Comparison between the computed u_1-velocity (resp., u_2-velocity) along the line $x_1 = 1/2$ (resp., $x_2 = 1/2$) (solid lines) and the results reported in Ghia, Ghia and Shin [105] (denoted by "+").

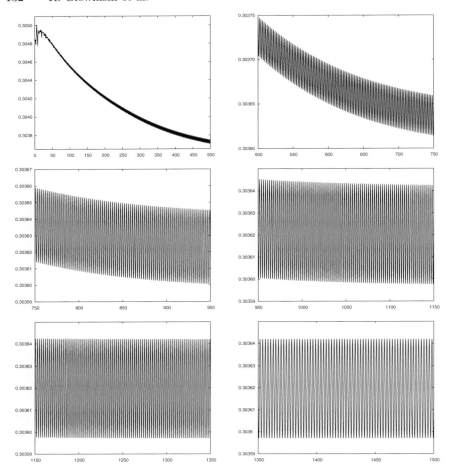

Fig. 6.5. The history of $\|\mathbf{u}_{\mathbf{h}}^n\|_2$ for the flow at $Re = 8500$ with $h_v = 1/256$ and $\triangle t = 0.0005$.

have taken $h_v = 1/128$ as mesh size for the velocity field for Reynolds numbers ($Re = 1/\nu$) up to 7500 and $h_v = 1/256$ for the range of Reynolds number from 5000 to 7500. Same mesh sizes were used in Ghia, Ghia and Shin [105]. For the time discretization we have taken $\triangle t = 0.0005$ and $Q_1 = 5$ in (195)-(197).

In the simulation when the relative change, $\|\mathbf{u}^n - \mathbf{u}^{n-1}\|_2/\|\mathbf{u}^n\|_2$, is less than 10^{-7}, $\mathbf{u}^n$ is taken as steady state. With $h_v = 1/128$, we computed the $Re = 100$ case with $\mathbf{u_0} = \mathbf{0}$ and then used the steady state of $Re = 100$ as the initial flow field condition for the $Re = 400$ case, and repeated this process up to the $Re = 8500$ case. We also used the steady state result at $Re = 5000$, obtained with $h_v = 1/128$, as the initial flow field condition for the case at

Fig. 6.6. One complete cycle of streamline contours at time interval of 2.27 at $Re = 8500$ with $h_v = 1/256$ (from left to right and from top to bottom).

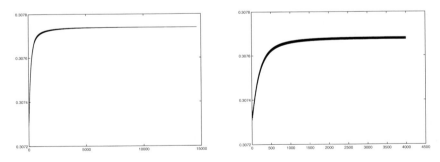

Fig. 6.7. The history of $\|\mathbf{u}_h^n\|_2$ for the flow at $Re = 8343$ (left) $Re = 8375$ (right) with $h_v = 1/128$ and $\triangle t = 0.0005$.

Re	ψ_{min}	location	ψ^*_{min}	location*
100	-0.10343	(0.617, 0.734)	-0.10342	(0.617, 0.734)
400	-0.11391	(0.555, 0.609)	-0.11391	(0.555, 0.605)
1000	-0.11917	(0.531, 0.562)	-0.11793	(0.531, 0.562)
3200	-0.12177	(0.516, 0.539)	-0.12038	(0.516, 0.547)
5000	-0.12122	(0.516, 0.535)	-0.11897	(0.512, 0.532)
7500	-0.12082	(0.516, 0.531)	-0.11998	(0.512, 0.532)

Table 6.1. Location at which the minimum ψ_{min} of the stream function is attained and minimal value of the stream function (ψ^*_{min} and location* are values taken from Ghia, Ghia and Shin [105]).

$Re = 5000$ with $h_v = 1/256$, and repeated this process up to the $Re = 8500$ case.

On Figures 6.1–6.3, we have visualized the streamlines, the vorticity contours and the isobars of the steady state solutions computed at $Re = 100$, 400, 1000, 3200, 5000, and 7500 respectively. Those values used to draw the streamlines and the vorticity contours are taken from Table III in Ghia, Ghia and Shin 1131. The values used to draw the isobars are from -0.1 to 0.1 with increment 0.01. In Figure 6.4, we have compared the computed u_1-velocity (resp., u_2-velocity) along the line $x_1 = 1/2$ (resp., $x_2 = 1/2$) with those results reported in Tables I and II in Ghia, Ghia and Shin [105]. In Table 6.1, we have the minimum of the stream function and the location at which the minimum is attained at $Re = 100$, 400, 1000, 3200, 5000, and 7500 respectively from the numerical experiments and Ghia, Ghia and Shin [105]. These results agree remarkably well with those obtained in Ghia, Ghia and Shin [105] using the stream function-vorticity formulation of the steady Navier-Stokes equations.

At $Re = 8500$, we are beyond a Hopf bifurcation point. In order to ensure that the computed periodic solution is not a numerical artifact, we have run simulations with three sets of mesh size and time step, namely $(h_v, \triangle t) = (1/128, 0.0005)$, $(h_v, \triangle t) = (1/128, 0.00025)$ and $(h_v, \triangle t) = (1/256, 0.0005)$. For the local time step in (195)-(197), we still have taken $Q_1 = 5$. For the case where $(h_v, \triangle t) = (1/128, 0.0005)$, we started from a steady state solution at $Re = 7500$ with $h_v = 1/128$ and run it till $t = 1500$. It took about 0.18 second per time step on a DEC personal workstation 500au. For the second case where $(h_v, \triangle t) = (1/128, 0.00025)$, we started from a solution of the previous case and run it till $t = 1650$. It took about 0.146 second per time step. For the third case where $(h_v, \triangle t) = (1/256, 0.0005)$, we started from a steady state solution at $Re = 7500$ with $h_v = 1/256$ and run it till $t = 1500$. It took about 0.76 second per time step.

The period of the Hopf bifurcated flow, at $Re = 8500$, is 2.24, 2.22, and 2.27 for the simulations done with $(h_v, \triangle t) = (1/128, 0.0005)$, $(h_v, \triangle t) = (1/128, 0.00025)$ and $(h_v, \triangle t) = (1/256, 0.0005)$, respectively. In Figure 6.5 the history of $\|\mathbf{u_h^n}\|_2$ for the flow at $Re = 8500$ with $(h_v, \triangle t) = (1/256, 0.0005)$ is presented. We can see, clearly, in Figure 6.5 that the solution reaches its asymptotic periodic state at $t = 1500$. In Figure 6.6, we have plotted a series of nine streamline contours for the flow at $Re = 8500$ with $(h_v, \triangle t) = (1/256, 0.0005)$ during a time interval of length 2.27 so that the nine plots make one complete period. We observe that there are persistent oscillations for all the secondary and tertiary vortices. The most significant changes during one period are the periodic appearance and disappearance of two tertiary vortices at the bottom left and at the top left. In Bruneau and Jouron [107], a transition to turbulence in the unit driven cavity flow was found for Reynolds number between 5000 and 10000 by solving the steady Navier-Stokes equations with a 512×512 grid. In Shen [108] due to the use of a regularized boundary condition $f(x) = 16x^2(1 - x)^2$, the critical Reynolds number for Hopf bifurcation is in the $(10,000, 10,500]$ range which is higher than the one for the less smooth boundary conditions used in Bruneau and Jouron [107] and Goyon [109], and the one in this article. In Goyon [109] the critical Reynolds number for the Hopf bifurcation is in the $(7,500, 10,000]$ range. A periodic solution was found at $Re = 10,000$ with period 2.41. Our results indicate that the critical Reynolds number is between 7500 and 8500. We then did use the flow field at $Re = 8500$ with $(h_v, \triangle t) = (1/128, 0.0005)$ as initial condition to roughly locate the critical Reynolds number for the Hopf bifurcation. At $Re = 8343$, we have obtained steady state and the history of $\|\mathbf{u_h^n}\|_2$ is shown in Figure 6.7. At $Re = 8375$, we have obtained Hopf bifurcated flow and the period is 2.235, whose the history of $\|\mathbf{u_h^n}\|_2$ is shown in Figure 6.7. Thus the critical Reynolds number (when using $(h_v, \triangle t) = (1/128, 0.0005)$) is between 8343 and 8375. Actually, using a spectral approximation, Auteri, Parolini and Quartapelle [110] have been able to locate the critical Reynolds number at $Re \approx 8020$.

6.2 A Three-dimensional Wall-driven Cavity Problem

In this section we consider a three-dimensional wall-driven cavity flow problem. We have $\Omega = (0, 1) \times (0, 1) \times (0, 1)$ and $\mathbf{g}(\mathbf{x}, t)$ defined by

$$\mathbf{g}(\mathbf{x}, t) = \begin{cases} (f(x_1)^2 f(x_2)^2 \theta(t), 0, 0)^T & on \ \{\mathbf{x} \mid \mathbf{x} = (x_1, x_2, 1)^T, 0 < x_1, x_2 < 1\}, \\ \mathbf{0} \ elsewhere \ on \ \partial\Omega \end{cases}$$

$$(400)$$

where $f(x)$ and $\theta(t)$ are defined in the previous section but with $a = 1/20$.

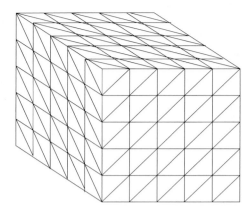

Fig. 6.8. An example of a regular tetrahedrization $\mathcal{T}_h$ of Ω.

The velocity $\mathbf{u}$ has been approximated by continuous piecewise affine vector-valued functions defined from a regular uniform "tetrahedrization" $\mathcal{T}_h$ with mesh size h_v (see Figure 6.8). The pressure p has been approximated by continuous piecewise affine functions, defined over a tetrahedrization $\mathcal{T}_{2h}$. Thus the mesh size for the pressure is $h_p = 2h_v$.

In the numerical experiments dedicated to steady flow computations, we have used two mesh sizes $h_v = 1/60$ and $1/80$ for $Re = 400$, 1000 and 1500. For the time discretization we have taken $\triangle t = 0.001$ and $Q_1 = 2$ in (195)-(197). Following the criterion used in Fujima, Tabata and Fukasawa [111], when the change in the simulation, $\|\mathbf{u^n} - \mathbf{u^{n-1}}\|_\infty / \triangle \mathbf{t}$, is less than 10^{-4}, $\mathbf{u^n}$ is taken as steady state. The initial condition of fluid field for $Re = 400$ (resp, 1000 and 1500) is the steady-state solution of $\mathbf{u}^n$ obtained at $Re = 100$ (resp., 400 and 1000) using the same mesh size and time step.

Figures 6.9 and 6.10 show the comparison of the our computational results at $Re = 400$ and 1000 with results obtained by Fujima, Tabata and Fukasawa [111] and Ku, Hirsh and Taylor [112], and Chiang, Sheu, Hwang [113]. All numerical results are in good agreement.

Velocity vectors of the steady flows obtained in the case of $Re = 400$, 1000, and 1500 are shown in Figures 6.11-6.13. Those vectors are projected orthogonally to the three planes, $x_2 = 0.5$, $x_1 = 0.5$, and $x_3 = 0.5$, and the length of the vectors has been doubled in the two later planes to observe the flow more clearly. We observe that the center of the primary vortex moves down as the Reynolds increases and secondary vortices appear in two lower corners, which is similar, in some sense, to what happens for the two-dimensional wall-driven cavity flow. At $x_1 = 0.5$, a pair of secondary vortices moves toward the lower corners when the Reynolds increases. Also another pair of vortices appears at the top corners at $Re = 1000$ and 1500.

At $Re = 3200$, the initial condition of fluid field is the steady-state solution of $\mathbf{u}^n$ obtained at $Re = 1500$ obtained with mesh size $h_v = 1/80$ and time step

$\triangle t = 0.001$. In Figure 6.14 the two- and three-minute averaged values of the 3D cavity central plane velocity profiles are compared with the experimental values by Prasad and Koseff [114]. The agreement is good. The vectors of velocity field at $t = 420$ are projected orthogonally to the three planes, $x_2 = 0.5$, $x_1 = 0.8125$, and $x_3 = 0.5$, and the length of the vectors has been doubled in the two later planes to observe the flow more clearly. A well developed pair of Taylor–Göertler–like vortices can be first observed at $t = 12$ (at least $t = 9$ there is no sign of it at $x = 0.8125$). The meandering of Taylor–Görtler–like vortices and its interaction with corner vortices is shown in Figures 6.16-6.17.

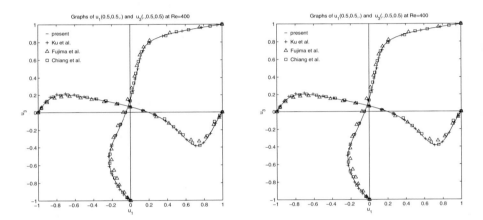

Fig. 6.9. 3D cavity central plane velocity profiles for $Re = 400$ obtained using $\triangle t = 0.001$ and mesh size $h_v = 1/60$ (left) and $1/80$ (right).

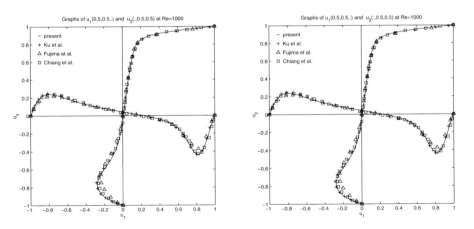

Fig. 6.10. 3D cavity central plane velocity profiles for $Re = 1000$ obtained using $\triangle t = 0.001$ and mesh size $h_v = 1/60$ (left) and $1/80$ (right).

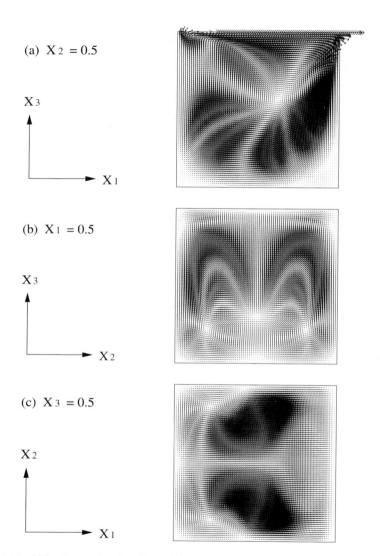

(a) $X_2 = 0.5$

X_3

X_1

(b) $X_1 = 0.5$

X_3

X_2

(c) $X_3 = 0.5$

X_2

X_1

Fig. 6.11. Velocity vector for $Re = 400$ at $t = 20.352$ on the planes (a) $x_2 = 0.5$, (b) $x_1 = 0.5$, and (c) $x_3 = 0.5$ obtained with mesh size $h_v = 1/80$.

(a) $X_2 = 0.5$

X_3

X_1

(b) $X_1 = 0.5$

X_3

X_2

(c) $X_3 = 0.5$

X_2

X_1

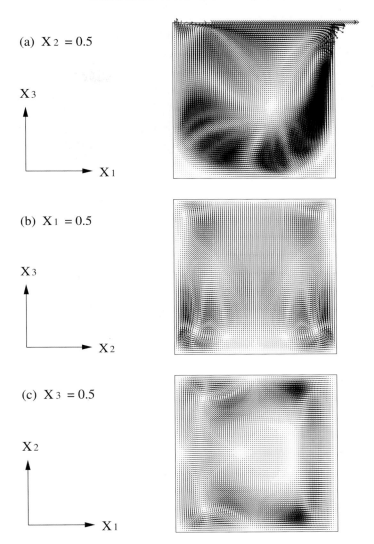

Fig. 6.12. Velocity vector for $Re = 1000$ at $t = 35.409$ on the planes (a) $x_2 = 0.5$, (b) $x_1 = 0.5$, and (c) $x_3 = 0.5$ obtained with mesh size $h_v = 1/80$.

(a) X$_2$ = 0.5

X$_3$

X$_1$

(b) X$_1$ = 0.5

X$_3$

X$_2$

(c) X$_3$ = 0.5

X$_2$

X$_1$

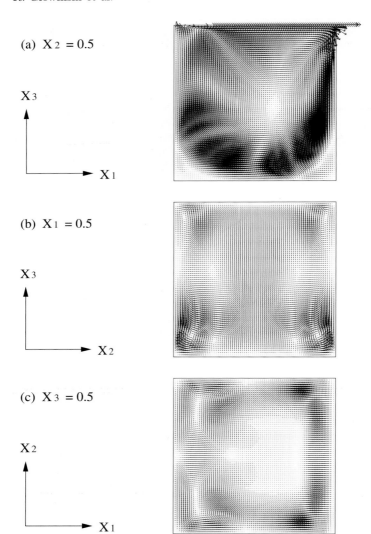

Fig. 6.13. Velocity vector for $Re = 1500$ at $t = 64.511$ on the planes (a) $x_2 = 0.5$, (b) $x_1 = 0.5$, and (c) $x_3 = 0.5$ obtained with mesh size $h_v = 1/80$.

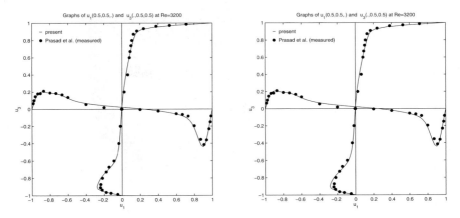

Fig. 6.14. 3D cavity central plane velocity profiles for $Re = 3200$. The simulation results are averaged values obtained from $t = 220$ second to $t = 340$ second (left) and $t = 220$ second to $t = 400$ second (right).

(a) $X_2 = 0.5$

X_3

X_1

(b) $X_1 = 0.8125$

X_3

X_2

(c) $X_3 = 0.5$

X_2

X_1

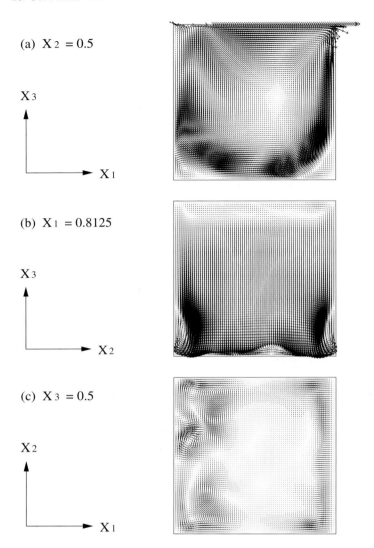

Fig. 6.15. Velocity vector for $Re = 3200$ at $t = 420$ on the planes (a) $x_2 = 0.5$, (b) $x_1 = 0.8125$, and (c) $x_3 = 0.5$ obtained with mesh size $h_v = 1/80$.

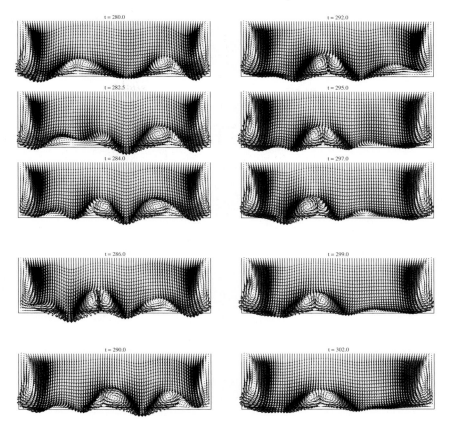

Fig. 6.16. Flow filed projected to the plane $x_1 = 0.8125$ near the downstream sidewall at $t =$280, 282.5, 284, 286, 290 92, 295, 297, 299 and 302 second (from top to bottom and then from left to right) for $Re = 3200$.

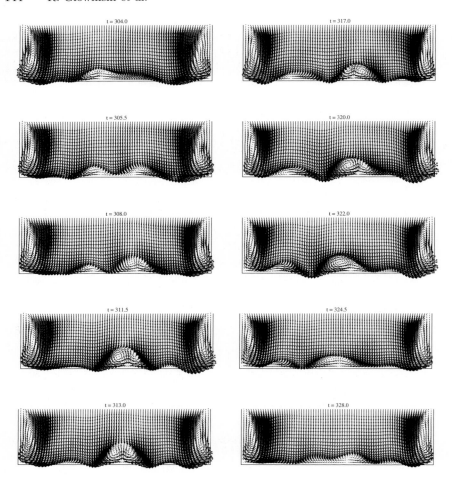

Fig. 6.17. Flow filed projected to the plane $x_1 = 0.8125$ near the downstream sidewall at t =304, 305.5, 308, 311.5, 313 317, 320, 322, 324.5, and 328 second (from top to bottom and then from left to right) for $Re = 3200$.

Acknowledgement

The authors acknowledge the support of NSF (grants DMS–9973318, CCR–9902035, and DMS–0209066), Texas Board of Higher Education (ARP grant 003652–0383–1999), and DOE/LASCI (grant R71700K–292–000–99). They warmly thank Professors V. Capasso and J. Periaux for inviting them to contribute to the EU funded 2003 Jacques–Louis Lions Summer School at Montecatini, Italy.

References

1. C. Canuto, M.Y. Hussaini, A. Quarteroni, and T.A. Zang, *Spectral Methods in Fluid Dynamics*, Springer-Verlag, New York, 1988.
2. M. Lesieur, *Turbulence in Fluids*, Kluwer, Dordrecht, 1990.
3. E. Guyon, J.P. Hulin, and L. Petit, *Hydrodynamique Physique*, Intereditions/Editions du CNRS, Paris, 1991.
4. R. Glowinski, Finite element methods for incompressible viscous flow, in *Handbook of Numerical Analysis*, **Vol. IX**, P.G. Ciarlet, J.-L. Lions eds., North-Holland, Amsterdam, 2003, 3–1176.
5. W. Prager, *Introduction to Mechanics of Continua*, Ginn and Company, Boston, MA, 1961.
6. G.K. Batchelor, *An Introduction to Fluid Mechanics*, Cambridge University Press, Cambridge, U.K., 1967.
7. A.J. Chorin and J.E. Marsden, *A Mathematical Introduction to Fluid Mechanics*, Springer-Verlag, New York, 1990.
8. R. Glowinski, *Numerical Methods for Nonlinear Variational Problems*, Springer–Verlag, New York, 1984.
9. M.O. Bristeau, R. Glowinski, B. Mantel, J. Periaux, P. Perrier, Numerical Methods for incompressible and compressible Navier-Stokes problems, in *Finite Element in Fluids*, **Vol. 6**, R.H. Gallagher, G. Carey, J.T. Oden, and O.C. Zienkiewicz eds., J. Wiley, Chicester, 1985, 1–40.
10. O. Pironneau, *Finite Element Methods for Fluids*, J. Wiley, Chicester, 1989.
11. J. Leray, Sur le mouvement d'un liquide visqueux emplissant l'espace, *Acta Mathematica*, **63**(1934), 193–248.
12. J. Leray, Essai sur les mouvements d'un liquide visqueux que limitent des parois, *J. Math. Pures et Appl.*, **13**(1934), 331–418.
13. E. Hopf, Uber die Anfangswertaufgabe fur die hydrodynamischen Grundgleichungen, *Math. Nachrichten*, **4**(1951), 213–231.
14. J. Leray, Aspects de la mécanique théorique des fluides, *La Vie des Sciences*, Comptes Rendus de l'Académie des Sciences, Paris, Série Générale, **11**(1994), 287–290.
15. J.L. Lions and G. Prodi, Un théorème d'existence et d'unicité dans les équations de Navier-Stokes en dimension 2, *C.R. Acad. Sci., Paris* **248**, 3519–3521.
16. J.L. Lions, *Equations Différentielles Opérationnelles et Problèmes aux Limites*, Springer–Verlag, Berlin, 1961.
17. J.L. Lions, *Quelques Méthodes de Résolution des Problèmes aux Limites Non Linéaires*, Dunod, Paris, 1969.
18. O. Ladysenskaya, *Theory and Numerical Analysis of the Navier-Stokes Equations*, Gordon and Breach, New York, NY, 1969.
19. R. Temam, *The Mathematical Theory of Viscous Incompressible Flow*, North-Holland, Amsterdam, 1977.
20. L. Tartar, *Topics in Nonlinear Analysis*, Publications Mathématiques d'Orsay, Université Paris-Sud, Départment de Mathématiques, Paris, 1978.
21. H.O. Kreiss and J. Lorenz, *Initial-Boundary Value Problems and the Navier-Stokes Equations*, Academic Press, Boston, MA, 1989.
22. P.L. Lions, *Mahtematical Topics in Fluid Mechanics, Vol I: Incompressible Models*, Oxford University, Oxford, UK, 1996.

23. M. Marion and R. Temam, *Navier-Stokes Equations*, in *Handbook of Numerical Analysis*, **Vol. VI**, P.G. Ciarlet, J.-L. Lions eds., North-Holland, Amsterdam, 1998, 503–689.

24. P.G. Ciarlet, *The Finite Element Method for Elliptic Problems*, North-Holland, Amsterdam, 1978.

25. F. Brezzi and M. Fortin, *Mixed and Hybrid Finite Element Methods*, Springer-Verlag, New York, NY, 1991.

26. S.C. Brenner and L.R. Scott *The Mathematical Theory of Finite Element Methods*, Springer-Verlag, New York, NY, 1994.

27. N.N. Yanenko, *The Method of Fractional Steps*, Springer-Verlag, Berlin, 1971.

28. G.I. Marchuk, *Methods of Numerical Mathematics*, Springer-Verlag, New York, NY, 1975.

29. G.I. Marchuk, Splitting and alternating direction methods. In Ciarlet, P.G., and Lions, J.L. (eds.) *Handbook of Numerical Analysis*, **Vol I**, North-Holland, Amsterdam, 1990, 197–462.

30. M. Crouzeix and A. Mignot, *Analyse Numérique des Equations Différentielles Ordiuaires*, Masson, Paris, 1984.

31. R. Glowinski and P. Le Tallec, *Augmented Lagrangians and Operator Splitting Methods in Nonlinear Mechanics*, SIAM, Philadelphia, PA, 1989.

32. R. Glowinski, Viscous flow simulation by finite element methods and related numerical techniques. In Murman, E.M., and Abarbanel, S.S. (eds.) *Progress and Supercomputing in Computational Fluid Dynamics*, Birkhauser, Boston, MA, 1985, 173–210.

33. R. Glowinski, Splitting methods for the numerical solution of the incompressible Navier-Stokes equations. In Balakrishnan, A.V., Dorodnitsyn, A.A., and Lions, J.L. (eds.) *Vistas in Applied Mathematics*, Optimization Software, New York, NY, 1986, 57–95.

34. G. Strang, On the construction and comparison of difference schemes, *SIAM J. Num. Anal.*, **5**(1968), 506–517.

35. J.T. Beale and A. Majda, Rates of convergence for viscous splitting of the Navier-Stokes equations, *Math. Comp.*, **37**(1981), 243–260.

36. R. Leveque and J. Oliger, Numerical methods based on additive splitting for hyperbolic partial differential equations, *Math. Comp.*, **37** (1983), 243–260.

37. P.A. Raviart and J.M. Thomas, *Introduction à l'Analyse Numérique des Equations aux Dérivées Partielles*, Masson, Paris, 1983.

38. A.J. Chorin, A numerical method for solving incompressible viscous flow problems, *J. Comp. Phys.*, **2** (1967), 12–26.

39. A.J. Chorin, Numerical solution of the Navier-Stokes equations, *Math. Comp.*, **23** (1968), 341–354.

40. R. Temam, Sur l'approximation des équations de Navier-Stokes par la méthode des pas fractionnaires (I), *Arch. Rat. Mech. Anal.*, **32** (1969), 135–153.

41. R. Temam, Sur l'approximation des équations de Navier-Stokes par la méthode des pas fractionnaires (II), *Arch. Rat. Mech. Anal.*, **33** (1969), 377–385.

42. A. Quarteroni and A. Valli, *Numerical Approximation of Partial Differential Equations*, Springer-Verlag, Berlin, 1994.

43. J. Daniel, *The Approximate Minimization of Functionals*, Prentice Hall, Englewood Cliffs, NJ, 1970.

44. E. Polak, *Computational Methods in Optimization*, Academic Press, New York, NY, 1971.

45. M.R. Hestenes and E.L. Stiefel, Methods of conjugate gradients for solving linear systems, *J. Res. Bureau National Standards*, Section B, **49** (1952), 409–436.

46. R.W. Freund, G.H. Golub, and N.M. Nachtigal, Iterative solution of linear systems, *Acta Numerica 1992*, Cambridge University Press, 1992, 57–100.

47. J. Nocedal, Theory of algorithms for unconstrained optimization, *Acta Numerica 1992*, Cambridge University Press, 1992, 199–242.

48. C.T. Kelley, *Iterative Methods for Linear and Nonlinear Equations*, SIAM, Philadelphia, PA, 1995.

49. Y. Saad, *Iterative Methods for Sparse Linear Systems*, PWS, Boston, MA, 1995.

50. G.H. Golub and D.P. O'Leary, Some history of the conjugate gradient and Lanczos algorithms: 1948-1976, *SIAM Review*, **31** (1989), 50–102.

51. E. Zeidler, *Nonlinear Functional Analysis and its Applications. Volume I: Fixed-Point Theorems*, Springer-Verlag, New York, NY, 1986.

52. I. Ekeland and R. Teman, *Convex Analysis and Variational Problems*, North-Holland, Amsterdam, 1976.

53. J.M. Ortega and W.C. Rheinboldt, *Iterative Solution of Nonlinear Equations in Several Variables*, Academic Press, New York, NY, 1970.

54. J.M. Ortega, and W.C. Rheinboldt, Local and global convergence of generalized linear iterations. In Ortega, J.M., and Rheinboldt, W.C. (eds.) *Numerical Solution of Nonlinear Problems*, SIAM, Philadelphia, PA, 1970.

55. J.M. Ortega and W.C. Rheinboldt, A general convergence result for unconstrained minimization methods, *SIAM J. Num. Anal.*, **9** (1972), 40–43.

56. M. Avriel, *Nonlinear Programming: Analysis and Methods*, Prentice-Hall, Englewood Cliffs, NJ, 1976.

57. M.J.D. Powell, Some convergence properties of the conjugate gradient method, *Math. Program.*, **11** (1976), 42–49.

58. M.J.D. Powell, Restart procedures of the conjugate gradient method, *Math. Program.*, **12** (1977), 148–162.

59. V. Girault and P.A. Raviart, *Finite Element Methods for Navier-Stokes Equations: Theory and Algorithms*, Springer-Verlag, Berlin, 1986.

60. J.B. Hiriart-Urruty and C. Lemarechal, *Convex Analysis and Minimization Algorithms*, Springer-Verlag, Berlin, 1993.

61. M. Crouzeix, Etude d'une méthode de linéarisation. Résolution numérique des équations de Stokes stationaires. In *Approximations et Méthodes Itératives de Résolution d'Inéquations Variationelles et de Problèmes Non Linéaires*, Cahiers de l'IRIA, **12**, 1974, 139-244.

62. M. Crouzeix, On an operator related to the convergence of Uzawa's algorithm for the Stokes equation. In Bristeau, M.O., Etgen, G., Fitzgibbon, W., Lions, J.L., Périaux, J., and Wheeler, M.F. (eds.) *Computational Science for the 21st Century*, Wiley, Chichester, 1997, 242–259.

63. J. Cahouet and J.P. Chabard, Some fast 3-D solvers for the generalized Stokes problem, *Int. J. Numer. Meth. in Fluids*, **8**, 1988, 269–295.

64. J.E. Dennis and R.B. Schnabel, A view of unconstrained optimization. In Newhauser, G.L., Rinnooy Kan, A.H.G., and Todd, M.J. (eds.) *Handbook in Operations Research and Management Science*, **Vol. 1**: *Optimization*, North-Holland, Amsterdam, 1989, 1–66.

65. F. Thomasset, *Implementation of Finite Element Methods for Navier-Stokes Equations*, Springer-Verlag, New York, NY, 1981.

66. R. Peyret and T.D. Taylor, *Computational Methods for Fluid Flow*, Springer-Verlag, New York, NY, 1982.

67. C. Cuvelier, A. Segal, and A. Van Steenhoven, *Finite Element Methods and Navier-Stokes Equations*, Reidel, Dordrecht, 1986.

68. M. Fortin, Finite element solution of the Navier-Stokes equations, *Acta Numerica 1993*, Cambridge University Press, 1993, 239–284.

69. M.D. Gunzburger, *Finite Element Methods for Viscous Incompressible Flows*, Academic Press, Boston, MA, 1989.

70. C.A.J. Fletcher, *Computational Techniques for Fluid Dynamics, Volume 1: Fundamental and General Techniques*, Springer-Verlag, Berlin, 1991.

71. C.A.J. Fletcher, *Computational Techniques for Fluid Dynamics, Volume 2: Specific Techniques for Different Flow Categories*, Springer-Verlag, Berlin, 1991.

72. M.D. Gunzburger and R.A. Nicolaides (eds.), *Incompressible Computational Fluid Dynamics,* Cambridge University Press, New York, NY, 1993.

73. L. Quartapelle, *Numerical Solution of the Incompressible Navier-Stokes Equations*, Birkhauser, Basel, 1993.

74. F.K. Hebeker, R. Rannacher, and G. Wittum (eds.), *Numerical Methods for the Navier-Stokes Equations*, Vieweg, Braunschweig/Wiesbaden, 1994.

75. P.M. Gresho and R.L. SANI, *Incompressible Flow and the Finite Element Method: Advection-Diffusion and Isothermal Laminar Flow,* J. Wiley, Chichester, 1998.

76. E. Fernandez-Cara and M.M. Beltran, The convergence of two numerical schemes for the Navier-Stokes equations, *Numerische Mathematik*, **55** (1989), 33–60.

77. P. Kloucek and F.S. Rys, On the stability of the fractional step-θ-scheme for the Navier-Stokes equations, *SIAM J. Num. Anal.*, **31** (1994), 1312–1335.

78. P. Hood and C. Taylor, A numerical solution of the Navier-Stokes equations using the finite element technique, *Computers and Fluids*, **1** (1973), 73–100.

79. M. Bercovier and O. Pironneau, Error estimates for finite element method solution of the Stokes problem in the primitive variables, *Numer. Math.*, **33** (1979), 211–224.

80. G. Strang and G. Fix, *An Analysis of the Finite Element Method*, Prentice Hall, Englewood Cliffs, NJ, 1973.

81. P.G. Ciarlet, Basic error estimates for elliptic problems. In Ciarlet, P.G., and Lions, J.L. (eds.) *Handbook of Numerical Analysis*, **Vol. II**, North-Holland, Amsterdam, 1991, 17–351.

82. R. Glowinski, Finite element methods for the numerical simulation of incompressible viscous flow. Introduction to the control of the Navier-Stokes equations. In Anderson, C.R., and Greengard, C. (eds.) *Vortex Dynamics and Vortex Methods*, Lecture in Applied Mathematics, Vol. 28, American Mathematical Society, Providence, RI, 1991, 219–301.

83. M.O. Bristeau, R. Glowinski, and J. Periaux, Numerical methods for the Navier-Stokes equations. Applications to the simulation of compressible and incompressible viscous flow, *Computer Physics Reports*, **6** (1987), 73–187.

84. E.J. Dean, R. Glowinski, and C.H. Li, Supercomputer solution of partial differential equation problems in Computational Fluid Dynamics and in Control, *Computer Physics Communications*, **53** (1989), 401–439.

85. R. Glowinski and O. Pironneau, Finite element methods for Navier-Stokes equations, *Annual Review of Fluid Mechanics*, **24** (1992), 167–204.
86. T.J.R. Hughes, L.P. Franca, and M. Balestra, A new finite element formulation for Computational Fluid Dynamics: V. Circumventing the Babaska-Brezzi Condition; A stable Petrov-Galerkin formulation of the Stokes problem accomodating equal-order interpolation, *Comp. Meth. Appl. Mech. Eng.*, **59** (1986), 85–100.
87. J. Douglas and J. Wang, An absolutely stabilized finite element method for the Stokes problem, *Math. Comp.*, **52** (1989), 495–508.
88. Z. Cai and J. Douglas, An analytic basis for multigrid methods for stabilized finite element methods for the Stokes problem. In Bristeau, M.O., Etgen, G., Fitzgibbon, W., Lions, J.L., Periaux, J., and Wheeler, M.F. (eds.) *Computational Science for the 21st Century*, Wiley, Chichester, 1997, 113–118.
89. R. Glowinski and C.H. Li, On the numerical implementation of the Hilbert Uniqueness Method for the exact boundary controllability of the wave equation, *C.R. Acad. Sc., Paris*, t. 311 (1990), Série I, 135-142.
90. Glowinski, R., C.H. Li, and J.L. Lions, A numerical approach to the exact boundary controllability of the wave equation (I) Dirichlet controls: Description of the numerical methods, *Japan J. Applied Math.*, **7** (1990), 1–76.
91. R. Glowinski and J.L. Lions, Exact and approximate controllability for distributed parameter systems, Part II, *Acta Numerica 1995*, Cambridge University Press, 1995, 159–333.
92. M. Crouzeix and P.A. Raviart, Conforming and nonconforming finite element methods for solving the stationary Stokes equations, *Revue Française d'Automatique, Informatique et Recherche Opérationnelle*, **R3** (1973), 33–76.
93. J.E. Roberts and J.M. Thomas, Mixed and hybrid methods. In Ciarlet, P.G., and Lions, J.L. (eds.) *Handbook of Numerical Analysis*, **Vol. II**, North-Holland, Amsterdam, 1991, 523–639.
94. R. Verfurth, Error estimates for a mixed finite element approximation of the Stokes problem, *Revue Française d'Automatique, Informatique et Recherche Opérationelle, Anal. Numer.*, **18** (1984), 175–182.
95. R. Glowinski, T.W. Pan, T.I. Hesla, and D.D. Joseph, A distributed Lagrange multiplier/fictitious domain method for particulate flow, *Int. J. Multiphase Flow*, **25** (1999), 755–794.
96. R. Glowinski, T.W. Pan, T.I. Hesla, D.D. Joseph, and J. Periaux , A distributed Lagrange multiplier/fictitious domain method for flows around moving rigid bodies: Application to particulate flow, *Int. J. Numer. Meth. in Fluids*, **30** (1999), 1043–1066.
97. R. Glowinski, T.W. Pan, T.I. Hesla, D.D. Joseph, and J. Periaux, A distributed Lagrange multiplier/fictitious domain method for the simulation of flow around moving rigid bodies: Application to particulate flow, *Comp. Meth. Appl. Mech. Eng.*, **184** (2000), 241–267.
98. R. Glowinski, T.W. Pan, T.I. Hesla, D.D. Joseph, and J. Periaux, A fictitious domain approach to the direct numerical simulation of incompressible viscous fluid flow past moving rigid bodies: Application to particulate flow, *J. Comp. Phys.*, **169** (2001), 363–426.
99. S. Turek, A comparative study of time-stepping techniques for the incompressible Navier-Stokes equations: from fully implicit non-linear schemes to semi-implicit projection methods, *Int. J. Num. Math. in Fluids*, **22**(1996), 987-1011.

100. E.J. Dean, R. Glowinski, A wave equation approach to the numerical solution of the Navier-Stokes equations for incompressible viscous flow, *C.R. Acad. Sci. Paris*, t. 325, Série I, (1997), 783–791.

101. E. Dean, R. Glowinski, T.-W. Pan, A wave equation approach to the numerical simulation of incompressible viscous fluid flow modeled by the Navier-Stokes equations, in *Mathematical and numerical aspects of wave propagation*, J. De Santo ed., SIAM, Philadelphia, 1998, 65–74.

102. R. Glowinski, L.H. Juarez, Finite element method and operator splitting for a time-dependent viscous incompressble free surface flow, *Computational Fluid Dynamics Journal*, **9**(2003), 459–468.

103. R. Glowinski, O. Pironneau, Finite Element Methods for Navier-Stokes Equations, *Annu. Rev. Fluid Mech.*, **24**(1992), 167–204.

104. C. Johnson, Streamline diffusion methods for problems in fluid mechanics, in *Finite Element in Fluids 6*, R. Gallagher ed., Wiley, 1986.

105. U. Ghia, K.N. Ghia, and C.T. Shin, High-Reynolds solutions for incompressible flow using Navier-Stokes equations and a multigrid method, *J. Comp. Phys.*, **48**(1982), 387–411.

106. R. Schreiber, H.B. Keller, Driven cavity flow by efficient numerical techniques, *J. Comp. Phys.*, **40**(1983), 310–333.

107. C.H. Bruneau and C. Jouron, Un nouveau schéma décentré pour le problème de la cavité entraînée, *C.R. Acad. Sci. Paris*, t. 307, Série I (1988), 359–362.

108. J. Shen, Hopf bifurcation of the unsteady regularized driven cavity flow, *J. Comp. Phys.*, **95**(1991), 228–245.

109. O. Goyon, High-Reynolds number solutions of Navier-Stokes equations using incremental unknowns, *Comput. Methods Appl. Mech. Engrg.*, **130** (1996), 319–335.

110. F. Auteri, N. Parolini, L. Quartapelle, Numerical investigation on the stability of singular driven cavity flow, *J. Comp. Phys.*, **183**(2002), 1–25.

111. S. Fujima, M. Tabata, Y. Fukasawa, Extension to three-dimensional problems of the upwind finite element scheme based on the choice up- and downwind points, *Comp. Meth. Appl. Mech. Eng.*, **112**(1994), 109–131.

112. H.C. Ku, R.S. Hirsh, T.D. Taylor, A pseudospectral method for solution of the three-dimensional incompressible Navier-Stokes equations, *J. Comp. Phys.*, **70**(1987), 439–462.

113. T.P. Chiang, W.H. Sheu, R.R. Hwang, Effect of Reynolds number on the eddy structure in a lid–driven cavity, *Int. J. Numer. Meth. in Fluids,* **26**(1998), 557–579.

114. A.K. Prasad and J.R. Koseff, Renolds number and end-wall efftects on a lid-driven cavity flow, *Phys. Fluids,* **A 1** (1989), 208–218.

Computational Techniques for the Verification and Control of Hybrid Systems

Claire J. Tomlin[1], Ian M. Mitchell, Alexandre M. Bayen[1], and Meeko K. M. Oishi[1]

[1] Hybrid Systems Laboratory, Department of Aeronautics and Astronautics
Stanford University, Stanford, CA 94305-4035
[2] Computer Science Department, University of British Columbia, Vancouver, BC, CANADA, V6T 1Z4

1 Introduction

Hybrid systems theory lies at the intersection of the two traditionally distinct fields of computer science verification and engineering control theory. It is loosely defined as the modeling and analysis of systems which involve the interaction of both discrete event systems (represented by finite automata) and continuous time dynamics (represented by differential equations). The goals of this research are in the design of verification techniques for hybrid systems, the development of a software toolkit for efficient application of these techniques, and the use of these tools in the analysis and control of large scale systems. In this paper, we present a summary of recent research results, and a detailed set of references, on the development of tools for the verification of hybrid systems, and on the application of these tools to some interesting examples.

The problem that has received much recent research attention has been the verification of the *safety* property of hybrid systems, which seeks a mathematically precise answer to the question: is a potentially unsafe configuration, or state, reachable from an initial configuration? For discrete systems, this problem has a long history in mathematics and computer science and may be solved by posing the system dynamics as a discrete game [1, 2]; in the continuous domain, control problems of the safety type have been addressed in the context of differential games [3]. For systems involving continuous dynamics, it is very difficult to compute and represent the set of states reachable from some initial set. In this lecture, we present recent solutions to the problem, including a method, based on the level set techniques of Osher and Sethian [4], which determines an implicit representation of the boundary of this *reachable set*. This method is based on the theorem, which is proved in [5] using two-person zero-sum game theory for continuous dynamical systems, that the solution to a particular Hamilton-Jacobi partial differential equation corresponds exactly to the boundary of the reachable set. In addition, we show that useful information for the control of such systems can be extracted from this boundary computation.

Much of the excitement in hybrid system research stems from the potential applications. With techniques such as the above, it is now possible to verify, and design safe, automated control schemes for low dimensional systems. We present two interesting examples in the verification of protocols for aircraft collision avoidance, and of mode switching logic in autopilots. We survey other applications that have been studied in this framework.

We conclude with a discussion of problem complexity and new directions that will enable treatment of problems of higher dimension.

The material in this paper is based on the hybrid system algorithm of [6], the level set implementation of [5, 7], the aircraft landing example of [8], and the interface analysis example of [9, 10]. It was presented as a lecture in the "Summer School Jacques Louis Lions", held during March 17-22 2003 in Montecatini, Italy.

2 Hybrid Model and Verification Methodology

2.1 Continuous, Discrete, and Hybrid Systems

Much of control theory is built around continuous-state models of system behavior. For example, the differential equation model given by

$$\dot{x} = f(x, u, d) \tag{1}$$

describes a system with *state* $x \in \mathbb{R}^n$ that evolves continuously in time according to the dynamical system $f(\cdot, \cdot, \cdot)$, a function of x, $u \in \mathcal{U} \subseteq \mathbb{R}^{n_u}$, $d \in \mathcal{D} \subseteq \mathbb{R}^{n_d}$. In general, u is used to represent parameters that can be controlled, called *control inputs*, and d represents *disturbance inputs*, which are parameters that cannot be controlled, such as the actions of another system in the environment. The initial state $x(0) = x_0$ is assumed to belong to a set $X_0 \subseteq \mathbb{R}^n$ of allowable initial conditions. A *trajectory* of (1) is represented as $(x(t), u(t), d(t))$, such that $x(0) \in X_0$, and $x(t)$ satisfies the differential equation (1) for control and disturbance input trajectories $u(t)$ and $d(t)$. We recommend [11, 12] as current references for continuous-state control systems.

Discrete-state models, such as finite automata, are also prevalent in control. The finite automaton given by

$$(Q, \Sigma, \mathsf{Init}, R) \tag{2}$$

models a system with is a finite set of *discrete state variables* Q, a set of input variables $\Sigma = \Sigma_u \cup \Sigma_d$ which is the union of *control actions* $\sigma_u \in \Sigma_u$ and *disturbance actions* $\sigma_d \in \Sigma_d$, a set of *initial states* $\mathsf{Init} \subseteq Q$, and a *transition relation* $R : Q \times \Sigma \to 2^Q$ which maps the state and input space to subsets of the state space (2^Q). A trajectory of (2) is a sequence of states and inputs, written as $(q(\cdot), \sigma(\cdot))$, where $q(0) \in$ Init and $q(i+1) \in R(q(i), \sigma(i))$ for index $i \in \mathbb{Z}$. The original work of Ramadge and Wonham [13] brought the use of discrete state systems to control, though parallels can be drawn between this work and that of Church, Büchi and Landweber [2, 14] who originally analyzed the von Neumann-Morgenstern [1] discrete games. A comprehensive reference for modeling and control of discrete state systems is [15].

Control theory is concerned with the design of a signal, either a continuous or discrete function of time, which when applied to the system causes the system state to exhibit desirable properties. These properties should hold despite possible disruptive action of the disturbance. A concrete example of a continuous-state control problem is in the control of an aircraft: here the state (position, orientation, velocity) of the aircraft evolves continuously over time in response to control inputs (throttle, control surfaces), as well as to disturbances (wind, hostile aircraft).

A *hybrid automaton* combines continuous-state and discrete-state dynamic systems, in order to model systems which evolve both continuously and according to discrete jumps. A hybrid automaton is defined to be a collection:

$$(S, \text{Init}, In, f, \text{Dom}, R) \tag{3}$$

where $S = Q \cup \mathbb{R}^n$ is the union of discrete and continuous states; Init $\subseteq S$ is a set of initial states; $In = (\Sigma_u \cup \Sigma_d) \cup (\mathcal{U} \cup \mathcal{D})$ is the union of actions and inputs; f is a function which takes state and input and maps to a new state, $f : S \times In \to S$; Dom $\subseteq S$ is a *domain*; and $R : S \times In \to 2^S$ is a *transition relation*.

The state of the hybrid automaton is represented as a pair (q, x), describing the discrete and continuous state of the system. The continuous-state control system is "indexed" by the mode and thus may change as the system changes modes. Dom describes, for each mode, the subset of the continuous state space within which the continuous state may exist, and R describes the transition logic of the system, which may depend on continuous state and input, as well as discrete state and action. A trajectory of this hybrid system is defined as the tuple: $((q(t), x(t)), (\sigma_u(t), \sigma_d(t)), (u(t), d(t)))$ in which $q(t) \in Q$ evolves according to discrete jumps, obeying the transition relation R; for fixed $q(t)$, $x(t)$ evolves continuously according to the control system $f(q(t), x(t), (\sigma_u(t), \sigma_d(t)), (u(t), d(t)))$. The introduction of disturbance parameters to both the control system defined by f and the reset relation defined by R will allow us to treat uncertainties, environmental disturbances, and actions of other systems.

This hybrid automaton model presented above allows for general nonlinear dynamics, and is a slight simplification of the model used in [6]. This model was developed from the early control work of [16, 17, 18, 19]. The emphasis of this work has been on extending the standard modeling, reachability and stability analyses, and controller design techniques to capture the interaction between the continuous and discrete dynamics. Other approaches to modeling hybrid systems involve extending finite automata to include simple continuous dynamics: these include timed automata [20], linear hybrid automata [21, 22, 23, 24], and hybrid input/output automata [25].

2.2 Safety Verification

Much of the research in hybrid systems has been motivated by the need to verify the behavior of safety critical system components. The problem of *safety verification* may be encoded as a condition on the region of operation in the system's state space: given a region of the state space which represents unsafe operation, *prove that the set of states from which the system can enter this unsafe region has empty intersection with the system's set of initial states.*

This problem may be posed as a property of the system's *reachable set* of states. There are two basic types of reachable sets. For a *forward reachable* set, we specify the initial conditions and seek to determine the set of all states that can be reached along trajectories that start in that set. Conversely, for a *backward reachable* set we specify a final or target set of states, and seek to determine the set of states from which trajectories start that can reach that target set. It is interesting to note that the forward and backward reachable sets are not simply time reversals of each other. The difference is illustrated

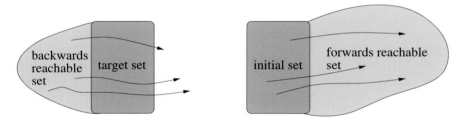

Fig. 1. Difference between backwards and forwards reachable sets.

in Figure 1 for generic target and initial sets, in which the arrows represent trajectories of the system. Figure 2 illustrates how a backwards reachable set may be used to verify system safety.

Powerful software tools for the automatic safety verification of discrete systems have existed for some time, such as Murϕ [26], PVS [27], SMV [28], and SPIN [29]. The verification of hybrid systems presents a more difficult challenge, primarily due to the uncountable number of distinct states in the continuous state space. In order to design and implement a methodology for hybrid system verification, we first need to be able to represent reachable sets of continuous systems, and to evolve these reachable sets according to the system's dynamics.

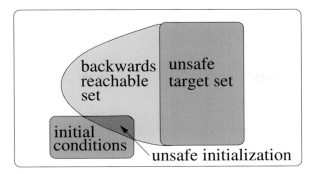

Fig. 2. Using the backwards reachable set to verify safety.

It comes as no surprise that the size and shape of the reachable set depends on the control and disturbance inputs in the system: control variables may be chosen so as to minimize the size of the backwards reachable set from an unsafe target, whereas the full range of disturbance variables must be taken into account in this computation. Thus, the methodology for safety verification has two components. The first involves computing the backward reachable set from an *a priori* specified unsafe target set; the second involves extracting from this computation the control law which must be used on the boundary of the backwards reachable set, in order to keep the system state out of this reachable set. Application of this methodology results in a system description with three simple modes (see Figure 3). Outside of the backwards reachable set, and away from its boundary, the system may use any control law it likes and it will remain safe (labeled as "safe" in Figure 3). When the system state touches the reachable set or unsafe target set boundary, the particular control law which is guaranteed to keep the system from entering the interior of the reachable set must be used. Inside the reachable set (labeled as "outside safe set" in Figure 3), there is no control law which will guarantee safety, however application of the particular optimal control law used to compute the boundary may still result in system becoming safe, if the disturbance is not playing optimally for itself.

In the following section, we first summarize different methods for computing reachable sets for continuous systems. We then provide an overview of our algorithm, which uses an implicit surface function representation of the reachable set, and a differential game theoretic method for its evolution. In the ensuing sections, we illustrate how this reachable set computation may be embedded as the key component in safety verification of hybrid systems.

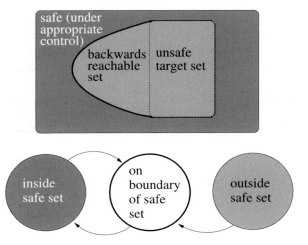

Fig. 3. A discrete abstraction with appropriate control information.

3 Verifying Continuous Systems

In this section, we review our methodology for computing reachable sets for continuous dynamic games. The backwards reachable set is the set of initial conditions giving rise to trajectories that lead to some target set. More formally, let $\mathcal{G}_0$ be the target set, $\mathcal{G}(\tau)$ be the backwards reachable set over finite horizon $\tau < \infty$, $x(\cdot)$ denote a trajectory of the system, and $x(\tau)$ be the state of that trajectory at time τ. Then $\mathcal{G}(\tau)$ is the set of $x(0)$ such that $x(s) \in \mathcal{G}_0$ for some $s \in [0, \tau]$. The choice of input values over time influences how a trajectory $x(t)$ evolves. For systems with inputs, the backwards reachable set $\mathcal{G}(\tau)$ is the set of $x(0)$ such that for every possible control input u there exists a disturbance input d that results in $x(s) \in \mathcal{G}_0$ for some $s \in [0, \tau]$ (where we abuse notation and refer interchangeably to the input signal over time and its instantaneous value).

The solution to the pursuit evasion game described in the previous section is a backwards reachable set. Let the target set be the collision set

$$\mathcal{G}_0 = \left\{ x \in \mathbb{R}^3 \,|\, \sqrt{x_1^2 + x_2^2} \leq d_0 \right\}. \tag{4}$$

Then $\mathcal{G}(\tau)$ is the set of initial configurations such that for any possible control input chosen by the evader, the pursuer can generate a disturbance input that leads to a collision within τ time units.

We use the very general implicit surface function representation for the reachable set: for example, consider the cylindrical target set (4) for the collision avoidance example. We represent this set as the zero sublevel set of a scalar function $\phi_0(x)$ defined over the state space

$$\phi_0(x) = \sqrt{x_1^2 + x_2^2} - d_0,$$
$$\mathcal{G}_0 = \left\{ x \in \mathbb{R}^3 | \phi_0(x) \le 0 \right\}.$$

Thus, a point x is inside $\mathcal{G}_0$ if $\phi_0(x)$ is negative, outside $\mathcal{G}_0$ if $\phi_0(x)$ is positive, and on the boundary of $\mathcal{G}_0$ if $\phi_0(x) = 0$. Constructing this signed distance function representation for $\mathcal{G}_0$ is straightforward for basic geometric shapes. Using negation, minimum, and maximum operators, we can construct functions $\mathcal{G}_0$ which are unions, intersections, and set differences. For example, if $\mathcal{G}_i$ is represented by $g_i(x)$, then, $\min[g_1(x), g_2(x)]$ represents $\mathcal{G}_1 \cup \mathcal{G}_2$, $\max[g_1(x), g_2(x)]$ represents $\mathcal{G}_1 \cap \mathcal{G}_2$, and $\max[g_1(x), -g_2(x)]$ represents $\mathcal{G}_1 \setminus \mathcal{G}_2$.

In [5] we proved that an implicit surface representation of the backwards reachable set can be found by solving a modified HJI PDE. Using $\nabla \phi$ to represent the gradient of ϕ, the modified HJI PDE is

$$\frac{\partial \phi(x,t)}{\partial t} + \min \left[0, H(x, \nabla \phi(x,t)) \right] = 0, \tag{5}$$

with Hamiltonian

$$H(x,p) = \max_{u \in \mathcal{U}} \min_{d \in \mathcal{D}} p \cdot f(x, u, d) \tag{6}$$

and terminal conditions

$$\phi(x,0) = \phi_0(x). \tag{7}$$

If $\mathcal{G}_0$ is the zero sublevel set of $\phi_0(x)$, then the zero sublevel set of the viscosity solution $\phi(x,t)$ to (5)–(7) specifies the backwards reachable set as

$$\mathcal{G}(\tau) = \left\{ x \in \mathbb{R}^3 | \phi(x, -\tau) \le 0 \right\}.$$

Notice that (5) is solved from time $t = 0$ backwards to some $t = -\tau \le 0$.

There are several interesting points to make about the HJI PDE (5)–(7). First, the $\min[0, H]$ formulation in (5) ensures that the reachable set only grows as τ increases. This formulation effectively "freezes" the system evolution when the state enters the target set, which enforces the property that a state which is labeled as "unsafe" cannot become "safe" at a future time. Second, we note that the $\max_u \min_d$ operation in computing the Hamiltonian (6) results in a solution which is not necessarily a "no regret", or saddle, solution to the differential game. By ordering the optimization so that the maximum occurs first, the control input u is effectively "playing" against an unknown disturbance – it is this order which produces a conservative solution, appropriate for the applicaton to system verification under uncertainty. Third, it is proven in [5] that out of many possible weak solutions, the viscosity solution [30] of (5)–(7) yields the reachable set boundary. The significance of this last point is that it enables us to draw from the well developed numerical schemes of the level set literature to compute accurate approximations of $\phi(x,t)$.

To compute numerical approximations of the viscosity solution to (5)–(7), we have developed a C++ implementation based on high resolution level set methods (an excellent introduction to these schemes can be found in [31]). We use a fifth order accurate weighted, essentially non-oscillatory (WENO) stencil [32, 33] to approximate $\nabla\phi(x,t)$, although we have also implemented a basic first order scheme for speed [4, 34]. We use the well studied Lax-Friedrichs (LF) approximation [35] to numerically compute Hamiltonian (6). Finally, we treat the time derivative in (5) with the method of lines and a second order total variation diminishing (TVD) Runge-Kutta scheme [36]. Numerical convergence of our algorithm is demonstrated and validated in [37, 5].

4 Verifying Hybrid Systems

In the previous section, we demonstrated the concept illustrated in Figure 3, in which the problem of verification of safety for continuous systems may be solved by a reachable set computation. This computation abstracts an uncountable number of states into the three classes: *inside safe set*, *boundary of safe set*, and *outside safe set*. We showed that this implicit surface function representation contains information which may be used for designing a safe control law. This safe control law could be used to filter any other control law as the system state approaches the reachable set boundary.

We now consider the problem of computing reachable sets for hybrid systems. Assuming that tools for discrete and continuous reachability are available, computing reachable sets for hybrid systems requires keeping track of the interplay between these discrete and continuous tools. Fundamentally, reachability analysis in discrete, continuous or hybrid systems seeks to partition states into two categories: those that are reachable from the initial conditions, and those that are not. Early work in this area focussed on decidable algorithms: it was shown that decidability results exist for timed and some classes of linear hybrid automata [38]. Software tools were designed to automatically compute reachable sets for these systems: Uppaal [39] and Kronos [40] for timed automata, and HyTech [41, 42] for linear hybrid automata. Some of these tools allow symbolic parameters in the model, and researchers began to study the problem of synthesizing values for these parameters in order to satisfy some kind of control objective, such as minimizing the size of the backwards reachable set. The procedure that we describe here was motivated by the work of [43, 44] for reachability computation and controller synthesis on timed automata, and that of [45] for controller synthesis on linear hybrid automata. Tools based on the analysis of piecewise linear systems, using mathematical programming tools such as CPLEX[46] have found success in several industrial applications.

Our hybrid system analysis algorithm [6] is built upon our implicit reachable set representation and level set implementation for continuous systems.

Thus, we are able to represent and analyze nonlinear hybrid systems, with generally shaped sets. In this sense, our work is related to that of the viability community [47, 48], which has extended concepts from viability to hybrid systems [49]; though the numerical techniques presented here differ from theirs. Other hybrid system reachability algorithms fall within this framework; the differences lie in their discrete and continuous reachability solvers and the types of initial conditions, inputs, invariants and guards that they admit. Tools such as **d/dt**, *Checkmate*, and *VeriSHIFT* have been designed using the different methods of continuous reachable set calculation surveyed in the previous section [50, 51, 52, 53, 54]: the complexity of these tools is essentially the complexity of the algorithm used to compute reachable sets in the corresponding continuous state space.

Methods for hybrid system verification listed above have found application in automotive control problems [46, 55], experimental industrial batch plants [56], vehicle collision avoidance problems [57, 58], as well as envelope protection problems [8, 59]. The problems that have been solved to date are generally of low dimension: to the best of our knowledge, the even the over-approximative methods to date have not been directly applied to systems of continuous dimension greater than 6. In the next section, we present results for envelope projection on nonlinear, hybrid systems with three continuous dimensions, representing the longitudinal dynamics of jet aircraft under hybrid control.

4.1 Computing Reachable Sets for Hybrid Systems

We describe the algorithm first with a picture, and then present the details of a few key components. The full details of the algorithm are in [6], with new implementation results presented in [7].

Consider the sequence of eight diagrams in Figure 4. We draw the hybrid automaton as a set of discrete states $\{q_1, \ldots, q_7\}$ with a transition logic represented by R (the arrows indicate the possible discrete state transitions, the dependence on continuous state and input variables is implied but not shown in the Figure). Associated to each discrete state q_i are the continuous dynamics $\dot{x} = f(q_i, x, (\sigma_u, \sigma_d), (u, d))$ and domain $\mathsf{Dom} \subseteq q_i \times \mathbb{R}^n$, neither of which are shown on the diagram. For illustrative purposes, we consider only one step of our algorithm applied in state q_1, from which there exist transitions to states q_2 and q_3 (shown in diagram 2). We initialize with the unsafe target sets (shown as sets in q_1 and q_2 in diagram 3), and sets which are known to be safe (shown as the "safe" set in q_3 in diagram 4). We augment the unsafe target set in q_1 with states from which there exists an uncontrolled transition to the unsafe set in q_2 (which is represented as a dashed arrow on diagram 5). Uncontrolled transitions may be caused by reset relations affected by disturbance actions. In the absence of other transitions out of state q_1, the set of states backwards reachable from the unsafe target set in q_1 may be computed using the reachable set algorithm of Section 3 on the dynamics

$\dot{x} = f(q_i, x(t), (\sigma_u(t), \sigma_d(t)), (u(t), d(t)))$ (diagram 6). However, there may exist regions of the state space in q_1 from which controllable transitions exist – these transitions could reset the system to a safe region in another discrete state. This is illustrated in diagram 7, with the region in which the system may "escape" to safety from q_1. Thus, the backwards reachable set of interest in this case is the set of states from which trajectories can reach the unsafe target set, without hitting this safe "escape" set first. We call this reachable set the *reach-avoid set*, and it is illustrated in diagram 8.

The algorithm illustrated above is implemented in the following way. The target set $\mathcal{G}_0 \subseteq Q \times \mathbb{R}^n$ can include different subsets of the continuous state space for each discrete mode:

$$\mathcal{G}_0 = \{(q, x) \in Q \times \mathbb{R}^n | g(q, x) \leq 0\} \tag{8}$$

for a level set function $g : Q \times \mathbb{R}^n \to \mathbb{R}$. We seek to construct the largest set of states for which the control, with action/input pair (σ_u, u) can guarantee that the safety property is met despite the disturbance action/input pair (σ_d, d).

For a given set $K \subseteq Q \times \mathbb{R}^n$, we define the *controllable predecessor* $\mathrm{Pre}_u(K)$ and the *uncontrollable predecessor* $\mathrm{Pre}_d(K^c)$ (where K^c refers to the complement of the set K in $Q \times \mathbb{R}^n$) by

$$\mathrm{Pre}_u(K) = \{(q, x) \in K : \exists (\sigma_u, u) \in \Sigma_{\mathbf{u}} \times \mathcal{U}$$
$$\forall (\sigma_d, d) \in \Sigma_{\mathbf{d}} \times \mathcal{D} \ R(q, x, \sigma_u, \sigma_d, u, d) \subseteq K\}$$
$$\mathrm{Pre}_d(K^c) = \{(q, x) \in K : \forall (\sigma_u, u) \in \Sigma_{\mathbf{u}} \times \mathcal{U}$$
$$\exists (\sigma_d, d) \in \Sigma_{\mathbf{d}} \times \mathcal{D} \ R(q, x, \sigma_u, \sigma_d, u, d) \cap K^c \neq \emptyset\} \cup K^c \tag{9}$$

Therefore $\mathrm{Pre}_u(K)$ contains all states in K for which controllable actions (σ_u, u) can force the state to remain in K for at least one step in the discrete evolution. $\mathrm{Pre}_d(K^c)$, on the other hand, contains all states in K^c, as well as all states from which uncontrollable actions (σ_d, d) may be able to force the state outside of K.

Consider two subsets $G \subseteq Q \times \mathbb{R}^n$ and $E \subseteq Q \times \mathbb{R}^n$ such that $G \cap E = \emptyset$. The reach-avoid operator is defined as:

$$\mathrm{Reach}(G, E) = \{(q, x) \in Q \times \mathbb{R}^n \mid \forall u \in \mathcal{U} \ \exists d \in \mathcal{D} \text{ and } t \geq 0 \text{ such that}$$
$$(q, x(t)) \in G \text{ and } (q, x(s)) \in \mathsf{Dom} \setminus E \text{ for } s \in [0, t]\} \tag{10}$$

where $(q, x(s))$ is the continuous state trajectory of $\dot{x}(s) = f(q, x(s), \sigma_u, \sigma_d, u(s), d(s))$ starting at (q, x).

Now, consider the following algorithm:

initialization: $W^0 = \mathcal{G}_0^c$, $W^{+1} = \emptyset$, $i = 0$
while $W^i \neq W^{i+1}$ **do**
$W^{i-1} = W^i \setminus \mathrm{Reach}\left(\mathrm{Pre}_d((W^i)^c), \mathrm{Pre}_u(W^i)\right)$
$i = i - 1$
end while

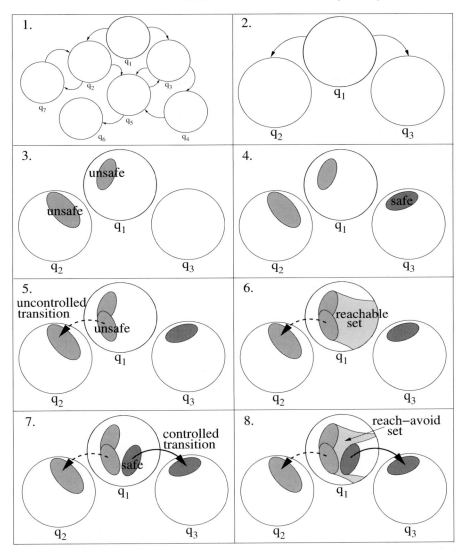

Fig. 4. An illustration of our algorithm for computing reachable sets for hybrid systems.

In the first step of this algorithm, we remove from $\mathcal{G}_0^c$ (the complement of $\mathcal{G}_0$), all states from which a disturbance forces the system either outside $\mathcal{G}_0^c$ or to states from which a disturbance action may cause transitions outside $\mathcal{G}_0^c$, without first touching the set of states from which there is a control action keeping the system inside $\mathcal{G}_0^c$. Since at each step, $W^{i-1} \subseteq W^i$, the set W^i decreases monotonically in size as i decreases. If the algorithm terminates, we denote the fixed point as W^*. The set W^* is used to verify the safety of the system. Recall once more from Figure 3: if the system starts inside W^*, then there exists a control law, extractable from our computational method, for which the system is guaranteed to be safe.

Returning to our pictorial description of the algorithm in Figure 4, and concentrating on the result of one step of the algorithm detailed in Figure 5, we note that, for iteration i: $\mathrm{Pre}_d((W^i)^c) = G_1 \cup G_2$, $E_1 \subset \mathrm{Pre}_u(W^i)$, and $\mathrm{Reach}\left(\mathrm{Pre}_d((W^i)^c), \mathrm{Pre}_u(W^i)\right) = G_3$.

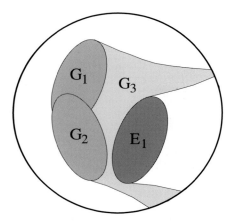

Fig. 5. Detail of the reach-avoid set from diagram 8 of Figure 4.

To implement this algorithm, we need to compute Pre_u, Pre_d, and Reach. The computation of Pre_u and Pre_d requires inversion of the transition relation R subject to the quantifiers $\exists$ and $\forall$; existence of this inverse can be guaranteed subject to conditions on the map R. In our examples, we perform this inversion by hand. The algorithm for computing $\mathrm{Reach}(G, E)$ is a direct modification of the reachable set calculation of Section 3, the details are presented in [7].

5 Flight Management System Example

In this section, we demonstrate our hybrid systems analysis on an interesting and current example, the landing of a civilian aircraft. This example is discussed in detail in [8] and [10]. In addition to the examples presented here, we have solved a range of multi-mode aircraft collision avoidance examples. Please refer to [7, 57] for these examples.

The autopilots of modern jets are highly automated systems which assist the pilot in constructing and flying four-dimensional trajectories, as well as altering these trajectories online in response to Air Traffic Control directives. The autopilot typically controls the throttle input and the vertical and lateral trajectories of the aircraft to automatically perform such functions as: acquiring a specified altitude and then leveling, holding a specified altitude, acquiring a specified vertical climb or descend rate, automatic vertical or lateral navigation between specified way points, or holding a specified throttle value. The combination of these throttle-vertical-lateral modes is referred to as the *flight mode* of the aircraft. A typical commercial autopilot has several hundred flight modes – it is interesting to note that these flight modes were designed to automate the way pilots fly aircraft manually: by controlling the lateral and vertical states of the aircraft to set points for fixed periods of time, pilots simplify the complex task of flying an aircraft. Those autopilot functions which are specific to aircraft landing are among the most safety critical, as reliable automation is necessary when there is little room for altitude deviations. Thus, the need for automation designs which guarantee safe operation of the aircraft has become paramount. Testing and simulation may overlook trajectories to unsafe states: "automation surprises" have been extensively studied [60] *after* the unsafe situation occurs, and "band-aids" are added to the design to ensure the same problem does not occur again. We believe that the computation of accurate reachable sets inside the aerodynamic flight envelope may be used to influence flight procedures and may help to prevent the occurrence of automation surprises.

5.1 Flap Deflection in a Landing Aircraft

In this example, we examine a landing aircraft, and we focus our attention on the flap setting choices available to the pilot. While flap extension and retraction are physically continuous operations, the pilot is presented with a button or lever with a set of discrete settings and the dynamic effect of deflecting flaps is assumed to be minor. Thus, we choose to model the flap setting as a discrete variable. The results in this section are taken from [37].

A simple point mass model for aircraft vertical navigation is used, which accounts for lift L, drag D, thrust T, and weight mg (see [61] and references therein). We model the nonlinear longitudinal dynamics

$$\begin{bmatrix} m\dot{V} \\ mV\dot{\gamma} \\ \dot{h} \end{bmatrix} = \begin{bmatrix} -D(\alpha, V) + T\cos\alpha - mg\sin\gamma \\ L(\alpha, V) + T\sin\alpha - mg\cos\gamma \\ V\sin\gamma \end{bmatrix} \tag{11}$$

in which the state $x = [V, \gamma, h] \in \mathbb{R}^3$ includes the aircraft's speed V, flight path angle γ, and altitude h. We assume the control input $u = [T, \alpha]$, with

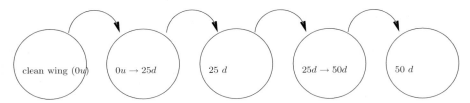

Fig. 6. Discrete transition diagram of flap deflection settings. Clean wing represents no deflection, 25 d represents a deflection of 25°, and 50 d, a deflection of 50°. The modes $0u \rightarrow 25d$ and $25d \rightarrow 50d$ are timed modes to reflect deflection time: if the pilot selects mode 25 d from clean wing, for example, the model will transition into an "intermediate" mode for 10 seconds, before entering 25 d. Thus, the transitions from clean wing to $0u \rightarrow 25d$ and from 25 d to $25d \rightarrow 50d$ are controlled transitions (σ_u) in our analysis, the others are uncontrolled transitions (σ_d).

aircraft thrust T and angle of attack α. The mass of the aircraft is denoted m. The functions $L(\alpha, V)$ and $D(\alpha, V)$ are modeled based on empirical data [62] and Prandtl's lifting line theory [63]:

$$L(\alpha, V) = \tfrac{1}{2}\rho SV^2 C_L(\alpha), \qquad D(\alpha, V) = \tfrac{1}{2}\rho SV^2 C_D(\alpha) \qquad (12)$$

where ρ is the density of air, S is wing area, and $C_L(\alpha)$ and $C_D(\alpha)$ are the dimensionless lift and drag coefficients.

In determining $C_L(\alpha)$ we will follow standard autoland procedure and assume that the aircraft switches between three fixed flap deflections $\delta = 0°$, $\delta = 25°$ and $\delta = 50°$ (with slats either extended or retracted), thus constituting a hybrid system with different nonlinear dynamics in each mode. This model is representative of current aircraft technology; for example, in civil jet cockpits the pilot uses a lever to select among four predefined flap deflection settings. We assume a linear form for the lift coefficient $C_L(\alpha) = h_\delta + 4.2\alpha$, where parameters $h_{0°} = 0.2$, $h_{25°} = 0.8$ and $h_{50°} = 1.25$ are determined from experimental data for a DC9-30 [62]. The value of α at which the vehicle stalls decreases with increasing flap deflection: $\alpha_{0°}^{\max} = 16°$, $\alpha_{25°}^{\max} = 13°$, $\alpha_{50°}^{\max} = 11°$; slat deflection adds 7° to the $\alpha^{\max}$ in each mode. The drag coefficient is computed from the lift coefficient as [63] $C_D(\alpha) = 0.041 + 0.045C_L^2(\alpha)$ and includes flap deflection, slat extension and gear deployment corrections. Thus, for a DC9-30 landing at sea level and for all $\alpha \in [-5°, \alpha_\delta^{\max}]$, the lift and drag terms in (11) are given by

$$L(\alpha, V) = 68.6\ (h_\delta + 4.2\alpha)V^2 \qquad D(\alpha, V) = (2.7 + 3.08\ (h_\delta + 4.2\alpha)^2)V^2$$

In our implementation, we consider three operational modes: $0u$, which represents $\delta = 0°$ with undeflected slats, $25d$, which represents $\delta = 25°$ with deflected slats, and $50d$, for $\delta = 50°$ with deflected slats.

Approximately 10 seconds are required for a 25° degree change in flap deflection. For our implementation, we define transition modes $0u \rightarrow 25d$ and

$25d \rightarrow 50d$ with timers, in which the aerodynamics are those of (11) with coefficients which interpolate those of the bounding operational modes. The corresponding discrete automaton is shown in Figure 6. Transition modes have only a timed switch at $t = t_{\text{delay}}$, so controlled switches will be separated by at least t_{delay} time units and the system is nonzeno. For the executions shown below, $t_{\text{delay}} = 10$ seconds.

The aircraft enters its final stage of landing close to 50 feet above ground level ([62, 64]). Restrictions on the flight path angle, aircraft velocity and touchdown (TD) speed are used to determine the initial safe set W_0:

$$
\begin{cases}
h \leq 0 & \text{landing or has landed} \\
V > V_\delta^{\text{stall}} & \text{faster than stall speed} \\
V < V^{\text{max}} & \text{slower than limit speed} \\
V \sin \gamma \geq \dot{z}_0 & \text{limited TD speed} \\
\gamma \leq 0 & \text{monotonic descent}
\end{cases}
\cup
\begin{cases}
h > 0 & \text{aircraft in the air} \\
V > V_\delta^{\text{stall}} & \text{faster than stall speed} \\
V < V^{\text{max}} & \text{slower than limit speed} \\
\gamma > -3° & \text{limited descent flight path} \\
\gamma \leq 0 & \text{monotonic descent}
\end{cases}
\tag{13}
$$

We again draw on numerical values for a DC9-30 [62]: stall speeds $V_{0u}^{\text{stall}} = 78$ m/s, $V_{25d}^{\text{stall}} = 61$ m/s, $V_{50d}^{\text{stall}} = 58$ m/s, maximal touchdown speed $\dot{h}_0 = 0.9144$ m/s, and maximal velocity $V^{\text{max}} = 83$ m/s. The aircraft's input range is restricted to a fixed thrust at 20% of its maximal value $T = 32KN$, and $\alpha \in [0°, \ 10°]$.

The results of our fixed point computation are shown in Figures 7 and 8. The interior of the surface shown in the first row of Figure 7 represents the initial envelopes W_0 for each of the $0u$, $25d$ and $50d$ modes. The second row of the figure shows the maximally controllable subset of the envelope for each mode individually, as determined by the reachable set computation for continuous systems. The clean wing configuration $0u$ becomes almost completely uncontrollable, while the remaining modes are partially controllable. The subset of the envelope that cannot be controlled in these high lift/high drag configurations can be divided into two components. For low speeds, the aircraft will tend to stall. For values of h near zero and low flight path angles γ, the aircraft cannot pull up in time to avoid landing gear damage at touchdown. The third row shows the results for the hybrid reachable set computation. Here, both modes $0u$ and $25d$ are almost completely controllable, since they can switch instantaneously to the fully deflected mode $50d$. However, no mode can control the states h near zero and low γ, because no mode can pull up in time to avoid landing gear damage. Figure 8 shows a slice through the reach and avoid sets for the hybrid analysis at a fixed altitude of $h = 5$m, for each of the $0u$, $25d$ and $50d$ modes. Here, the grey-scale represents the following: dark grey is the subset of the initial escape set that is also safe in the current mode, mid-grey is the initial escape set, light grey is the known unsafe set, and white is the computed reach set, or those states from which the system can neither remain in the same mode nor switch to safety.

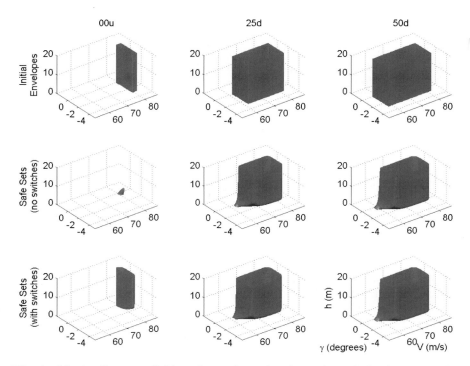

Fig. 7. Maximally controllable safe envelopes for the multimode landing example. From left to right the columns represent modes $0u$, $25d$ and $50d$.

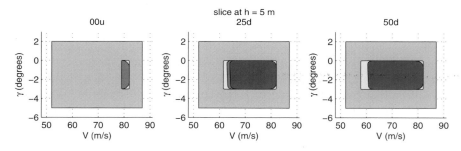

Fig. 8. Slices through the reach and avoid sets for the hybrid analysis at a fixed altitude of $h = 5$m. From left to right the columns represent modes $0u$, $25d$ and $50d$.

5.2 Take Off / Go Around Interface Analysis

We now examine another aircraft landing example with the goal of using hybrid system verification in order to prove desirable qualities about the pilot's display. Naturally, only a subset of all information about the aircraft is displayed to the pilot – but how much information is enough? When the pilot does not have the required information at his disposal, and becomes confused by the cockpit automation, automation surprises and mode confusion can occur. Automation surprises are contributing factors in many aircraft incidents, commonly regarded as indicators of future aircraft accidents. Currently, extensive flight simulation and testing are used to validate autopilot systems and their displays. However, discovering design errors as early as possible in the design process is important for aircraft manufacturers as well as pilots, and hybrid verification tools can aid in this process. The results in this section are taken from [9], which uses the same form of longitudinal dynamic model (11) as the previous section, with new parameters for a large commercial aircraft [9].

In modeling $C_L(\alpha)$ and $C_D(\alpha)$ as in (12), we define $C_L(\alpha) = C_{L_0} + C_{L_\alpha}\alpha$ and $C_D(\alpha) = C_{D_0} + KC_L^2(\alpha)$. The constants C_{L_0}, C_{D_0}, and K represent a particular aircraft configuration, as indicated in Table 1. $C_{L_\alpha} = 5.105$ in all modes. The aircraft has mass $m = 190000$ kg, wing surface area $S = 427.80$ m/s^2, and maximum thrust $T_{\max} = 686700$ N.

Table 1. Aerodynamic constants for autoland modes indexed by $\dot{x} = f_i(x, u)$.

i	C_{L_0}	C_{D_0}	K	Flaps Setting	Landing Gear
1	0.4225	0.024847	0.04831	Flaps-20	Down
2	0.7043	0.025151	0.04831	Flaps-25	Down
3	0.8212	0.025455	0.04831	Flaps-30	Down
4	0.4225	0.019704	0.04589	Flaps-20	Up
5	0.7043	0.020009	0.04589	Flaps-25	Up
6	0.8212	0.020313	0.04589	Flaps-30	Up

The model for this example also varies from the previous example in that we directly account for the user's actions in the hybrid system. We assume that the pilot operates the aircraft according to strict procedure, shown in Figure 9. During landing, if for any reason the pilot or air traffic controller deems the landing unacceptable (debris on the runway, a potential conflict with another aircraft, or severe wind shear near the runway, for example), the pilot must initiate a go-around maneuver. A go-around can be initiated at any time after the glideslope has been captured and before the aircraft touches down. Pushing the go-around button engages a sequence of events designed to make the aircraft climb as quickly as possible to a preset missed-approach altitude, $h_{\mathrm{alt}} = 2500$ feet.

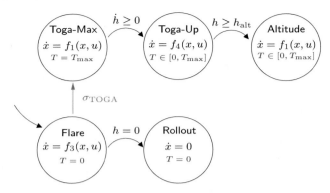

Fig. 9. Hybrid procedural automaton $H_{\text{procedure}}$. The dynamics $f_i(x, u) = f(q_i, x, u)$ differ in the values of aerodynamic coefficients affecting lift and drag.

The initial state of the procedural model $H_{\text{procedural}}$ (Figure 9) is Flare, with flaps at Flaps-30 and thrust fixed at idle. When a pilot initiates a go-around maneuver (often called a "TOGA" due to the "Take-Off/Go-Around" indicator on the pilot display), the pilot changes the flaps to Flaps-20 and the autothrottle forces the thrust to $T_{\max}$ (Toga-Max). When the aircraft obtains a positive rate of climb, the pilot raises the landing gear, and the autothrottle allows $T \in [0, T_{\max}]$ (Toga-Up). The aircraft continues to climb to the missed approach altitude, h_{alt}, then automatically switches into an altitude-holding mode, Altitude, to prepare for the next approach (with the landing gear down). If a go-around is not initiated from Flare, the aircraft switches to Rollout when it lands. (We do not model the aircraft's behavior after touchdown.)

Although go-arounds are unpredictable and may be required at any time during the autoland prior to touchdown, we model σ_{TOGA} as a controlled transition because the pilot must initiate the go-around for it to occur. Certain events occur simultaneously: changing the flaps to Flaps-30 and event σ_{TOGA}, raising the landing gear and $\dot{h} \geq 0$, and lowering the landing gear and $h \geq h_{\text{alt}}$.

Each mode in the procedural automaton is subject to state and input bounds, due to constraints arising from aircraft aerodynamics and desired aircraft behavior. These bounds, shown in Table 2, form the boundary of the initial envelope W_0. Bounds on V and α are determined by stall speeds and structural limitations for each flap setting. Bounds on γ and T are determined by the desired maneuver [65]. Additionally, at touchdown, $\theta \in [0°, 12.9°]$ to prevent a tail strike, and $\dot{h} \geq -1.829$ m/s to prevent damage to the landing gear.

We separate the hybrid procedural model (Figure 9) across the user-controlled switch σ_{TOGA}, into two hybrid subsystems: H_F and H_T. H_F encompasses Flare and Rollout, $H_{\underset{T}{\circ}}$ encompasses Toga-Max, Toga-Up, and Altitude. Computationally, automatic transitions are smoothly accomplished by concatenating modes across automatic transitions, so that the change in dy-

Table 2. State bounds for autoland modes of $H_{\mathrm{procedural}}$.

Mode	V [m/s]	γ [degrees]	α [degrees]
Flare	$[55.57,\ 87.46]$	$[-6.0°, 0.0°]$	$[-9°, 15°]$
Toga-Max	$[63.79, 97.74]$	$[-6.0°, 0.0°]$	$[-8°, 12°]$
Toga-Up	$[63.79, 97.74]$	$[0.0°, 13.3°]$	$[-8°, 12°]$
Altitude	$[63.79, 97.74]$	$[-0.7°, 0.7°]$	$[-8°, 12°]$

namics across the switching surface is modeled as another nonlinearity in the dynamics. Additionally, we assume in $H_{\mathring{T}}$ that if the aircraft leaves the top of the computational domain ($h = 20$ m) without exceeding its flight envelope, it is capable of reaching Altitude mode, which we consider to be completely safe.

The initial flight envelopes for $H_{\mathbf{F}}$ and $H_{\mathbf{T}}$, $(W_{\mathbf{F}})_0$ and $(W_{\mathbf{T}})_0$, are determined by state bounds on each mode given in Table 2. We perform the reachable set computation on $H_{\mathbf{F}}$ and $H_{\mathbf{T}}$ separately to obtain the safe flight envelopes $W_{\mathbf{F}}$ and $W_{\mathbf{T}}$. Figure 10 shows $W_{\mathbf{F}}$, and Figure 11 shows $W_{\mathbf{T}}$ in Toga-Up and Toga-Max modes. (Note that the boundary of $W_{\mathbf{F}}$ along $\gamma = 0$ corresponds with the transition boundary of $W_{\mathbf{T}}$ between Toga-Up and Toga-Max, $\dot{h} = 0$.)

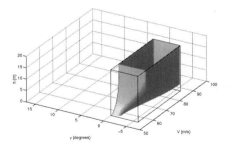

Fig. 10. Safe region W_F; the outer box is $(W_F)_0$.

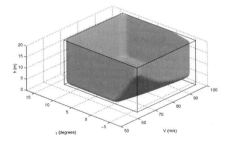

Fig. 11. Safe region W_T: the outer box is $(W_T)_0$.

Figure 12 shows the continuous region $W_{\mathbf{F}} \cap W_{\mathbf{T}}$ from which we can guarantee both a safe landing and a safe go-around. Notice that this set is smaller than $W_{\mathbf{F}}$, the region from which a safe landing is possible: the pilot is further restricted in executing a go-around. There are states from which a safe landing is possible, but a safe go-around is not.

Verification within a hybrid framework allows us to account for the inherently complicated dynamics underlying the simple, discrete representations displayed to the pilot. In this example, in order to safely supervise the system, the pilot should have enough information to know before entering a go-around maneuver whether or not the aircraft will remain safe: thus the

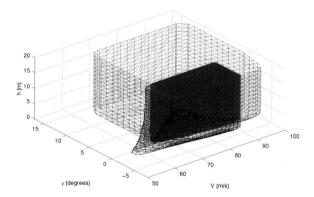

Fig. 12. The solid shape is the safe region $W_F \cap W_T$, from which safe landing and safe go-around is possible. The meshes depict $W_\mathbf{F}$ and W_T.

pilot could respond to this information by increasing speed, decreasing ascent rate, or decreasing angle of attack.

Further details on how hybrid system verification is used to verify information contained in user-interfaces can be found in [9].

6 Summary

We have presented a method and algorithm for hybrid systems analysis, specifically, for the verification of safety properties of hybrid systems. We have also given a brief summary of other available methods. All techniques rely on the ability to compute reachable sets of hybrid systems, and they differ mainly in the assumptions made about the representation of sets, and evolution of the continuous state dynamics. We have described and demonstrated our algorithm, which represents a set implicitly as the zero sublevel set of a given function, and computes its evolution through the hybrid dynamics using a combination of constrained level set methods and discrete mappings through transition functions.

Many directions for further work are available, and we are pursuing several of them. Our algorithm is currently constrained by computational complexity: examples with four continuous dimensions take several days to run on our standard desktop computers, five dimensions takes weeks. We are working on a variant of our algorithm which first projects the high dimensional target into a set of lower dimensional subspaces of the state space, computes the reachable sets of these projections (quickly, as they are in low dimensions), and then "backprojects" these sets to form, in high dimensions, an overapproximation of the actual reachable set. The actual reachable set need never be computed, and overapproximations of unsafe sets can be used to verify safety, as we showed in the last lecture. We are also developing methods to compute tight polyhedral overapproximations of the reachable set for general nonlinear hybrid systems [66] – these methods scale well in continuous dimension, as they do not require gridding the state space.

Acknowledgments

We would like to thank John Lygeros and Shankar Sastry, who are coauthors of the hybrid system algorithm and provided valuable insight into the more recent viscosity solution proofs. We thank Ron Fedkiw and Stan Osher for their help and discussions regarding level set methods; the rendering software that we used in Section 3 was written by Ron Fedkiw.

References

1. J. von Neumann and O. Morgenstern, *Theory of Games and Economic Behavior*. Princeton University Press, 1947.
2. A. Church, "Logic, arithmetic, and automata," in *Proceedings of the International Congress of Mathematicians*, pp. 23–35, 1962.
3. R. Isaacs, *Differential Games*. John Wiley, 1967.
4. S. Osher and J. A. Sethian, "Fronts propagating with curvature-dependent speed: Algorithms based on Hamilton-Jacobi formulations," *Journal of Computational Physics*, vol. 79, pp. 12–49, 1988.
5. I. Mitchell, A. M. Bayen, and C. J. Tomlin, "Computing reachable sets for continuous dynamic games using level set methods," *IEEE Transactions on Automatic Control*, December 2001. Submitted.
6. C. J. Tomlin, J. Lygeros, and S. Sastry, "A game theoretic approach to controller design for hybrid systems," *Proceedings of the IEEE*, vol. 88, no. 7, pp. 949–970, July 2000.
7. I. Mitchell, *Application of Level Set Methods to Control and Reachability Problems in Continuous and Hybrid Systems*. PhD thesis, Scientific Computing and Computational Mathematics, Stanford University, August 2002.
8. A. M. Bayen, I. Mitchell, M. Oishi, and C. J. Tomlin, "Automatic envelope protection and cockpit interface analysis of an autoland system using hybrid system theory." Submitted to the AIAA Journal of Guidance, Control, and Dynamics, December 2002.
9. M. Oishi, I. Mitchell, A. M. Bayen, C. J. Tomlin, and A. Degani, "Hybrid verification of an interface for an automatic landing," in *Proceedings of the IEEE Conference on Decision and Control*, (Las Vegas, NV), December 2002.
10. M. Oishi, C. J. Tomlin, and A. Degani, "Verification of user interfaces for hybrid systems." Submitted as a NASA Technical Memorandum, Ames Research Center, November 2002.
11. S. S. Sastry, *Nonlinear Systems: Analysis, Stability and Control*. New York: Springer Verlag, 1999.
12. J. Doyle, B. Francis, and A. Tannenbaum, *Feedback Control Theory*. New York: Macmillan, 1992.

13. P. J. G. Ramadge and W. M. Wonham, "The control of discrete event dynamical systems," *Proceedings of the IEEE*, vol. Vol.77, no. 1, pp. 81–98, 1989.
14. J. R. Büchi and L. H. Landweber, "Solving sequential conditions by finite-state operators," in *Proceedings of the American Mathematical Society*, pp. 295–311, 1969.
15. C. Cassandras and S. Lafortune, *Introduction to Discrete Event Systems*. Boston: Kluwer, 1999.
16. R. Brockett, "Hybrid models for motion control systems," in *Perspectives in Control* (H. Trentelman and J. Willems, eds.), pp. 29–54, Boston: Birkhauser, 1993.
17. M. S. Branicky, *Control of Hybrid Systems*. PhD thesis, Department of Electrical Engineering and Computer Sciences, Massachusetts Institute of Technology, 1994.
18. J. Lygeros, *Hierarchical, Hybrid Control of Large Scale Systems*. PhD thesis, Department of Electrical Engineering and Computer Sciences, University of California at Berkeley, 1996.
19. A. Nerode and W. Kohn, "Models for hybrid systems: Automata, topologies, controllability, observability," in *Hybrid Systems* (R. L. Grossman, A. Nerode, A. P. Ravn, and H. Rischel, eds.), LNCS 736, pp. 317–356, New York: Springer Verlag, 1993.
20. R. Alur and D. Dill, "A theory of timed automata," *Theoretical Computer Science*, vol. 126, pp. 183–235, 1994.
21. R. Alur, C. Courcoubetis, T. Henzinger, P.-H. Ho, X. Nicollin, A. Olivero, J. Sifakis, and S. Yovine, "The algorithmic analysis of hybrid systems," in *Proceedings of the 11th International Conference on Analysis and Optimization of Systems: Discrete-event Systems* (G. Cohen and J.-P. Quadrat, eds.), no. 199 in LNCS, pp. 331–351, Springer Verlag, 1994.
22. T. Henzinger, "The theory of hybrid automata," in *Proceedings of the 11th Annual Symposium on Logic in Computer Science*, pp. 278–292, IEEE Computer Society Press, 1996.
23. A. Puri and P. Varaiya, "Decidability of hybrid systems with rectangular differential inclusions," in *CAV94: Computer-Aided Verification*, no. 818 in LNCS, pp. 95–104, Stanford, CA: Springer Verlag, 1995.
24. O. Shakernia, G. J. Pappas, and S. S. Sastry, "Decidable controller synthesis for classes of linear systems," in *Hybrid Systems: Computation and Control* (B. Krogh and N. Lynch, eds.), LNCS 1790, pp. 407–420, Springer Verlag, 2000.
25. N. Lynch, R. Segala, and F. Vaandraager, "Hybrid I/O automata," 2002. Submitted. Also, Technical Report MIT-LCS-TR-827b, MIT Laboratory for Computer Science, Cambridge, MA 02139.
26. D. L. Dill, "The Murφ verification system," in *Conference on Computer-Aided Verification*, LNCS, pp. 390–393, Springer-Verlag, July 1996.
27. S. Owre, J. M. Rushby, and N. Shankar, "PVS: A prototype verification system," in *11th International Conference on Automated Deduction (CADE)* (D. Kapur, ed.), vol. 607 of *Lecture Notes in Artificial Intelligence*, (Saratoga, NY), pp. 748–752, Springer-Verlag, June 1992.
28. J. Burch, E. M. Clarke, K. McMillan, D. Dill, and L. Hwang, "Symbolic model checking: 10^{20} states and beyond," *Information and Computation*, vol. 98, pp. 142–170, June 1992.

29. G. Holzmann, "The model checker Spin," *IEEE Transactions on Software Engineering*, vol. 23, pp. 279–295, May 1997. Special issue on Formal Methods in Software Practice.

30. M. G. Crandall, L. C. Evans, and P.-L. Lions, "Some properties of viscosity solutions of Hamilton-Jacobi equations," *Transactions of the American Mathematical Society*, vol. 282, no. 2, pp. 487–502, 1984.

31. S. Osher and R. Fedkiw, *The Level Set Method and Dynamic Implicit Surfaces.* Springer-Verlag, 2002.

32. G.-S. Jiang and D. Peng, "Weighted ENO schemes for Hamilton-Jacobi equations," *SIAM Journal on Scientific Computing*, vol. 21, pp. 2126–2143, 2000.

33. S. Osher and C.-W. Shu, "High-order essentially nonoscillatory schemes for Hamilton-Jacobi equations," *SIAM Journal of Numerical Analysis*, vol. 28, no. 4, pp. 907–922, 1991.

34. J. A. Sethian, *Level Set Methods and Fast Marching Methods.* New York: Cambridge University Press, 1999.

35. M. G. Crandall and P.-L. Lions, "Two approximations of solutions of Hamilton-Jacobi equations," *Mathematics of Computation*, vol. 43, no. 167, pp. 1–19, 1984.

36. C.-W. Shu and S. Osher, "Efficient implementation of essentially non-oscillatory shock-capturing schemes," *Journal of Computational Physics*, vol. 77, pp. 439–471, 1988.

37. I. Mitchell, A. M. Bayen, and C. J. Tomlin, "Validating a Hamilton-Jacobi approximation to hybrid system reachable sets," in *Hybrid Systems: Computation and Control* (M. D. D. Benedetto and A. Sangiovanni-Vincentelli, eds.), LNCS 2034, pp. 418–432, Springer Verlag, 2001.

38. T. A. Henzinger, P. W. Kopke, A. Puri, and P. Varaiya, "What's decidable about hybrid automata," in *Proceedings of the 27th Annual ACM Symposium on Theory of Computing*, 1995.

39. K. Larsen, P. Pettersson, and W. Yi, "Uppaal in a nutshell," *Software Tools for Technology Transfer*, vol. 1, 1997.

40. S. Yovine, "Kronos: A verification tool for real-time systems," *Software Tools for Technology Transfer*, vol. 1, pp. 123–133, 1997.

41. R. Alur, C. Courcoubetis, T. A. Henzinger, and P.-H. Ho, "Hybrid automata: An algorithmic approach to the specification and verification of hybrid systems," in *Hybrid Systems* (R. L. Grossman, A. Nerode, A. P. Ravn, and H. Rischel, eds.), LNCS, pp. 366–392, New York: Springer Verlag, 1993.

42. T. A. Henzinger, P. Ho, and H. Wong-Toi, "HyTech: A model checker for hybrid systems," *Software Tools for Technology Transfer*, vol. 1, pp. 110–122, 1997.

43. O. Maler, A. Pnueli, and J. Sifakis, "On the synthesis of discrete controllers for timed systems," in *STACS 95: Theoretical Aspects of Computer Science* (E. W. Mayr and C. Puech, eds.), no. 900 in LNCS, pp. 229–242, Munich: Springer Verlag, 1995.

44. E. Asarin, O. Maler, and A. Pnueli, "Symbolic controller synthesis for discrete and timed systems," in *Proceedings of Hybrid Systems II, Volume 999 of LNCS* (P. Antsaklis, W. Kohn, A. Nerode, and S. Sastry, eds.), Cambridge: Springer Verlag, 1995.

45. H. Wong-Toi, "The synthesis of controllers for linear hybrid automata," in *Proceedings of the IEEE Conference on Decision and Control*, (San Diego, CA), 1997.

46. A. Bemporad and M. Morari, "Verification of hybrid systems via mathematical programming," in *Hybrid Systems: Computation and Control* (F. Vaandrager and J. H. van Schuppen, eds.), no. 1569 in LNCS, pp. 30–45, Berlin: Springer Verlag, 1999.

47. P. Cardaliaguet, M. Quincampoix, and P. Saint-Pierre, "Set-valued numerical analysis for optimal control and differential games," in *Stochastic and Differential Games: Theory and Numerical Methods* (M. Bardi, T. Parthasarathy, and T. E. S. Raghavan, eds.), vol. 4 of *Annals of International Society of Dynamic Games*, Birkhäuser, 1999.

48. J.-P. Aubin, J. Lygeros, M. Quincampoix, S. Sastry, and N. Seube, "Impulse differential inclusions: A viability approach to hybrid systems," *IEEE Transactions on Automatic Control*, vol. 47, no. 1, pp. 2–20, 2002.

49. A. M. Bayen, E. Crück, and C. J. Tomlin, "Guaranteed overapproximation of unsafe sets for continuous and hybrid systems: Solving the Hamilton-Jacobi equation using viability techniques," in *Hybrid Systems: Computation and Control* (C. J. Tomlin and M. R. Greenstreet, eds.), LNCS 2289, pp. 90–104, Springer Verlag, 2002.

50. T. Dang, *Vérification et synthèse des systèmes hybrides*. PhD thesis, Institut National Polytechnique de Grenoble (Verimag), 2000.

51. E. Asarin, O. Bournez, T. Dang, and O. Maler, "Approximate reachability analysis of piecewise-linear dynamical systems," in *Hybrid Systems: Computation and Control* (B. Krogh and N. Lynch, eds.), LNCS 1790, pp. 21–31, Springer Verlag, 2000.

52. A. Chutinan and B. H. Krogh, "Verification of infinite-state dynamic systems using approximate quotient transition systems," *IEEE Transactions on Automatic Control*, vol. 46, no. 9, pp. 1401–1410, 2001.

53. O. Botchkarev and S. Tripakis, "Verification of hybrid systems with linear differential inclusions using ellipsoidal approximations," in *Hybrid Systems: Computation and Control* (B. Krogh and N. Lynch, eds.), LNCS 1790, pp. 73–88, Springer Verlag, 2000.

54. R. Vidal, S. Schaffert, J. Lygeros, and S. S. Sastry, "Controlled invariance of discrete time systems," in *Hybrid Systems: Computation and Control* (B. Krogh and N. Lynch, eds.), LNCS 1790, pp. 437–450, Springer Verlag, 2000.

55. A. Balluchi, M. D. Benedetto, C. Pinello, C. Rossi, and A. Sangionvanni-Vincentelli, "Hybrid control for automotive engine management: The cut-off case," in *Hybrid Systems: Computation and Control* (T. Henzinger and S. Sastry, eds.), no. 1386 in LNCS, pp. 13–32, New York: Springer Verlag, 1998.

56. Esprit, "Verification of Hybrid Systems: Results of a European Union Esprit Project," in *European Journal of Control* (O. Maler, ed.), Vol. 7, Issue 4, 2001.

57. C. J. Tomlin, I. Mitchell, and R. Ghosh, "Safety verification of conflict resolution maneuvers," *IEEE Transactions on Intelligent Transportation Systems*, vol. 2, no. 2, pp. 110–120, 2001. June.

58. M. Oishi, C. J. Tomlin, V. Gopal, and D. Godbole, "Addressing multiobjective control: Safety and performance through constrained optimization," in *Hybrid Systems: Computation and Control* (M. D. D. Benedetto and A. Sangiovanni-Vincentelli, eds.), LNCS 2034, pp. 459–472, Springer Verlag, 2001.

59. T. Dang and O. Maler, "Reachability analysis via face lifting," in *Hybrid Systems: Computation and Control* (S. Sastry and T. Henzinger, eds.), no. 1386 in LNCS, pp. 96–109, Springer Verlag, 1998.

60. A. Degani, *Modeling Human-Machine Systems: On Modes, Error, and Patterns of Interaction.* PhD thesis, Department of Industrial and Systems Engineering, Georgia Institute of Technology, 1996.

61. C. J. Tomlin, *Hybrid Control of Air Traffic Management Systems.* PhD thesis, Department of Electrical Engineering, University of California, Berkeley, 1998.

62. I. M. Kroo, *Aircraft Design: Synthesis and Analysis.* Stanford, California: Desktop Aeronautics Inc., 1999.

63. J. Anderson, *Fundamentals of Aerodynamics.* New York: McGraw Hill Inc., 1991.

64. United States Federal Aviation Administration, *Federal Aviation Regulations,* 1990. Section 25.125 (landing).

65. T. Lambregts, "Automatic flight control: Concepts and methods." FAA National Resource Specialist, Advanced Controls, 1995.

66. I. Hwang, D. Stipanovic, and C. Tomlin, "Polyhedral reachable sets for continuous and hybrid systems." Submitted to the 2003 American Control Conference, August 2002.

Part II

Case Studies

Data Assimilation Methods for an Oceanographic Problem

Didier Auroux[1] and Jacques Blum[2]

[1] Laboratoire J.A. Dieudonné, Université de Nice Sophia-Antipolis, Parc Valrose, F-06108 Nice Cedex 2, France. `auroux@math.unice.fr`
[2] Laboratoire J.A. Dieudonné, Université de Nice Sophia-Antipolis, Parc Valrose, F-06108 Nice Cedex 2, France. `jblum@math.unice.fr`

1 Introduction

The dynamics of the oceans play a major role in the knowledge of our environment and especially in the Earth's climate. Over the past twenty years, the new satellite techniques for observing the oceans, and especially the use of altimeter measurements, have greatly improved our knowledge of the oceans by allowing synoptic monitoring of the surface. The measurements of the sea-surface height have clearly demonstrated the feasibility and the usefulness of satellite altimetry. It was with the availability of Topex/Poseidon data since 1992, that the oceanographic community began intensive exploitation of this new observational source. It has already given incomparable information to study the general circulation of the ocean, to estimate the energy levels of the upper ocean, and to examine the local dynamics of different regions of particular interest, such as the Gulf Stream area, the Kuroshio extension, the Antarctic circumpolar current and the tropical oceans.

At the interface between the two major components of oceanographic science, i.e. observations and models, lies the domain of so-called data assimilation (DA). DA covers all the mathematical and numerical techniques which allow us to blend as optimally as possible all the sources of information coming from theory, models and other types of data. Clearly these techniques may not only apply in oceanography but also to other environmental disciplines. DA allows us to recreate the time-space structure of a system from a set of information which has, in general, a large disparity in nature, in space-time distribution and in accuracy. There are two main categories of DA methods: variational methods based on the optimal control theory [Lio68] and statistical methods based on the theory of optimal statistical estimation. The prototype of the first class which is actually of interest here is the optimal control method which was first introduced in meteorology (see [Lew85], [LeD86], [Tal87]) and more recently for the ocean (see [Tha88], [She90], [Moo91], [Sch93], [Nec94], [Luo98]). The prototype of statistical methods is the Kalman filter whose introduction in oceanography dates back roughly a decade (see, for example, [Ghi89] and [Ghi91]). The Kalman filter was extended to nonlinear cases ([Jaz70], [Gel74]) but it has been mostly applied in oceanography to quasi-linear situations of the tropical oceans ([Gou92], [Fuk95], [Fuk93], [Can96], [Ver99]). We also refer to the recent book of Bennett [Ben02] on inverse methods, both for oceanography and meteorology.

All DA techniques encounter major difficulties in practice for computing reasons: memory size and computing costs. The full Kalman filter would, in principle, require the manipulation of $(N \times N)$ matrices where N is the state vector dimension which is typically 10^7 or 10^8 in an oceanic problem. The optimal control adjoint method often requires several hundred iterations of the minimization process to converge, thus implying an equivalent number of model runs.

In this paper, we first focus our interest on the use of the variational adjoint method in a relatively simple ocean model in order to try to reconstruct the four-dimensional ocean system from altimetric surface observations of the ocean. The variational method uses the strong constraint hypothesis, i.e. the ocean circulation model is assumed to be exact. The assimilation process is carried out by an identification of the initial state of the dynamical system which minimizes a cost function. This cost function is the mean-square difference between the observations and the corresponding model variables. The functional will be minimized using a numerical unconstrained optimization method such as the limited memory BFGS algorithm (see [Gil89]). The gradient vector is obtained analytically from the adjoint state, which can be interpreted as the Lagrange multiplier of the model equations. We then use a dual method, which consists in considering the model as a weak constraint. The use of an observation vector as a Lagrange multiplier for this constraint allows us to consider the minimization problem in a dual way. The dual cost function, measuring the difference between the data and the model state corresponding to a vector of the observation space, is minimized in the observation space, still using the BFGS algorithm.

In section 2, we introduce the physical model used for the theorical and numerical results. The primal and dual methods applied to our ocean model are introduced in sections 3 and 4 respectively. Some numerical results are given in section 5. A few conclusions will be given in section 6.

2 Physical Model

2.1 Quasi-geostrophy

The system which governs the behaviour of the ocean is called the primitive equation system, constituted by the conservation laws of mass, momentum (Navier-Stokes equations), temperature and salinity. Most large-scale geophysical flows are based on the geostrophic equilibrium between the rotational effect due to the Coriolis force and the horizontal pressure gradient.

We will use here a simplified quasi-geostrophic ocean model. This model arises from the primitive equations, assuming first that the rotational effect (Coriolis force) is much stronger than the inertial effect. This can be quantified by the fact that the ratio between the characteristic time of the rotation of the Earth and the inertial time is small. This ratio is called the Rossby number. The quasi-geostrophic model also assumes that the size of the ocean is small compared to the size of the Earth, and that this ratio is close to the Rossby number. Quasi-geostrophy finally assumes that the depth of the basin is small compared to its width (the ocean is supposed to be a thin layer of the Earth). In the case of the Atlantic Ocean, all these assumptions are not valid, but it has been shown that this approximate model reproduces quite well the ocean circulations at intermediate latitudes, such as the Gulf Stream.

The thermodynamic effects are neglected, and we also assume that the forcing is due to the wind at the surface of the ocean and that the dissipation is essentially due to bottom and lateral friction.

2.2 Equations of the Model

The ocean is supposed to be stratified in n layers, each of them having a constant fluid density [Hol78]. The quasi-geostrophic model is obtained by making a first order expansion of the Navier-Stokes equation with respect to the Rossby number [Ped79]. The model system is then composed of n coupled equations resulting from the conservation law of the potential vorticity. The equations can be written as :

$$\frac{D_1\left(\theta_1(\Psi) + f\right)}{Dt} - \beta\Delta^2\Psi_1 = F_1 \qquad \text{in } \Omega \times]0, T[, \tag{1}$$

at the surface layer $(k = 1)$,

$$\frac{D_k\left(\theta_k(\Psi) + f\right)}{Dt} - \beta\Delta^2\Psi_k = 0 \qquad \text{in } \Omega \times]0, T[, \tag{2}$$

at intermediate layers $(k = 2, \ldots, n - 1)$, and

$$\frac{D_n\left(\theta_n(\Psi) + f\right)}{Dt} + \alpha\Delta\Psi_n - \beta\Delta^2\Psi_n = 0 \qquad \text{in } \Omega \times]0, T[, \tag{3}$$

at the bottom layer $(k = n)$, where

- $\Omega \subset \mathbb{R}^2$ is the circulation basin and $]0, T[$ is the time interval,

- n is the number of layers,

- Ψ_k is the stream function at layer k, Ψ is the vector $(\Psi_1, \ldots, \Psi_n)^T$,

- θ_k is the sum of the dynamical and thermal vorticity at layer k :

$$\theta_k(\Psi) = \Delta\Psi_k - (W\Psi)_k,$$

with $-(W\Psi)_k = \dfrac{f_0^2\rho}{H_kg}\left(\dfrac{\Psi_{k+1} - \Psi_k}{\rho_{k+1} - \rho_k} - \dfrac{\Psi_k - \Psi_{k-1}}{\rho_k - \rho_{k-1}}\right).$

- f is the Coriolis force (f_0 is the Coriolis force at the reference latitude of the ocean).
  In the β-plane approximation, the Coriolis force varies linearly with respect to the latitude.
- g represents the gravity, ρ_k the fluid density at layer k (and ρ the average fluid density), and H_k the depth of the layer k,

- $\dfrac{D_k}{Dt}$ is the Lagrangian particular derivative : $\dfrac{D_k}{Dt} = \dfrac{\partial}{\partial t} + J(\Psi_k, .)$,

  where J is the Jacobian operator $J(f,g) = \dfrac{\partial f}{\partial x}\dfrac{\partial g}{\partial y} - \dfrac{\partial f}{\partial y}\dfrac{\partial g}{\partial x}$,

- $\Delta\Psi_n$ represents the bottom friction dissipation, $\Delta^2\Psi_k$ represents the lateral friction dissipation,

- and F_1 is the forcing term, the wind stress applied to the ocean surface.

2.3 Boundary Conditions

The tridiagonal matrix W (used to couple the stream functions at different layers) can be diagonalized :

$$W = P.\mathrm{diag}(\lambda_1, \ldots, \lambda_n).P^{-1}, \tag{4}$$

where $0 = \lambda_1 < \lambda_2 \leq \cdots \leq \lambda_n$ are the eigenvalues, and P is the transformation matrix. We can then define the mode vector of the stream functions $\Phi = (\Phi_1, \ldots, \Phi_n)^T$:

$$\Phi = P^{-1}\Psi.$$

The first mode Φ_1 corresponds to the eigenvalue 0 and is called the barotropic mode. The next modes ares the baroclinic modes. The boundary conditions result from the mass conservation law (Holland 1978), and can then be written as :

$$\Phi_1 = 0 \qquad \text{in } \partial\Omega\times]0, T[,$$

$$\int_\Omega \Phi_k(t)d\sigma = 0 \qquad \forall t \in [0, T], \quad \forall k \geq 2, \tag{5}$$

and

$$\Delta\Psi_k(t) = 0 \qquad \text{in } \partial\Omega\times]0, T[, \quad \forall k. \tag{6}$$

The initial conditions $\Psi_k(0)$ complete the equations of the direct model.

3 Primal Variational Method

We suppose that the data we want to assimilate come from satellite measurements of the sea-surface height, which is directly related to the upper layer stream function Ψ_1 by $h = \dfrac{f_0}{g}\Psi_1$. Thus, we assume that we have an observational stream function Ψ_1^{obs}. These observations are only available at times t_i, $i = 1\ldots N$, over the data assimilation period $[0, T]$, and are also discrete in space. We consider then that the vector $\Psi_1^{obs}(t_i)$ represents the observations of the ocean surface available at time t_i.

The control vector u (which has to be determined) is the initial state of the stream functions at all layers $(\Psi_k(0))_{k=1\ldots n}$.

3.1 Cost Function

We can define a cost function

$$
\begin{aligned}
\mathcal{J}(u) = \frac{1}{2}\sum_{i=1}^{N}\langle R_i^{-1}\left(H_i\Psi_1(t_i) - \Psi_1^{obs}(t_i)\right), H_i\Psi_1(t_i) - \Psi_1^{obs}(t_i)\rangle \\
+ \frac{1}{2}\langle P_0^{-1}u, u\rangle,
\end{aligned}
\tag{7}
$$

where P_0 and R_i are covariance matrices, H_i are (linear) observation operators connecting observations Ψ_1^{obs} and model solutions Ψ_1, and $\langle \, . \, , \, . \, \rangle$ is the canonical real scalar product.

The first part of the cost function quantifies the difference between the observations and the state function, and the second part is a regularisation term. The inverse problem which consists in the minimization of $\mathcal{J}$ is then well-posed.

3.2 Adjoint Model

In order to minimize the cost function, we need its gradient $\nabla\mathcal{J}$. Because of the large dimension of the model state vector (more than 10^6), it is not possible to compute directly the gradient by using finite difference methods. The gradient vector of the functional is then obtained by solving backwards in time the adjoint model ([LeD86]). The quasi-geostrophic adjoint equations are :

$$
\begin{aligned}
\frac{\partial \theta_1^T(\Lambda)}{\partial t} - \Delta J(\Psi_1, \Lambda_1) - (W^T J(\Psi, \Lambda))_1 - J\left(\Lambda_1, \theta_1(\Psi) + f\right) \\
-\beta\Delta^2\Lambda_1 = E_1
\end{aligned}
\tag{8}
$$

at the surface layer,

$$\frac{\partial \theta_k^T(\Lambda)}{\partial t} - \Delta J(\Psi_k, \Lambda_k) - (W^T J(\Psi, \Lambda))_k - J(\Lambda_k, \theta_k(\Psi) + f) \tag{9}$$
$$- \beta \Delta^2 \Lambda_k = 0$$

at the intermediate layers, and

$$\frac{\partial \theta_n^T(\Lambda)}{\partial t} - \Delta J(\Psi_n, \Lambda_n) - (W^T J(\Psi, \Lambda))_n - J(\Lambda_n, \theta_n(\Psi) + f) \tag{10}$$
$$+ \alpha \Delta \Lambda_n - \beta \Delta^2 \Lambda_k = 0$$

at the bottom layer, in $\Omega \times]0, T[$, where

- $\Lambda_1, \dots, \Lambda_n$ is the adjoint vector,

- $\theta_k^T(\Lambda) = -\Delta \Lambda_k + (W^T \Lambda)_k$ is the *vorticity* corresponding to the adjoint state,

- and E_1 is the derivative of $\mathcal{J}$ with respect to Ψ_k :

$$E_1(t) = \sum_{i=1}^N R_i^{-1}(H_i \Psi_1(t) - \Psi_1^{obs}(t)) \, \delta(t - t_i).$$

If we denote by $\chi = (\chi_1, \dots, \chi_n)^T$ the modal adjoint vector :

$$\chi = P^T \Lambda,$$

the space boundary conditions satisfied by the adjoint state Λ are :

$$\chi_1 = 0 \qquad \text{in } \partial\Omega \times]0, T[,$$
$$\int_\Omega \chi_k(t) d\sigma = 0 \qquad \forall t \in [0, T], \quad \forall k \geq 2, \tag{11}$$

and

$$\Delta \Lambda_k(t) = 0 \qquad \text{in } \partial\Omega \times]0, T[, \quad \forall k. \tag{12}$$

The gradient of the first part of $\mathcal{J}$ is obtained by solving equations (8-12) with a final condition of nullity of the adjoint state. The gradient of the second part of $\mathcal{J}$ is obtained directly by deriving it with respect to u, and we obtain :

$$\nabla \mathcal{J} = H(-\Delta + W)H^{-1} \begin{pmatrix} \Lambda_1(0) \\ \vdots \\ \Lambda_n(0) \end{pmatrix} + P_0^{-1} u \tag{13}$$

where H is the diagonal matrix with the layers' depths H_k on the diagonal.

3.3 Minimization Process

The numerical minimization of the cost function $\mathcal{J}$ can be realized using a quasi-Newton method. The Newton class of minimization algorithms is based on an iterative process, using at iteration k the descent direction $d_k = -H_k^{-1}.\nabla\mathcal{J}(x_k)$, where $H_k = \nabla^2\mathcal{J}(x_k)$ is the Hessian matrix of the cost function. The direct computation of the Hessian matrix is impossible (its dimension being too large), but it is possible to evaluate it, using the second order adjoint equations. However, the inversion of H is nearly impossible. The quasi-Newton algorithms consist in replacing the inverse of the Hessian matrix by a succession of matrices $(W_k)_{k\in\mathbb{N}}$ which are symmetric positive definite approximations to H^{-1}.
The BFGS algorithm ([Bro69]) uses the following update formula :

$$W_{k+1} = U(W_k, s_k, \eta_k) := \left(I - \frac{s_k \otimes \eta_k}{\langle \eta_k, s_k \rangle}\right)W_k\left(I - \frac{\eta_k \otimes s_k}{\langle \eta_k, s_k \rangle}\right) + \frac{s_k \otimes s_k}{\langle \eta_k, s_k \rangle}$$

with $s_k = x_{k+1} - x_k$, $\eta_k = \nabla\mathcal{J}(x_{k+1}) - \nabla\mathcal{J}(x_k)$ and $a \otimes b : c \mapsto \langle b, c \rangle a$. The disadvantage of this formula is the need to store all pairs (s_k, η_k).
The L-BFGS algorithm ([Liu89]) is a limited memory version of the previous algorithm. Only the last M pairs are stored, M being often equal to 5. The update formula is then :

$$W_k = U(W_{k-1}, s_{k-1}, \eta_{k-1}), \quad 1 \le k \le M,$$

and

$$\begin{cases} W_k^0 = D_k, \\ W_k^{i+1} = U(W_k^i, s_{k-M+i}, \eta_{k-M+i}), \quad i = 0 \dots M - 1, \\ W_k = W_k^M, \end{cases}$$

for $k \ge M+1$, where D_k is a diagonal matrix. The update formula for D_k is :

$$D_{k+1}^{(i)} = \left(\frac{1}{D_k^{(i)}} + \frac{\eta_k^{(i)^2}}{\langle \eta_k, s_k \rangle} - \frac{s_k^{(i)^2}}{(D_k^{(i)})^2 \langle D_k^{-1} s_k, s_k \rangle}\right)^{-1}.$$

4 Dual Method

4.1 General Description

The primal method has many disadvantages. First, the minimization process is often stopped before convergence to the minimum, because of the size of the state vector. Moreover, it is also impossible to take into account a model error : in the previous section, we have supposed that the model and the equations were perfect. This is obviously not the case (for example, not all parameters are well known). The only solution to incorporate the model error into the minimization process is to add corrective terms to the model, consider them as part of the control vector, and add a third term to the cost function. This is not computationally realistic because the size of the control vector would be multiplied by the number of time steps. Therefore, it is not possible to take into account in a straightforward way the model error in the primal variational approach.

A new approach to data assimilation problems has been recently introduced ([Amo95], [Ben92], [Cou97]). Rather than minimizing a cost function on the state space, the dual method consists in working in the observation space (which is smaller than the state space).

4.2 Dual Algorithm

Instead of solving first the direct equations and then the adjoint equations in the primal variational approach, the dual method consists in solving first the adjoint equations in order to use the information contained in the observation vector, and then the direct equations in order to reconstruct a trajectory. The dual algorithm for the quasi-geostrophic model can be constructed as follows :

- Let m be an observation vector that can be directly related to Ψ_1 (assume that m is a vector containing an observation of a part of the ocean surface at different times t_i),
- Solve the adjoint equations (with a final condition equal to zero) :

$$
\frac{\partial \theta_1^T(\Lambda)}{\partial t} - \Delta J(\Psi_1, \Lambda_1) - (W^T J(\Psi, \Lambda))_1 - J(\Lambda_1, \theta_1(\Psi) + f) \\
- \beta \Delta^2 \Lambda_1 = \tilde{E}_1(m),
$$

$$
\frac{\partial \theta_k^T(\Lambda)}{\partial t} - \Delta J(\Psi_k, \Lambda_k) - (W^T J(\Psi, \Lambda))_k - J(\Lambda_k, \theta_k(\Psi) + f) \\
- \beta \Delta^2 \Lambda_k = 0, \quad 1 < k < n, \tag{14}
$$

$$
\frac{\partial \theta_n^T(\Lambda)}{\partial t} - \Delta J(\Psi_n, \Lambda_n) - (W^T J(\Psi, \Lambda))_n - J(\Lambda_n, \theta_n(\Psi) + f) \\
+ \alpha \Delta \Lambda_n - \beta \Delta^2 \Lambda_k = 0,
$$

where

$$
\tilde{E}_1(m)(t) = \sum_{i=1}^{N} H_i^T R_i^{-1}(m(t) - \Psi_1^{obs}(t)) \, \delta(t - t_i).
$$

- Solve the direct equations

$$\frac{D_1\left(\theta_1(\Psi) + f\right)}{Dt} - \beta\Delta^2\Psi_1 = F_1 + (Q\Lambda)_1,$$

$$\frac{D_k\left(\theta_k(\Psi) + f\right)}{Dt} - \beta\Delta^2\Psi_k = (Q\Lambda)_k, \qquad\qquad 1 < k < n \qquad (15)$$

$$\frac{D_n\left(\theta_n(\Psi) + f\right)}{Dt} + \alpha\Delta\Psi_n - \beta\Delta^2\Psi_n = (Q\Lambda)_n,$$

with the initial conditions

$$\Psi_k(0) = \Psi_k^e(0) + (P_0\Lambda(0))_k,$$

where Q and P_0 are statistical preconditioning matrices, and $\Psi_k^e(0)$ is an *a priori* estimation of $\Psi_k(0)$.

- Define the operator $\mathcal{D} : (\mathcal{D}m)(t) = \sum_{i=1}^{N} H_i\Psi_1(t_i)\,\delta(t - t_i).$

We can then define the dual cost function as follows :

$$\mathcal{J}_\mathcal{D}(m) = \frac{1}{2}\langle\mathcal{D}m, m\rangle - \langle\Psi_1^{obs}, m\rangle. \qquad (16)$$

$\mathcal{J}_\mathcal{D}$ measures the difference between $\mathcal{D}m$ and Ψ_1^{obs}, i.e. between the trace (in the observation space) of a solution of the direct model and the observation vector.

As $\mathcal{D}$ is a linear symmetric positive definite operator, the gradient is obviously given by

$$\nabla\mathcal{J}_\mathcal{D}(m) = \mathcal{D}m - \Psi_1^{obs}. \qquad (17)$$

It is therefore easy to perform the minimization of $\mathcal{J}_\mathcal{D}$, given its gradient, simply by using a quasi-Newton method such as a BFGS algorithm. Once the minimum has been found, it is easy to reconstruct the corresponding trajectory in the state space by solving (14-15).

We can observe that the minimization of the dual cost function takes place over a smaller space than the minimization of the primary one. Moreover, this method also takes into account the model error, which was numerically impossible in the classical approach.

5 Numerical Results

5.1 Model Parameters

The numerical experiments are performed for a square three-layered ocean. The basin has horizontal dimensions of 4000 km × 4000 km and its depth is 5 km. The layers' depths are 300 meters for the surface layer, 700 meters for the intermediate layer, and 4000 meters for the bottom layer. The ocean is discretized by a Cartesian mesh of $200 \times 200 \times 3$ grid zones. The time step is 1.5 hour. The initial conditions are chosen equal to zero for a six-year ocean spin-up phase, the final state of which being then the initial state for the data assimilation period. Then the assimilation period starts (time $t = 0$) with this initial condition ($\Psi_k(0)$), and lasts 5 days (time $t = T$), i.e. 80 time steps. The numerical method used to minimize the cost functions is a limited memory BFGS quasi-Newton method. The M1QN3 code by Gilbert and Lemaréchal ([Gil89]) is used for our experiments.

The experimental approach consists in performing *twin experiments* with simulated data. First, a reference experiment is run and the corresponding data are extracted. This reference trajectory will be further called the exact solution. Experimental surface data are supposed to be obtained on every fifth gridpoint of the model, with a time sampling of 7.5 hours (every 5 time steps). Simulated surface data are then noised with a blank Gaussian distribution, and provided as observations for the cost function. The first guess of the assimilation experiments is chosen as the reference state of the ocean one year before the assimilation period. The results of the identification process are then compared to the reference experiment.

5.2 Exact Solution, Noised Observations

Figure 1 represents the stream function Ψ_1 at the surface layer, at the beginning and at the end of the assimilation period. These fields will be useful to measure the identification of the initial state, and also the reconstruction of the stream function at the final time. One can observe the turbulent structure

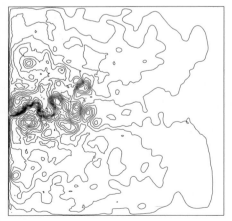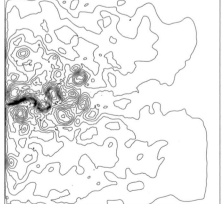

Fig. 1. Exact solution at the beginning (a), resp. the end (b), of the assimilation period.

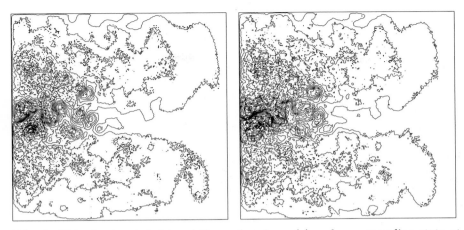

Fig. 2. Noised extracted data at the surface layer (a) and corresponding state at the end of the assimilation period (b)

of the ocean, with a main current simulating a Gulf Stream type configuration.

The first part of Fig. 2 represents the noised data extracted from the reference run, still at the surface layer. The second part of this figure is the corresponding state after a model run using the noised data as initial condition. This experiment clearly shows the importance of data assimilation. The model will indeed not smooth the trajectory, and it is not possible to obtain good predictions by simply integrating the model with observation data as initial conditions.

5.3 Primal Method

The initial estimated vector to start the minimization process is chosen to be the reference state of the ocean one year before the assimilation period. The minimization process is stopped after 40 iterations, each iteration consisting of one integration of the forward direct model (in order to compute $\mathcal{J}$) and one integration of the backward adjoint model (in order to compute $\nabla\mathcal{J}$). The result of the minimization is shown on Fig. 3-a. The direct model is then integrated over the assimilation period, using the computed minimizer as initial condition, and the corresponding state of the ocean at the end of the assimilation period is shown on Fig. 3-b.

We can notice that the stream function of the solution at time $t = 0$ at the surface layer is comparable to the exact solution at the same time, but to a lesser extent at time $t = T$. This can be explained by the fact that the primal

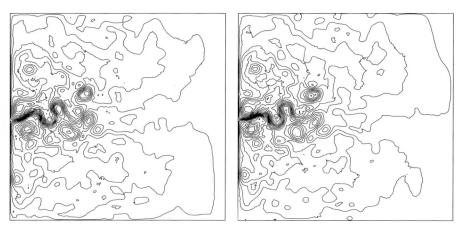

Fig. 3. Result of the minimization of the primal cost function. Solution at the beginning (a) and the end (b) of the assimilation period

algorithm gives more importance to the state at $t = 0$ than to any other time, as it is the control vector.

5.4 Dual Method

The initial estimated vector is the same. The minimization process is still stopped after 40 iterations, each iteration consisting now of one integration of the backward adjoint equations and one integration of the direct equations (in order to compute $\mathcal{J}_D$ and $\nabla\mathcal{J}_D$).

The result of the minimization process is shown on Fig. 4-a at time $t = 0$ and Fig. 4-b at time $t = T$.

The stream function appears to be less smooth than in the primal case. This is due to the fact that the observations are noised and the dual algorithm works over the observation space. The corresponding state at the end of the assimilation period is closer to the exact solution than in the primal case. The dual algorithm looks indeed for a global acceptable solution : the control vector is a set of *observations* all over the assimilation period.

5.5 Comparison between the two Methods

Figure 5 represents the root mean square (RMS) error over the entire assimilation period between the exact solution and an identified solution, using either the observations, the primal solution, or the dual one.

The RMS error at time t for one of these solutions is :

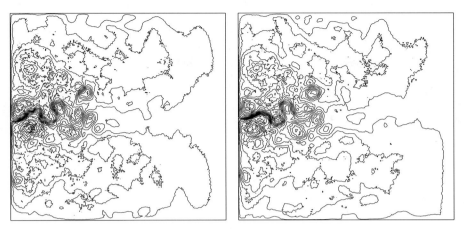

Fig. 4. Result of the minimization of the dual cost function. Solution at the beginning (a) and the end (b) of the assimilation period

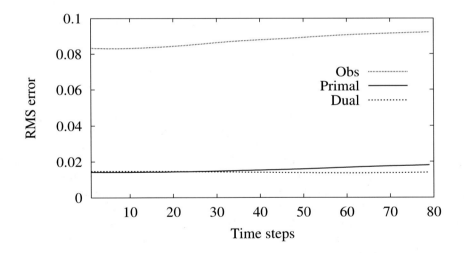

Fig. 5. RMS errors of the different methods versus number of time steps over the assimilation period, using as initial conditions: the first noised observation (thin line), the primal algorithm (bold line) and the dual algorithm (dot line).

$$rms(t) = \frac{\int_{\Omega} \left[\Psi_1^{\text{sol}}(t) - \Psi_1^{\text{exact}}(t) \right]^2 d\sigma}{\int_{\Omega} \left[\Psi_1^{\text{exact}}(t) \right]^2 d\sigma}.$$

The error reaches nearly 10% in the case of the trajectory resulting from the observations, and increases in time. This is due to the inherent non-linearities of the model.

In the case of the two data assimilation methods, the RMS error is clearly smaller (by a factor of more than 5). This proves once again the usefulness of data assimilation, which makes it possible to reconstruct a trajectory with less than 2% RMS error using noised observations with a 10% RMS error. The RMS error of the primal method tends to increase in time. This is in agreement with the remark concerning the resemblance between the stream function of the primal solution and that of the exact solution at initial time, and the loss of this similarity at final time. The RMS error of the dual solution is a little larger, but it tends to remain constant (and even decrease a little bit sometimes) in time. Moreover, as the dimension of the control vector u in the primal variational approach is 121203 (201 × 201 × 3), versus 28577 in the dual one (41 × 41 × 17), the minimization of the dual cost function is faster than for the primal approach.

6 Conclusion

As a matter of fact, the oceanic circulation model is governed by complex equations and behave as certain typical characteristics of the turbulent flow. Besides, in practice, the observation data are of various nature and should be combined together in the same functional to be minimized. In the present work, synthetic data are sampled using the whole surface layer. Generally, in the framework of the realistic oceanic data assimilation, the data are available only along ground tracks for time intervals corresponding to the satellite repeat period. Therefore, the optimal initial state would not be as well estimated because of the relatively small number of observations and their heterogeneous spatial distribution. Also, we notice that the performances of these methods have been assessed with a quasi-geostrophic model. It is necessary to apply them to a more complicated model such as the primitive equation model.

The dual data assimilation method is promising taking into account the computing time which is smaller than the primal optimal control method because of the smaller dimension of the observation space. Moreover, it enables to introduce an error in the model and not to consider the equations of the fluid as a strong constraint. The non linear character of the equations remains a problem for the proof of convergence of the dual method.

It remains a promising step towards operational oceanography.

References

[Amo95] Amodei, L.: Solution approchée pour un problème d'assimilation de données météorologiques avec prise en compte de l'erreur modèle. C.R. Acad. Sci. Paris, **321 (II)**, 1087–1094 (1995)

[Ben92] Bennett, A.F.: Inverse methods in physical oceanography. Cambridge University Press, Cambridge (1992)

[Ben02] Bennett, A.F.: Inverse Modeling of the Ocean and Atmosphere. Cambridge University Press, Cambridge (2002)

[Bro69] Broyden, C.G.: A new double-rank minimization algorithm. Notices American Math. Soc., **16**, 670 (1969)

[Can96] Cane, M.A., Kaplan, A., Miller, R.N., Tang, B., Hackert, E.C., Busalacchi, A.J.: Mapping tropical Pacific sea level: data assimilation via a reduced state Kalman filter. J. Geophys. Res., **101**(C10), 22599–22617 (1996)

[Cou97] Courtier, P.: Dual formulation of four-dimensional variational assimilation. Q. J. R. Meteorol. Soc., **123**, 2449–2461 (1997)

[Fuk93] Fukumori, I., Benveniste, J., Wunsch, C., Haidvogel, D.B.: Assimilation of sea surface topography into an ocean circulation model using a steady state smoother. J. Phys. Oceanogr., **23**, 1831–1855 (1993)

[Fuk95] Fukumori, I.: Assimilation of Topex sea level measurements with a reduced-gravity, shallow water model of the tropical Pacific ocean. J. Geophys. Res., **100**(C12), 25027–25039 (1995)

[Gel74] Gelb, A.: Applied Optimal Estimation. MA: MIT Press, Cambridge (1974)

[Ghi89] Ghil, M.: Meteorological data assimilation for oceanographers. Part I: description and theoretical framework. Dyn. Atmos. Oceans, **13**, 171–218 (1989)

[Ghi91] Ghil, M., Manalotte-Rizzoli, P.: Data assimilation in meteorology and oceanography. Adv. Geophys., **23**, 141–265 (1991)

[Gil89] Gilbert, J.-Ch., Lemaréchal, C.: Some numerical experiments with variable storage quasi-Newton algorithms. Math. Prog., **45**, 407–435 (1989)

[Gou92] Gourdeau, L., Arnault, S., Ménard, Y., Merle, J.: Geosat sea-level assimilation in a tropical Atlantic model using Kalman filter. Ocean. Acta, **15**, 567–574 (1992)

[Hol78] Holland, W.R.: The role of mesoscale eddies in the general circulation of the ocean. J. Phys. Ocean., **8-3**, 363–392 (1978)

[Jaz70] Jazwinski, A.H.: Stochastic Processes and Filtering Theory. Academic, New York (1970)

[LeD86] Le Dimet, F.-X., Talagrand, O.: Variational algorithms for analysis and assimilation of meteorogical observations : theoretical aspects. Tellus, **38A**, 97–110 (1986)

[Lew85] Lewis, J.M., Derber, J.C.: The use of adjoint equations to solve a variational adjustment problem with convective constraints. Tellus, **37A**, 309–322 (1985)

[Lio68] Lions, J.L.: Contrôle optimal de systèmes gouvernés par des équations aux dérivées partielles. Dunod (1968)

[Liu89] Liu, D.C., Nocedal, J.: On the limited memory BFGS method for large scale optimization. Math. Prog., **45**, 503–528 (1989)

[Luo98] Luong, B., Blum, J., Verron, J.: A variational method for the resolution of a data assimilation problem in oceanography. Inverse Problems, **14**, 979–997 (1998)

[Moo91] Moore, A.M.: Data assimilation in a quasigeostrophic open-ocean model of the Gulf-Stream region using the adjoint model. J. Phys. Oceanogr., **21**, 398–427 (1991)

[Nec94] Nechaev, V., Yaremchuk, M.I.: Application of the adjoint technique to processing of a standard section data set: world ocean circulation experiment section S4 along 67S in the Pacific ocean. J. Geophys. Res., **100**(C1), 865–879 (1994)

[Ped79] Pedlosky, J.: Geophysical fluid dynamics. Springer-Verlag, New-York (1979)

[Sch93] Schröter, J., Seiler, U., Wenzel, M.: Variational assimilation of Geosat data into an eddy-resolving model of the Gulf Stream area. J. Phys. Oceanogr., **23**, 925–953 (1993)

[She90] Sheinbaum, J., Anderson, D.L.T.: Variational assimilation of XBT data. Part I. J. Phys. Oceanogr., **20**, 672–688 (1990)

[Tal87] Talagrand, O., Courtier, P.: Variational assimilation of meteorological observations with the adjoint vorticity equation. Part I: Theory. Q. J. R. Meteorol. Soc., **113**, 1311–1328 (1987)

[Tha88] Thacker, W.C., Long, R.B.: Fitting dynamics to data. J. Geophys. Res., **93**, 1227–1240 (1988)

[Ver99] Verron, J., Gourdeau, L., Pham, D.T., Murtugudde, R., Busalacchi, A.J.: An extended Kalman filter to assimilate satellite altimeter data into a nonlinear numerical model of the tropical Pacific Ocean: method and validation. J. Geophys. Res., **104**, 5441–5458 (1999)

Ant Colonies: a Nature inspired Paradigm for the Mathematical Modelling of Self-Organizing Systems.

Vincenzo Capasso, Daniela Morale

MIRIAM & Department of Mathematics , University of Milano, Via C. Saldini, 50, Milano, Italy,{capasso,morale}@mat.unimi.it

1 Introduction

A widespread interest has arisen in recent years regarding systems of many individuals (cells, insects, human beings, etc.) that exhibit a collective behavior, such as swarming, schooling, herding etc [24]. Among those a very interesting impact on models of distributed computing, and optimization of highly complex systems derives from the so called Ant Colonies (AC), so that the corresponding paradigm of optimization is known as ACO [11]. For a presentation of applications to ACO's we refer also to [18] and the references therein. A nice popular presentation at the same topic may be found in [4, 5]

Attention to AC's has a long tradition in manuals and treatises on organization of groups/armies [29]:"*A clear and familiar example of natural organization is the ant colony... the ant colony is totally information centered ... the fundamental activities of an information-centered organization are gathering processing using and giving out information...*" [Sun Tzu, 500 BC,][29] A remarkable aspect of these global organizations is that individuals move altogether in a coordinated (though random) fashion even though interaction among them via relevant senses (sight, smell, hearing, etc) are typically limited to much shorter distances than the size of the group.

Over the past couple of decades, a large amount of literature has been devoted to the mathematical modelling of self-organizing populations, based on the concepts of short range/long range "social interaction" among different individuals of a biological population.

A fruitful approach suggested since long by various authors [9, 23, 24] is based on the modelling of the "movement" of each individual "particle" embedded in the total population of similar particles (the so called individual based model - IBM).

Often a multiple scale approach is preferable: the global behavior of the population is described, at the macroscopic scale, by a continuum density whose evolution in terms of integro-differential equations is derived by a limiting process from the empirical distribution associated with a large number of particles. The large knowledge available on nonlinear PDE's often helps in providing the qualitative behavior of a global "swarm".

In this report we follow the above scheme starting from the mathematical modelling of $N \in \mathbb{N}$ social individuals as a system of N stochastic differential equations (i.e. ODE's subject to Wiener noise).

This allows the use of methods of stochastic calculus for obtaining evolution equations for the global behavior of the system via Itô formulas. More informative models would allow, as noise, more general processes such as Levy processes that include both Wiener and Poisson jump processes [8, 27].

We present few applications: an aggregation model for the modelling of armies of ants; a model of vasculogenesis based on chemotaxis; a model of price herding in economics, in order to show the variety of different fields that may be modelled by an AC paradigm.

2 Particles as Social Individuals

From a Lagrangian point of view, the state of a system of N particles may be described as a (stochastic) process $\{X_N^k(t)\}_{t \in \mathbb{R}_+}$ in $(\mathbb{R}^d, \mathcal{B}_{\mathbb{R}^d})$ on a suitable probability space $(\Omega, \mathcal{F}, P)$, where $X_N^k(t)$ models the state of the k-th particle out of N.

In another way, the state of the k-th particle may be modelled as a Dirac-random measure in $\mathcal{M}(\mathbb{R}^d)$

$$\epsilon_{X_N^k(t)}. \tag{1}$$

The localizing measure (1) is defined as follows

$$\epsilon_{X_N^k(t)}(B) = \begin{cases} 1 & \text{if } X_N^k(t) \in B \\ 0 & \text{if } X_N^k(t) \notin B \end{cases} \qquad \forall B \in \mathbb{R}^d, \tag{2}$$

and for any sufficiently smooth $f : \mathbb{R}^d \to \mathbb{R}$

$$\int_{\mathbb{R}^d} f(y)\epsilon_{X_N^k(t)}(dy) = f\left(\epsilon_{X_N^k(t)}\right).$$

From an Eulerian point of view, the collective behavior of the discrete (in the number of particles) system, may be given in terms of the spatial distribution of the system at time t, expressed in term of an empirical measure

$$X_N(t) = \frac{1}{N} \sum_{k=1}^{N} \epsilon_{X_N^k(t)} \in \mathcal{M}(\mathbb{R}^d), \tag{3}$$

such that $\forall B \in \mathcal{B}(\mathbb{R}^d)$,

$$(X_N(t))(B) = \frac{1}{N}\sum_{k=1}^{N}\epsilon_{X_N^k(t)}(B) = \frac{\sharp\text{particles in } B \text{ at time } t}{N}.$$

This means that $X_N(t)$ measures the spatial relative frequency of the particles. The whole history of the system is described by the following empirical process X_N

$$X_N : t \in \mathbb{R}_+ \rightarrow X_N(t) \in \mathcal{M}(\mathbb{R}^d).$$

Given a description of the state of the system, one is interested in its dynamics, i.e. the time evolution of the state (1) of the single particle and of the empirical distribution (3). Here we are interested in those systems where each particle acts as a social individual. This means that the change of the state $X_N^k(t)$ during the time interval $(t, t+dt]$ in addition to some individual factors and random factors, depends significatively on the mutual interaction among individuals. Collective behavior may be related to individual decisions, and it is assumed that the "motion" of an individual is a combined result of both population-independent and population-dependent decisions. This idea comes out from the observation of natural populations, but it may be used in many other fields [11, 21, 1] as we will see in the applications.

The motion of cells, grouping of animals, self organization of molecules cannot be purely random. Pure random dispersal of individuals may explain phenomena of spatial homogeneization of a population; organization must be due to relevant mechanisms of interaction among individuals. There is always an element of choice in location. So interaction contrasts diffusion and on the other way round, dispersion contrasts possible crowding effect.

This leads to a system of N stochastic differential equations. The source of stochasticity could be either due to the randomness in the environment or given by other factors, so that we may model the velocity as following

$$dX_N^k(t) = h_N(X_N(t), B_t, t)\,dt, \quad k = 1, 2, \ldots, N, \qquad (4)$$

where $h_N : \mathcal{M}(\mathbb{R}^d) \times \mathcal{M}(\mathbb{R}^d) \times R_+ \rightarrow \mathbb{R}$ is generally a smooth function. The random perturbing function B_t models a random forcing factor. A simple way to model randomness is to consider an independent additive noise, acting on each particle; so that (4) becomes

$$dX_N^k(t) = \left[f_N^k(t) + F_N\left[X_N(t)\right]\left(X_N^k(t)\right)\right]dt$$
$$+ \sigma(X_N(t), t)dW^k(t), \quad k = 1, \ldots, \Lambda_N(t); \qquad (5)$$

the functional F_N defined on $\mathcal{M}(\mathbb{R}^d)$, i.e. the deterministic drift term depending on the empirical measure (3), describes the interaction of the k-th particle with other particles in the system; the function $f_N^k : \mathbb{R}_+ \rightarrow \mathbb{R}$ describes the individual dynamics which may depend only on time or on the state of the particle itself; $\{W^k, k = 1, \dots\}$ is a family of independent standard Wiener processes.

The number of particles may be either constant over time, say $\Lambda_N(t) = N, \forall t \in \mathbb{R}_+$, or a dynamical variable itself described e.g. by a birth and death process whose intensities may be strongly coupled with the particle population. In this case N represents only a scaling parameter [10].

The functional form of the interaction term F_N is related to the mathematical modelling of the interaction among individuals.

It is clear that the limit dynamics of the population will depend on the particular mathematical structure of the interaction term.

3 Modelling Interaction

Here let us consider in (5) $\Lambda_N(t) = N, \forall t \in \mathbb{R}_+$. We want to discuss the expression of the functional $F_N[X_N(t)]$.
In the system there are three main scales (cf. Fig. 1):

a. the macroscopic scale regarding the typical volume occupied by the total population;
b. the microscopic scale, regarding the typical volume of each individual;
c. the mesoscale, which is in between the previous two; the typical distance is much larger with respect to the distance between particles, but it is much smaller with respect the whole space.

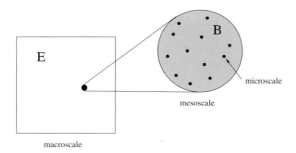

Fig. 1. Different scales.

If we consider a system of N particles located in $\mathbb{R}^d$, in the macroscopic space-time coordinates the typical distance between neighboring particles is is $O(N^{-1/d})$ and the order of the size of the whole space $O(1)$.

The interaction among particles is mathematically modelled by an interaction potential depending on the distance between two particles. The range of the potential kernel gives the "size" of the interaction.

We may distinguish three main type of interactions:

a. *McKean-Vlasov interaction*: any particle interacts with $O(N)$ other particles; collective long-range forces are predominant and the particles are weakly interacting; the range of the interaction is not restricted to a small neighborhood: any particle or individual can interact with essentially any other members of the population, i.e. the range gets very large in comparison with the typical distance between neighboring particles, and its strength decreases fast, like $1/N$.

b. *hydrodynamic interaction*: any particle interacts with $O(1)$ other particles in a very small neighborhood with volume $O(1/N)$. The interaction gets short-ranged and rather strong for large N.

c. *moderate interaction* : any particle interacts with many $O\left(N/\alpha(N)\right)$ other particles is a small volume $O\left(1/\alpha(N)\right)$ where both $\alpha(N)$ and $(N/\alpha(N))$ tend to infinity as $N \to \infty$;

As said above, the interaction is described via a potential kernel. The three types of interaction may be obtained from a reference function V_1 by means of an appropriate rescaling. In the rescaling fundamental roles are played by two parameters: the total number of particles N and a parameter $\beta \in [0, 1]$.

Suppose the particles are located in $\mathbb{R}^d$, and assume that for a system of size N the interaction between two particles located in x and y respectively is determined by

$$\frac{1}{N} V_N(x - y), \tag{6}$$

where

$$V_N(z) = N^\beta V_1(N^{\beta/d} z), \tag{7}$$

for some sufficiently regular function V_1 and a scaling exponent $\beta \in [0, 1]$. So in (5)

$$F_N[X_N(t)](X_N^k(t)) = (X_N(t) * V_N)(X_N^k(t))$$

$$= \sum_{i=1}^{N} \frac{1}{N} V_N\left(X_N^i(t) - X_N^k(t)\right)$$

$$= \sum_{i=1}^{N} N^{1-\beta} V_1\left(N^{\beta/d}\left(X_N^i(t) - X_N^k(t)\right)\right) \tag{8}$$

We have

a'. *McKean-Vlasov interaction*, if $\beta = 0$;
b'. *hydrodynamic interaction*, if $\beta = 1$;
c'. *moderate interaction*, if $\beta \in (0, 1)$.

Obviously, it would be possible also to consider interacting particle systems rescaled by $\beta > 1$. This means that the range of interaction decreases much faster than the typical distance between neighboring particles. So most of the time the particles do not approach sufficiently close to feel the interaction. The interactions under study here are the McKean-Vlasov and the moderate interaction.

Observe that for the moderate interaction c., the function $\alpha(N) = N^\beta$, so that any particle can react with other particles located in a volume of order $O(N^{-\beta})$ and the interaction range (the so called mesoscale) is sufficiently small with respect to the macroscale (order of the size of the whole space) and large enough with respect to the microscale (typical distance between particles). As a consequence each small interval at the mesoscale, when N is sufficiently large, may contain a large number of particles. So from a mathematical point of view laws of large numbers may be applied.

4 Diffusion

In this model randomness may be due to both external sources and "social" reasons. The external sources are, for example, unpredictable irregularities of the environment. On the other hand the innate need of interaction with similars is a social reason. As a consequence, randomness is modeled by a multidimensional Brownian motion $\underline{W}_t$; the coefficient of $d\underline{W}_t$ is a matrix function depending upon the distribution of particles or some environmental parameters. In the example presented here, we take into account just the intrinsic stochasticity due to the need of each particle of interacting with others. Hence, we consider just one Brownian motion dW_t and the variance of each particle σ_N depending on the number of particles, not on their distribution, i.e. in (5)

$$\sigma(X_N(t), t) = \sigma_N, \forall t.$$

We could interpret this as a rough approximation of the model by considering all the stochasticities (also the ones due to the environment) modeled by

$$\sigma_N dW_t. \tag{9}$$

If σ_N expresses the intrinsic randomness of each individual due to its need of socializing, it might be allowed to consider σ_N reducing as N increases, and in particular to assume it vanishing. Indeed if the number of particles is large, the mean free path of each particle may reduce up to a limiting value that may eventually be zero

$$\lim_{N \to \infty} \sigma_N = \sigma_\infty \geq 0. \tag{10}$$

5 Time Evolution at the Macroscopic Scale

System (5) and (8) model the time evolution of the system from the Lagrangian point of view. By an Eulerian approach, one has to find the time evolution for the empirical spatial distribution (3) of the particles.

A fundamental tool for the limiting procedure is Itô's formula which establishes the time evolution of some function $f(X_N^k(t), t)$, $f \in C_b^2(\mathbb{R}^d \times \mathbb{R}_+)$, of the trajectory $\{X_N^k(t), t \in \mathbb{R}_+\}$ of the individual particle given the sde (5) [8, 17]:

$$
\begin{aligned}
f(X_N^k(t), t) = {}& f(X_N^k(0), 0) \\
&+ \int_0^t \left\{ f_N^k(t) + F_N[X_N(s)](X_N^k(s)) \right\} \nabla f(X_N^k(s), s) ds \\
&+ \int_0^t \left[\frac{\partial}{\partial s} f(X_N^k(s), s) + \frac{\sigma_N^2}{2} \Delta f(X_N^k(s), s) \right] ds \\
&+ \sigma_N \int_0^t \nabla f(X_N^k(s), s) dW_s.
\end{aligned}
\tag{11}
$$

From (11), we get the following weak formulation of the time evolution of $X_N(t)$, for any $f \in C_b^{2,1}(\mathbb{R}^d \times [0, \infty))$:

$$
\begin{aligned}
\langle X_N(t), f(\cdot, t) \rangle = {}& \langle X_N(0), f(\cdot, 0) \rangle \\
&+ \int_0^t \langle X_N(t), \left\{ f_N^k(t) + F_N[X_N(s)](X_N^k(s)) \right\} \cdot \nabla f(\cdot, s) \rangle ds \\
&+ \int_0^t \left\langle X_N(t), \frac{\sigma_N^2}{2} \Delta f(\cdot, s) + \frac{\partial}{\partial s} f(\cdot, s) \right\rangle ds \\
&+ \frac{\sigma_N}{N} \int_0^t \sum_k \nabla f(X_N^k(s), s) dW^k(s).
\end{aligned}
\tag{12}
$$

In (12) we have introduced the following quantity

$$
g_N(x, t) = (X_N(t) * V_N)(x).
\tag{13}
$$

Furthermore, we have denoted the integration of any (sufficiently smooth) function $f : \mathbb{R}^d \longrightarrow \mathbb{R}$ with respect of a measure μ on $(\mathbb{R}^d, \mathcal{B}_{\mathbb{R}^d})$ as

$$
\langle \mu, f \rangle = \int f(x) \mu(dx).
\tag{14}
$$

Let observe that the only explicit dependence on the stochasticity in (12) is in the last term

$$M_N(f,t) = \frac{\sigma_N}{N} \int_0^t \sum_k \nabla f(X_N^k(s), s) dW^k(s).$$ (15)

It is a zero mean martingale with respect to the natural filtration of the process $\{X_N(t), t \in \mathbb{R}_+\}$.

We may apply Doob's inequality [17] to get an upper bound for the second variation of (15)

$$E\left[\sup_{t \leq T} |M_N(f,t)|\right]^2 \leq \frac{4\sigma_N^2 \|\nabla f\|_\infty^2 T}{N}.$$ (16)

This means that as N increases to ∞, (16) decreases to zero. This will be important later on. Just observe by now that when $N \to \infty$, $M_N(t)$ vanishes in probability, i.e. the random part of (12) vanishes and the dynamics of X_N gets deterministic.

6 Individual Based Models. Applications

In the next we consider some examples of individual based models. In different context the concept of interacting individuals may be applied and the dynamics of the whole system may be described by (5). We differentiate the drift part, because of the different interactions.

6.1 An Example in Ecology: an Army of Ants. An Aggregation Model

System (5) may well describe the collective behaviour of individuals in herds, swarms, etc. [12, 15, 16, 24, 28]. Now we consider a specific model studied to describe the aggregative behavior of an army of ants [3].

Aggregation is due to "social" forces induced by the interaction of each individual with other individuals in the population which belong to a suitable neighborhood.

We assume a short range repulsion among individuals that prevents their accumulation in a single point in space [19].

Hence we consider the following basic biological assumptions :

(i) particles tend to aggregate subject to their interaction within a range of size $R > 0$ (finite or not). This corresponds to the assumption that each particle is capable of perceiving the others only within a suitable sensory range; in other words each particle has a limited knowledge of the spatial distribution of its neighbors.

(ii) particles are subject to repulsion when they come "too close" to each other.

(iii)particles are subject to random dispersal with a diffusion coefficient σ_N possibly depending upon the total number of particles.

We may express assumptions (i) and (ii) by introducing in the drift term F_N in (5) two additive components: $F_{N,1}$, responsible of aggregation, and $F_{N,2}$, responsible of repulsion, such that

$$F_N = F_{N,1} + F_{N,2},$$

both of the type (8).

The Aggregation Term $F_{N,1}$

Aggregation range depends on the range of visibility, due to the environment. This means that each particle acts upon the knowledge of the neighboring spatial distribution of the population within a limited range of the same size.

So in (8) there is not any rescaling, i.e. $\beta = 0$. So the aggregation is modeled by a McKean-Vlasov interaction kernel.

$$G : \mathbb{R}^d \longrightarrow \mathbb{R}_+$$

having a support confined to the ball centered at $0 \in \mathbb{R}^d$ and radius $R \in \bar{\mathbb{R}}_+$ as the range of sensitivity for aggregation. A "generalized" gradient operator is obtained as follows. Given a measure μ on $\mathbb{R}^d$ we define the function

$$[\nabla G * \mu](x) = \int_{\mathbb{R}^d} \nabla G(x - y)\mu(dy), \quad x \in \mathbb{R}^d$$

as the classical convolution of the gradient of the kernel G with the measure μ. Furthermore, G is such that

$$G(x) = \widehat{G}(|x|), \tag{17}$$

with $\widehat{G}$ decreasing function in $\mathbb{R}_+$.

We assume that the aggregation term $F_{N,1}$ depends on such a generalized gradient of $X_N(t)$ at $X_N^k(t)$:

$$F_{N,1}[X_N(t)]\left(X_N^k(t)\right) = [\nabla G * X_N(t)]\left(X_N^k(t)\right). \tag{18}$$

Each individual feels this generalized gradient of the measure $X_N(t)$ with respect to the kernel G; the positive sign for $F_{N,1}$ and (17) expresses a force of attraction of the particle in the direction of increasing concentration of individuals.

We emphasize the great generality included in this definition of generalized gradient of a measure μ on $\mathbb{R}^d$. By using particular shapes of G, one may include angular ranges of sensitivity, asymmetries, etc. at a finite distance [16].

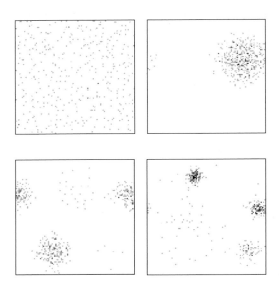

Fig. 2. Different aggregation profiles: aggregation is stronger as the support of G gets smaller.

The Repulsion Term $F_{N,2}$

As far as repulsion is concerned we proceed in a similar way by introducing a convolution kernel

$$V_N : \mathbb{R}^d \longrightarrow \mathbb{R}_+,$$

which describe a moderate interaction between particles, i.e. the $F_{N,2}$ is of the type (8), with $\beta \in (0,1)$.

We define

$$F_{N,2}[X_N(t)](X_N^k(t)) = -\left(\nabla V_N * X_N(t)\right)\left(X_N^k(t)\right)$$

$$= -\tfrac{1}{N} \sum_{m=1}^{N} \nabla V_N(X_N^k(t) - X_N^m(t)). \tag{19}$$

Each individual feels the gradient of the population in a small neighborhood; the negative sign for $F_{N,2}$ expresses a drift towards decreasing concentration of individuals. In this case the range of the repulsion kernel decreases to zero as the size N of the population increases to infinity.

So the dynamics of the system of N ants is modelled by the following system of SDEs

$$dX_N^k(t) = \left[\left(X_N(t) * \nabla G\right)\left(X_N^k(t)\right) - \left(X_N(t) * \nabla V_N\right)\left(X_N^k(t)\right)\right] dt$$

$$+ \sigma(X_N(t), t) dW^k(t), \quad k = 1, \dots, N. \tag{20}$$

We may see from Fig. 2 that the model well describes an aggregation phenomenon. The aggregating force gets stronger as the interaction range of G gets smaller.

By (12), (18) and (19) we have the following equation for the empirical measure:

$$\langle X_N(t), f(\cdot, t)\rangle = \langle X_N(0), f(\cdot, 0)\rangle$$
$$+ \int_0^t \langle X_N(t), (X_N(s) * \nabla G) \cdot \nabla f(\cdot, s)\rangle ds$$
$$- \int_0^t \langle X_N(t), \nabla g_N(\cdot, s) \cdot \nabla f(\cdot, s)\rangle ds$$
$$+ \int_0^t \left\langle X_N(t), \frac{\sigma_N^2}{2}\triangle f(\cdot, s) + \frac{\partial}{\partial s}f(\cdot, s)\right\rangle ds$$
$$+ \frac{\sigma_N}{N}\int_0^t \sum_k \nabla f(X_N^k(s), s) dW^k(s). \tag{21}$$

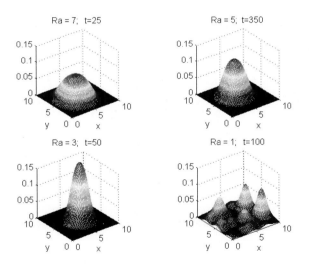

Fig. 3. Different aggregation profiles for the empirical distribution of particles.

In Fig. 3 we see the density distribution of particles. The aggregation is qualitatively similar to the one highlighted in the density profiles of a particular family of ants. Indeed, in Fig. 4 four profiles of the section transversal to the

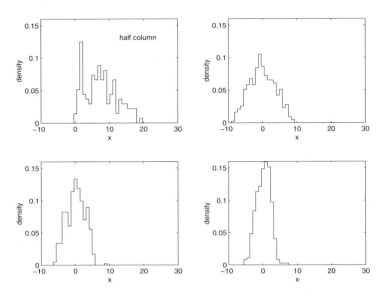

Fig. 4. "Field"-experiments: the pictures show the density profile in a cross section of an army of *Polyergus* rufescens on different type of terrain: the number of obstacles is increasing, from a concrete plane terrain at the upper left (where only one half of the army is represented) to a high grass field at the bottom right. The width of the distribution gets narrower, from about 40 cm to 10 cm.

direction of the motion of an army of ants are shown. The theoretical hypothesis, confirmed by model (21) is that the aggregation depends on the range of visibility of the environment: more the obstacles are, lower is the visibility, stronger is the aggregation.

6.2 An Example in Medicine: Vasculogenesis

In this section we consider a stochastic particle model for describing the first step of the process of formation of vascular type networks.

From a mathematical point of view, we introduce here a new element: a multiple scale approach. Indeed, a spatial point process for the seeding of new cells, and a system of stochastic differential equation for the movement of cells are strongly coupled with the faster dynamic of as chemotactic factor, the evolution is described by a PDE.

Vasculogenesis is a typical example of a self-organization phenomenon mediated by systems of underlying fields. In particular a process of cellular organization occurs at a microscopic scale, while diffusion of chemoattractants occurs at a macroscopic scale. Indeed soluble mediators (VGEF,...) are on one side released by the cells themselves and on the other side play the role of attractors by chemotaxis.

Endothelial cells are cultured on a protein matrix (Matrigel), which is plunged into a liquid. By means of a pipette, a number N of cell is released; they "swim" across the liquid and seed on the matrigel. Then they start to migrate and aggregate, building up a network, composed, at a macroscopic level, by nodes and cords.

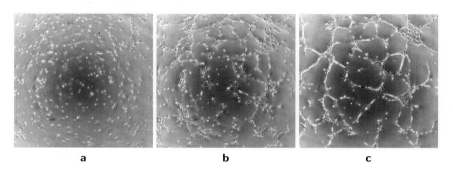

Fig. 5. The process of formation of endothelial cells networks (Courtesy by F. Bussolino *et al.*, Candiolo Institute, Torino, Italy)

The whole process is observed through videomicroscopy. It is composed by three main steps:

1. endothelial cells randomly seed on the matrigel surface, and start moving in different directions, interact and adhere to their neighbors. Their form is essentially symmetric and one may assume each cell to sit at a single point of the space (Fig. 5, a);
2. the network undergoes a slight modification: neighbor cells stretch out and stick (Fig. 5, b);
3. individual cells fold up to form tubes (capillaries), so that a three dimensional structure is created (Fig. 5, c).

We describe the "seeding" and the aggregation processes for the cells, i.e. the formation of the so called "pre-pattern" of the network (Fig. 5, a). It seems natural to consider a stochastic component in both processes, since endothelial cells by themselves exhibit a random movement.

Let be $X_N^k(t), v_N^k(t)$ the position and the velocity of the k-th cell. In this case (5) becomes

$$dX_N^k(t) = v_N^k(t)dt$$
$$dv_N^k(t) = F_N[X_N(t), C(\cdot, t)](X_N^k(t), v_N^k(t))dt + \sigma dW^k(t), \qquad (22)$$

where $\{W^k, k = 1, \ldots, N\}$ is a family of independent standard Wiener processes. The drift in (22) depends also on $C(x, t)$, the concentration of the chemoattractor released by the cells.

We consider the following assumptions:

a. a microscale for the particle dynamics and a macroscale for the chemoat-
    tractor, due to the fact that the latter evolves much faster than the move-
    ment of the particle
b. each cell goes towards higher concentration of the chemical substrate, the
    interaction occurring via a nonlocal gradient associated via a suitable ker-
    nel G; furthermore cells do not overlap.
c. the chemical underlying field C, produced by the particles, diffuses and
    degrades.

By hypotheses a. and b., the drift term may be modelled as the following

$$F_N[X_N(t), C(\cdot, t)](X_N^k(t), v_N^k(t)) = \left[\alpha_k \frac{v_N^k(t)}{|v_N^k(t)|} + [C(t, \cdot) * \nabla G](X_N^k(t)) dt \right.$$

$$\left. - [X_N(t, \cdot) * \nabla V](X_N^k(t))] \right.$$

$$\times \left(1 - N \int \delta_{A_N^k(t)}(x) X_N(t)(dx) \right) \quad (23)$$

In (23) we have introduced the concept of preferred direction or inertia, by
means of the term $f^k(t) = \alpha_k v_N^k(t)/|v_N^k(t)|$: cells do not turn very often; the
choice of the new direction is the output of the balance between the weight
of the last direction and the attraction weight. Besides they cannot occupy a
spatial position where another cell sits. This exclusion principle is modelled
by the term $1 - \sum_{j=1}^k \delta_{A_N^k(t)}(X_N^j(t))$.

Furthermore the aggregative component is given by the underlying field
$C(t, x)$ of the chemoattractor. So this is not a population-dependent model.

By hypothesis c., we consider the following PDE for the density C

$$\frac{\partial C}{\partial t}(t, x) = \Delta C(t, x) + \eta \sum_{j=1}^k \delta_{X_N^k(t)}(x) - 1/\gamma C(t, x), \quad (24)$$

where η is the rate of release and γ is the characteristic time of the soluble
factor.

Initial conditions

Since, in vitro experiments, chemotaxis occurs also in the vertical direction,
while endothelial cells reach the matrigel, we do not consider an initial uniform distribution of the N cells. A first group of $n << N$ cells is uniformly
distributed on the matrigel; the remaining $N-n$ cells are allocated as a spatial
point process with the same rate of the chemotaxis process

$$[C(t, \cdot) * \nabla G](x) dx dt \left(1 - \sum_{j=1}^k \delta_{X_N^k(t)}(x) \right).$$

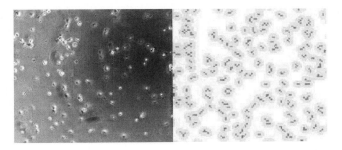

Fig. 6. Left: EC pre-pattern; Right: simulations; the darker grey points represent the endothelial cells; the lighter grey is the chemotactic substance.

In Fig. 6.2, we may qualitatively compare the equilibrium structure obtained by simulation with the distribution of cells in the first step of the in vitro experiment (right). We may see a pre-pattern has been created, which exhibits aggregating clusters with holes.

6.3 An Example from Economics: Price Herding

In this section we consider a cooperative particle system in a completely different setting. We show how we may apply the basic ideas used above.

Let us consider a market formed by N sellers competing for the trade of similar goods. Let be $X_N^k(t)$ the price of a good of the k-th seller. The questions we want to answer are the following:

i)   how may different prices influence each other?
ii) how may we introduce a concept of cooperation?
iii) which are the basic aspects of the dynamics of a system of prices?

We have to take into account general effects like inflation and random fluctuations of prices due to unexpected events, as well as possible interactions among prices themselves.

We consider the following assumptions on the rates of change of prices:

a) in general, prices of similar products follow a process of imitation and tend "to aggregate" themselves;
b) the influence between prices which are significantly different is lower;
c) sellers do not influence each other in the same way; in general, their "importance" is given from the share of their sales;
d) prices are subject to inflation;
e) rates are subject to Wiener noise.

We may then model the evolution of prices, in the framework of (5) as follows:

$$\frac{dX_N^k(t)}{X_N^k(t)} = \left[\tilde{f}_N^k(t) + \tilde{F}_N[X_N(t)] \right] dt + \sigma_N^k \, dW^k(t), \quad k = 1, \ldots, N, \quad (25)$$

From assumption d) the drift term $\tilde{f}_N^k(t)$ in (25) becomes

$$\tilde{f}_N^k(t) = \beta_k \alpha(t), \tag{26}$$

where $\alpha(t) \in \mathbb{R}$ is the inflation function and $\beta_k \in \mathbb{R}_+$ represents the sensibility of the k-th price, i.e. the k-th seller to the inflation.

Assumption a) and b) introduce in the model the concept of interaction, well described by (8), with $\beta = 0$. In particular from assumption b), we choose the following form for the interaction kernel V_N:

$$V_N(x) = \nabla K_\epsilon(x),$$
$$K_\epsilon(x) = \frac{1}{\sqrt{2\pi\epsilon^2}} \exp\left(-\left(\frac{x}{2\epsilon}\right)^2 \right), \quad x \in \mathbb{R}^d. \tag{27}$$

Assumption c) states that the response force of the k-th price to the interaction with the j-th prices depends on the ratio of the share of the two sellers. So we introduce a weight for the interaction kernel. Let $I_k(t), k = 1, \ldots, N$ represent the share of sales of the k-th seller at time t. We assume that $I_k(t) > 0 \, \forall \, t \geq t_0, \, \forall \, i = 1, \ldots, N$.

For the interaction of the k-th price with the j-th price we choose the following weight:

$$\frac{I_j(t)}{I_k(t)}$$

so that the final system of SDEs for the rates is, for $\beta_{j,k}, \gamma_{j,k} \in \mathbb{R}_+$.

$$\frac{dX_N^k(t)}{X_N^k(t)} = \left[\beta_k \alpha(t) + \frac{1}{N} \sum_{j=1}^{N} \beta_{j,k} \left(\frac{I_j(t)}{I_k(t)} \right)^{\gamma_{j,k}} \nabla K_\epsilon(X_N^j(t) - X_N^k(t)) \right] dt$$

$$+ \sigma_N^k \, dW^k(t), \quad k = 1, \ldots, N. \tag{28}$$

For $n = 1$, we may think that the evolution of price $X(t)$ is uniquely determined by inflation. In fact, (28) reduces to the linear equation

$$\frac{dX(t)}{X(t)} = \alpha \, dt + \epsilon dW(t), \tag{29}$$

where $\alpha(t) \in \mathbb{R}$ represents the inflation.

In order to validate the model (28), we have compared the simulation results with real time series for a family of car prices (Fig. 7, right) . We have collected some data relative to Italian car market in the 90's. These data concern the most important vehicle characteristics (price, mass, dimensions, max speed, acceleration, power, sale period) as well as information about quantities of sales and inflation. We have selected a number of trademarks. They are representative of the whole Italian car trading, since they cover about the 90-percent of the total quantity of sales. All these data have been ordered in time series.

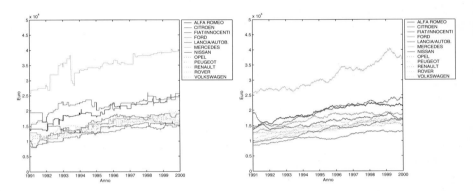

Fig. 7. Left: Real price time series from 1/1/1991 until 12/31/1999 (Euro). The general increase is principally due to inflation. The largest part of them constitutes a band included between 8000 − 17000 Euro, which seems to follow Fiat's price. On the contrary, Mercedes, Alfa Romeo and Lancia evolve independently (Courtesy of Quattroruote Pubb.); Right: Simulated price evolution.

7 Dynamics for Large Populations

In the previous sections we have always considered a large, but finite number N of particles.

An interesting question is how the dynamics (12) changes as the population becomes large and dense, so that densities may be considered. Mathematically it is possible to consider continuum models only when it is possible to perform a law of large numbers.

We may conjecture that a limiting measure valued deterministic process $\{X_\infty(t), t \in \mathbb{R}_+\}$ exists whose evolution equation (in weak form) should be

$$\langle X_\infty(t), f(\cdot, t) \rangle = \langle X_\infty(0), f(\cdot, 0) \rangle$$

$$+ \int_0^t \langle X_\infty(s), F[(X_\infty(s)](\cdot) \nabla f(\cdot, s) \rangle ds$$

$$+ \int_0^t \left\langle X_\infty(s), \frac{\sigma_\infty^2}{2} \triangle f(\cdot, s) + \frac{\partial}{\partial s} f(\cdot, s) \right\rangle ds. \qquad (30)$$

for $\sigma_\infty^2 \geq 0$.

Actually various nontrivial mathematical problems arise in connection with the existence of a limiting measure valued process $\{X_\infty(t), t \in \mathbb{R}_+\}$, above all for the case $\beta \in (0, 1]$ in (8). Indeed, for $\beta > 0$ the support of V_N tends to localize at 0 for N tending to infinity. The space variation of V_N and worse of ∇V_N become larger and larger in a neighborhood of 0. So we need some regularity on V_N, claiming also small variations for small time intervals. Here the concept of mesoscale becomes important: in order to perform a law of large numbers a neighborhood at the mesoscale should contain a sufficiently large number of particles; roughly speaking, this is equivalent to the fact that V_N converges to a Delta-dirac function sufficiently slow. From the mathematical point of view this is equivalent to properties of relative compactness for the process X_N [2, 20, 26].

The typical programme for the proof of the convergence includes the following [20, 25]

a) prove existence of a deterministic limiting measure values process $\{X_\infty(t), t \in \mathbb{R}_+\}$.

b) prove absolute continuity of the limiting measure with respect to the usual Lebesgue measure on $\mathbb{R}^d$

c) provide an evolution equation for the density $\rho(x, t)$

We will not go into further details; but here we confine ourselves to a formal convergence procedure for the model (21).

Let us now suppose that indeed the empirical process $\{X_N(t), t \in \mathbb{R}_+\}$ tends, as $N \to \infty$, to a deterministic process $\{X(t), t \in \mathbb{R}_+\}$, which is for any $t \in \mathbb{R}_+$ absolutely continuous with respect to the Lebesgue measure on $\mathbb{R}^d$, with density $\rho(x, t)$:

$$\lim_{N \to +\infty} \langle X_N(t), f(\cdot, t) \rangle = \langle X(t), f(\cdot, t) \rangle$$

$$= \int f(x, t) \rho(x, t) dx, t \geq 0.$$

As a formal consequence we get

$$\lim_{N \to +\infty} g_N(x,t) = \lim_{N \to +\infty} (X_N(t) * V_N)(x) = \rho(x,t),$$

$$\lim_{N \to +\infty} \nabla g_N(x,t) = \nabla \rho(x,t),$$

$$\lim_{N \to +\infty} (X_N(t) * \nabla G)(x) = (X(t) * \nabla G(x))$$

$$= \int \nabla G(x-y)\rho(y,t)dy.$$

Hence, by applying the above limits, from (21) we get

$$\int_{\mathbb{R}^d} f(x,t)\rho(x,t)dx = \int_{\mathbb{R}^d} f(x,0)\rho(x,0)dx$$

$$+ \int_0^t ds \int_{\mathbb{R}^d} dx [(\nabla G * \rho(\cdot,s))(x) - \nabla \rho(x,s)] \cdot \nabla f(x,s)\rho(x,s)$$

$$+ \int_0^t ds \int_{\mathbb{R}^d} dx \left[\frac{\partial}{\partial s} f(x,s)\rho(x,s) + \frac{\sigma_\infty^2}{2} \triangle f(x,s)\rho(x,s) \right], \qquad (31)$$

where σ_∞ is defined in (10).

We recognize that (31) is a weak version of the following equation for the spatial density $\rho(x,t)$:

$$\frac{\partial}{\partial t}\rho(x,t) = \frac{\sigma_\infty^2}{2} \triangle \rho(x,t) + \nabla \cdot (\rho(x,t)\nabla \rho(x,t))$$

$$- \nabla \cdot [\rho(x,t)(\nabla G * \rho(\cdot,t))(x)], \quad x \in \mathbb{R}^d, t \ge 0,$$

$$\rho(x,0) = \rho_0(x), \quad x \in \mathbb{R}^d. \qquad (32)$$

In the degenerate case, i.e. if in the (10) the equality holds, (32) becomes

$$\frac{\partial}{\partial t}\rho(x,t) = \nabla \cdot (\rho(x,t)\nabla \rho(x,t)) - \nabla \cdot [\rho(x,t)(\nabla G * \rho(\cdot,t))(x)],$$

$$x \in \mathbb{R}^d, t \ge 0,$$

$$\rho(x,0) = \rho_0(x), \quad x \in \mathbb{R}^d, \qquad (33)$$

By this limiting procedure, we see how when a particle system becomes large, one may describe it via the density. In this case the individuality is lost and in particular, and the system becomes deterministic, even though a memory of the individual randomness is in the diffusion term in (32).

Acknowledgement

With regards to vasculogenesis it is a pleasure to acknoledge Prof. Bussolino (Candiolo Institute, Torino) for providing pictures and data and Prof. Luigi Preziosi (Politecnico of Torino) for relevant discussions. For the analysis of the price herding it is a pleasure to acknowledge the magazine QUATTRORUOTE (Editoriale Domus) for providing data about car prices on the Italian market. This work has been performed within a research contract with an great international company, in collaboration with Fabio Sioli and AnnaMaria Bianchi.

References

1. Bianchi A., Capasso, V., Morale D., Sioli, F. 'Mathematical Models for Price Herding. Application to prediction of Multivariate Time Series". MIRIAM REport 07/2003 (confidential), University of Milano, 2003
2. Billingsley P.,"Convergence of Probability Measures ", (Wiley series in Probability and Mathematical Statistics. Tracts on probability and statistics). New York : John Wiley & Sons, 1968.
3. Boi, S., Capasso. V, Morale, D. "Modeling the aggregative behavior of ants of the species *Polyergus rufescens*.", Nonlinear Analysis: Real World Applications, I, 2000, p. 163-176.
4. Bonabeau E., Théraulaz G. "Swarm Smart", Scientific American, March 2000, p. 73-79
5. E. Bonabeau E., M. Dorigo, Théraulaz G. "Inspiration for Optimization from Social Insect Behavior", *Nature*,**406**, 2000, p. 39 - 42
6. Bussolino, F. *et al.* Hepatocyte growth factor is a potent angiogenic factor which stimulates endothelial cell motility and growth. *J. Cell Biol.* **119**, 1992, 629-641.
7. Carmeliet P. Mechanisms of angiogenesis and arteriogenesis *Nature Medicine* **403**, 2000, 207 - 211
8. V. Capasso, D. Bakstein "Introduction to the Theory and Applications of Continuous-time Stochastic Processes.", Birkhäuser. In press.
9. Durrett, R., Levin, S.A., " The importance of being discrete (and spatial)." Theor. Pop. Biol., **46**, 1994, 363-394.
10. V. Capasso, D. Morale, C. Salani "Polymer Crystallization Processes via Many Particle Systems". In *Mathematical Modelling for Polymer Processing. Polymerization, Crystallization, Manufacturing.* (V.Capasso Ed.), Springer-Verlag, chapter 6, p. 243-259 2003
11. A.Colorni, M.Dorigo, F.Maffioli, V.Maniezzo, G.Righini, M.Trubian. *Heuristics from nature for hard combinatorial optimization problems.* International Transaction in Operational Research, 3, 1, pag. 1-21, 1995.
12. Durrett, R., Levin, S.A., " The importance of being discrete (and spatial)." Theor. Pop. Biol., **46**, 1994, 363-394.
13. Folkman, J. & Haudenschild, C. Angiogenesis *in vitro. Nature* **288**, 551-556 (1980).
14. Gamba, A., Ambrosi, D., Coniglio, A., de Candia, A., Di Talia, S., Giraudo, E., Serini, G., Preziosi & L., Bussolino, F. Percolation, morphogenesis, and Burgers dynamics in blood vessels formation. Submitted to *Phys. Rev. Lett.*

15. Gueron, S., Levin, S.A., "The dynamics of group formation. Math. Biosci., **128**, 1995, 243-264.
16. Gueron, S., Levin, S.A., Rubenstein, D.I., "The dynamics of herds: from individuals to aggregations, J. Theor. Biol., **182**, 1996, 85-98.
17. Friedman A., "Stochastic Differential Equations and Applications", Vols. I and II, Academic Press, London, 1975.
18. Maniezzo V. , "An Ant System Heuristic for the Two-Dimensional Finite Bin Packing Problem: preliminary results", This Volume.
19. Morale, D., Capasso, V. Oelschlaeger K. "An interacting particle system modelling aggregation behavior: from individuals to populations.", J. Mathematical Biology, 2004. In press.
20. Morale, D., Capasso, V. Oelschläger K. "A rigorous derivation of the mean-field nonlinear integro-differential equation for a population of aggregating individuals subject to stochastic fluctuations", Preprint 98-38 (SFB 359), IWR, Universität Heidelberg, Juni 1998.
21. Morale D., A stochastic particle model for vasculogenesis: a multiple scale approach In *Mathematical Modelling & Computing in Biology and Medicine. 5th Conference of the ESMTB 2002*, (V. Capasso, Ed.), ESCULAPIO, Bologna, p. 616-622 , 2003.
22. Morale D. "Laws of Large Numbers" for interacting particle systems: from discrete to continuum. An aggregation model., PhD Thesis, University of MIlano, 1999.
23. Okubo, A., S. Levin "Diffusion and ecological problems : Modern Perspectives", Springer-Verlag, 2002.
24. Okubo, A. "Dynamical aspects of animal grouping: swarms, school, flocks and herds." Adv. BioPhys., **22** (1986), 1-94.
25. Oelschläger, K., "Many-particle systems and the continuum description of their dynamics", Habilitationsschrift, Faculty of Mathematics, University of Heidelberg, Germany, 1989.
26. Oelschläger K. "A law of large number for moderately interacting diffusion processes", Z. Wahrsch. Verw. Geb., **69** (1985), 279-322.
27. Protter P., "Stochastic Integration and Differential Equations", Springer Verlag, Berlin; 1990
28. Skellam, J.G. "Random dispersal in theoretical populations." Biometrika, **38**, 1951, 196-218.
29. Krause, D.G. "Sun Tzu: The Art of War for Executives.", Nicholas Brealey Pub., London, 1996.

Distribution Theoretic Approach to Multi-phase Flow

Hideo Kawarada[1], Eiichi Baba[3], Mitsumasa Okada[3], and H. Suito[2]

[1] Faculty of Distribution and Logistics Systems, Ryutsu Keizai University,
Hirahata 120, Ryuugasaki, Ibaraki, Japan
kawarada@rku.ac.jp
[2] Environmental and Mathematical Sciences Department, Okayama University,
Japan
[3] Hiroshima University, Japan

1 Introduction

We discuss theoretical study to clarify phenomena involving multi-phase flow arising near a tidal flat. Hereafter, we state the reason why an applied mathematics group of Chiba University including the author started this study.

In 1994, an environment study group of Hiroshima University started a study of tidal flat. Based on fieldwork, they found the importance of fluid flow to understand the biological activities inside tidal flat and sandy beach. They found that number of bacteria inside seabed has a strong correlation to the amount of silt (fine sand less than 50 micron). Larger slope gives smaller amount of silt. They noticed the importance of appropriate slope for a tidal flat where sound biological activities are made inside seabed.

On the other hand, independently in 1992, an applied mathematics group of Chiba University was studying on a sliding problem of two different media. In 1993, they extended the work to explain a penetration problem from one side to the other side in different phases. Then, in 1994, the study was extended to a study of wave motion on the sloping sandy beach. At this moment this mathematics group did not notice yet their potential to explore coastal environmental problems.

The environmental experiment group needed rationalization of their experimental results by theoretical standpoint. Then, the flow phenomenon observed by the experiment was informed to Chiba University mathematics group by one of the environment study group in 1998, and this experimental results attracted and inspired the mathematics group to recognize a large potential to explore the flow phenomenon inside seabed, and then they extended their analysis to a unified treatment of fluid flow in various regions such as air, water, wetted sand and dry sand. It was found that mathematically simulated flow behavior inside sloping beach well explained observed phenomena.

Thus, the flow phenomena observed by the environment study group have been rationalized.

The environment study group continued work further and found that stranded spilled oil over sloping beach prevent the infiltration of fresh seawaters into seabed. This brings shortage of oxygen. Therefore, benthos suffers from surviving there. The mathematics group extended their work to simulating behavior and decomposition of stranded spilled oil based on unified Navier-Stokes equations with Bingham fluid model for oil. Multi-phase flow analyses were made. In this study, decomposition of spilled oil into soluble components by bacteria is simulated as a chemical reaction. Furthermore, in order to represent phenomena of gas nucleus generation accompanied by oil decomposition, compressible fluid phase was added to the obtained multi-phase formulation.

2 The Strategy for the Research

The strategy to be applied is;

1. To grasp the global image of the concerned phenomena based on physical evidence.
   - Conservation of mass and momentum for multi-phase flow in 3.2 and 3.3.
2. To represent the local structure by use of mathematical concept.
   - Description of interfacial interactions due to Signorini type boundary conditions and reactivity condition in 3.4 [3][4][5].
3. To justify effectiveness of mathematical concept through mathematical analysis.
   - Distribution theoretic approach to multi-phase flow in 3.5 [6].
4. To construct mathematical model describing the concerned phenomena according to the procedures stated in 1, 2 and 3.
   - Unified flow equations for total flow system in 3.6.
5. To select numerical method suited for solving mathematical model.
   - Finite difference method with time-independent mesh in 4.1.
6. To develop solution algorithm for obtaining discrete solutions.
   - Anti-smearing device for numerical free surface in 3.7.
   - Modification of density distribution in [15].
7. To estimate an error between a solution of mathematical model and discrete one.
   - Error estimation of finite element problem issued from fictitious domain techniques [16].
8. To visualize numerical results in the form of animation.
   - Various visualization techniques to understand unsteady complex phenomena [7].
9. To evaluate mathematical model by comparing to physical evidence.

- Decription of model evaluations in [12], [14] and [18].
10. To predict natural phenomena related with unrealizable physical experiments.
    - Visualization of oil decomposition phenomena in [17].

3 Mathematical Model

3.1 Notations

x_i	: Cartesian coordinates$(i = 1, 2, 3)$
u_i	: Velocity vector$(i = 1, 2, 3)$
u_n	: Normal component of velocity on boundaries
u_T	: Tangential component of velocity on boundaries
p	: Pressure
ρ	: Density
η	: Yield value of Bigham fluid
t	: Time
D_{ij}	: Rate of strain tensor
D_Π	: An invariant of rate of strain tensor
Ω_α	: domain
$\Omega = \underset{\alpha}{\cup} \Omega_\alpha$	: Total domain
$\Gamma_{\alpha,\beta}$	: Boundary between α-phase and β-phase
ν^α	: Kinematic viscosity of α-phase
μ^α	: Viscosity of α-phase
χ^α	: Characteristic function representing a domainα
σ_{ij}^α	: Stress tensor of α-phase
ρ^α	: Density of α-phase
c	: Coefficient of registance force receiving from sand
$[A]_{\alpha,\beta}$	: $\left(A^\alpha - A^\beta\right)\big|_{\Gamma_{\alpha,\beta}}$

3.2 Conservation of Mass for Multi-phase Flow

$$\frac{\partial \rho}{\partial t} + \frac{\partial}{\partial x_j}(\rho u_j) = 0, \ \text{in } \Omega. \tag{1}$$

Here, $\rho = \sum_{l=1}^{6} \rho^{(l)} \chi^{(l)}, \quad u_i = \sum_{l=1}^{6} u_i^{(l)} \chi^{(l)}$ and l means $\{a, w, f, as, ws, fs\}$.
Fluids for each phase are assumed to be incompressible, i.e., $\rho^{(l)} = $const.

- Incompressibility conditions for constant density phases

$$\frac{\partial u_j^{(l)}}{\partial x_j} = 0 \ \text{in } \Omega_l, \ \text{for each } l, \ t > 0 \tag{2}$$

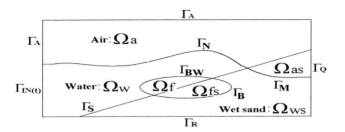

Fig. 1. Geometry

• Motion equations of interfaces between different phases

$$\frac{\partial \chi^{(l)}}{\partial t} + u_j^{(l)} \frac{\partial \chi^{(l)}}{\partial x_j} = 0 \ \text{ in } \Omega, \ \text{ for each } l, \ t > 0 \tag{3}$$

3.3 Conservation of Momentum for Multi-phase Flow

• Navier-Stokes equations for water and air phases

$$\frac{\partial u_i}{\partial t} + u_j \frac{\partial u_i}{\partial x_j} = -\frac{1}{\rho_w} \frac{\partial p}{\partial x_i} + \nu_w \frac{\partial D_{ij}}{\partial x_j} + K_i. \tag{4}$$

• Bingham equations for oil phase

$$\frac{\partial u_i}{\partial t} + u_j \frac{\partial u_i}{\partial x_j} = -\frac{1}{\rho_f} \frac{\partial p}{\partial x_i} + \frac{1}{\rho_f} \frac{\partial}{\partial x_j} \left\{ \left(\mu_f + \frac{\tau}{\sqrt{4D_{\Pi} + \varepsilon_b}} \right) D_{ij} \right\} + K_i, \tag{5}$$

where $\tau = \frac{\eta}{\rho_f}$ and $D_{\Pi} = \frac{1}{2} D_{ij} D_{ij}$.

• **Unification** of the above mentioned equations

$$\frac{\partial u_i}{\partial t} + u_j \frac{\partial u_i}{\partial x_j} = -\frac{1}{\rho} \frac{\partial p}{\partial x_i} + K_i \tag{6}$$

$$+\frac{1}{\rho} \frac{\partial}{\partial x_j} \left[\left\{ \mu_{air} \chi_{air} + \mu_w \chi_w + \left(\mu_f + \frac{\tau}{\sqrt{4D_{\Pi} + \varepsilon_b}} \right) \chi_f \right\} D_{ij} \right]. \tag{7}$$

3.4 Interfacial Interactions

Simplified Coulomb Law for Friction

Description of adhesion, sliding and infiltration phenomena of oil in sand using **Signorini** type boundary conditions.

$$\begin{cases} ||[\sigma_T]|| \leq g_T, \\ g_T \cdot |u_T| + [\sigma_T] \cdot u_T = 0, \end{cases} \quad \text{on } \Gamma_B. \tag{8}$$

$$\begin{cases} ||[\sigma_T]|| < g_T \longmapsto u_T = 0 & \text{(Adhesion)} \\ ||[\sigma_T]|| = g_T \longmapsto u_T = 0 \text{ or } u_T \neq 0 \text{ (Sliding)} \end{cases} \tag{9}$$

where g_T is the friction coefficient.

$$-[\sigma_T] = g_T \cdot \partial\left(|u_T|\right) \qquad \text{on } \Gamma_B. \tag{10}$$

Regularization;

$$-\left[\sigma_{T_j}\right] = g_T \cdot \frac{u_{T_j}}{\sqrt{|u_T|^2 + \varepsilon_g^2}} \quad \text{on } \Gamma_B, \quad (j = 1, 2). \tag{11}$$

Reactivity Condition

Description of oil decomposition by bacteria using **reactivity condition**. Here, k_f and k_w mean the rate of consumption for oil and the rate of production for water, respectively. Therefore, k_f is a negative constant and k_w is a positive constant. $\chi^{(b)}$ is characteristic function of Ω_b.

The noseparation of oil and water phases shows;

$$u_j^{(fs)} + k_f n_j^{(fs)} \chi^{(b)} = u_j^{(ws)} + k_w n_j^{(ws)} \chi^{(b)} \quad \text{on } \Gamma_B, \quad (j = 1, 2, 3), \tag{12}$$

Reactivity condition;

$$\left[u \cdot n^{(ws)}\right] = u^{(ws)} n^{(ws)} - u^{(fs)} n^{(ws)} = -(k_f + k_w) \chi^{(b)} \text{ on } \Gamma_B, \tag{13}$$

$$\left[u \cdot T^{(fs)}\right] = u^{(fs)} T^{(fs)} - u^{(ws)} T^{(fs)} = 0 \qquad \text{on } \Gamma_B. \tag{14}$$

3.5 Distribution Theoretic Approach to Multi-phase Flow

Let us note that the interfacial interactions stated in 3.5 are represented by the jump boundary conditions for the Dirichlet type (the reactivity condition) and the Neumann type (the frictional condition). In order to introduce such jump conditions into an unified model for multi-phase flow system, we have developed distribution theoretic approach to multi-phase flow. For simplicity, we discuss the case of steady Stokes problem defined in $\Omega \subset R^3$. Ω is divided into subdomains Ω_1 and Ω_2, whose interface is denoted by Γ. (See figure 2.) The two phase Stokes problem (S_1) with jump boundary conditions on Γ is defined as follows.

$$(S_1) \begin{cases} -\triangle u^{(1)} + \nabla p^{(1)} = f^{(1)} & \text{in } \Omega_1, \\ \operatorname{div} u^{(1)} = 0 & \text{in } \Omega_1, \\ -\triangle u^{(2)} + \nabla p^{(2)} = f^{(2)} & \text{in } \Omega_2, \\ \operatorname{div} u^{(2)} = 0 & \text{in } \Omega_2, \\ [\sigma_n] = \sigma_n^{(1)} - \sigma_n^{(2)} = a & \text{on } \Gamma, \\ [\sigma_{T_j}] = \sigma_{T_j}^{(1)} - \sigma_{T_j}^{(2)} = b_j \; (j=1,2) & \text{on } \Gamma, \\ [u_n] = u_n^{(1)} - u_n^{(2)} = c & \text{on } \Gamma, \\ [u_{T_j}] = u_{T_j}^{(1)} - u_{T_j}^{(2)} = d_j \; (j=1,2) & \text{on } \Gamma, \\ u^{(1)} = 0, & \text{on } \partial\Omega, \end{cases} \tag{15}$$

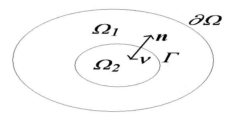

Fig. 2. Geometry for Stokes problem

Let χ be the characteristic function of Ω_2 in Ω and assume a, b_j, c and d_j be constants. Then (S_1) is settled into a single equation (S_2) in the following way;

$$(S_2) \begin{cases} -\triangle u + \nabla p + a \cdot \nabla\chi + b_1 \cdot \Sigma_1\chi + b_2 \cdot \Sigma_2\chi \\ \quad + c \cdot \boldsymbol{n} \cdot \triangle\chi + d_1 \cdot T_1 \cdot \triangle\chi + d_2 \cdot T_2 \cdot \triangle\chi = f, \text{ in } \Omega, \\ \operatorname{div}\boldsymbol{u} = 0, & \text{in } \Omega\backslash\Gamma, \\ u = 0, & \text{on } \partial\Omega, \end{cases} \tag{16}$$

where

$$f = (1 - \chi)f^{(1)} + \chi f^{(2)}, \tag{17}$$

and

$$\nabla \chi = \left(\frac{\partial \chi}{\partial x_1}, \frac{\partial \chi}{\partial x_2}, \frac{\partial \chi}{\partial x_3} \right), \tag{18}$$

$$\Sigma_1 \chi = \left(\frac{\partial \chi}{\partial x_2}, -\frac{\partial \chi}{\partial x_1}, 0 \right), \tag{19}$$

$$\Sigma_2 \chi = \left(\frac{\partial \chi}{\partial x_1} \frac{\partial \chi}{\partial x_3}, \frac{\partial \chi}{\partial x_2} \frac{\partial \chi}{\partial x_3}, \left(\frac{\partial \chi}{\partial x_3} \right)^2 - |\nabla \chi|^2 \right), \tag{20}$$

$$\boldsymbol{n} = -\frac{\nabla \chi}{|\nabla \chi|}, \qquad T_1 = \frac{\Sigma_1 \chi}{|\Sigma_1 \chi|}, \qquad T_2 = \frac{\Sigma_2 \chi}{|\Sigma_2 \chi|}. \tag{21}$$

Here, $\frac{\partial \chi}{\partial x_j}$ ($j = 1, 2, 3$) is a distribution supported on Γ, multiplication and division of them are defined by convolution of distribution with support. Summarizing the above mentioned facts, we have

Theorem 1

   (S_1) *is equivalent to* (S_2).

Remark 1

   The surface distribution of the dipole moment with the normal direction to the surface brings about the jump of the normal component of velocity for the fluid flowing across the interface.

Proof of Theorem 1

The case of the jump condition for the Dirichlet type

   Let us consider the problem $(Pr)_c$ to generate $[u_n] = c$. For simplicity, let us put $c = 1$.

$$(Pr)_c \begin{cases} \triangle u - \nabla p = \tilde{n} \triangle \chi & \text{in } \Omega, \\ \mathrm{div} u = 0 & \text{in } \Omega, \\ u = 0 & \text{on } \partial \Omega. \end{cases} \tag{22}$$

where $\triangle \chi = \frac{\partial}{\partial \nu} \delta(\Gamma)$ is a dipole moment, $\Gamma = \partial \Omega_2$ is smooth and $\tilde{n}$ is smoothly extended into R^3 so as to satisfy $\left. \frac{\partial \tilde{n}_j}{\partial \nu} \right|_\Gamma = 0$ ($j = 1, 2, 3$) and $\triangle \tilde{n} = 0$ in Ω_2.

Step 1. Potential flow is defined to satisfy

$$(Pr)_{c1} \begin{cases} \triangle U - \nabla P = \tilde{n} \cdot \triangle \chi & \text{in } R^3, \\ \mathrm{div}\, U = 0 & \text{in } R^3. \end{cases} \tag{23}$$

   We define fundamental solution: $\boldsymbol{E}^k = \left\{ E_j^k \right\}_{j=1}^3$ ($k = 1, 2, 3$) s.t.

$$\begin{cases} \triangle \boldsymbol{E}^k - \nabla q^k = \delta(x-y) \cdot \boldsymbol{e}_k & \text{in } R^3, \\ \operatorname{div} \boldsymbol{E}^k = 0 & \text{in } R^3, \end{cases} \tag{24}$$

where $\boldsymbol{e}^k$ is an unit vector of kth axis ($k = 1, 2, 3$) and $x = (x_1, x_2, x_3)$, $y = (y_1, y_2, y_3)$.

$$\begin{cases} E_j^k(x, y) = -\dfrac{\delta_j^k}{4\pi |x-y|} + \dfrac{(x_k - y_k)(x_j - y_j)}{|x-y|^3} \\ \qquad\quad = \delta_j^k E(x-y) + F_j^k(x-y) \in L_y^1(R^3), \\ q^k(x, y) = \dfrac{\partial}{\partial x_k} E(x-y) \in L_y^1(R^3). \end{cases} \tag{25}$$

Potential $\{U, P\}$ is represented in the following way;

$$\begin{cases} \boldsymbol{U}(x) = \boldsymbol{E} * \tilde{n}\triangle\chi = \boldsymbol{E} * \triangle(\tilde{n}\chi) & \text{in } \mathcal{D}'(R^3), \\ \boldsymbol{P}(x) = \boldsymbol{q} * \tilde{n}\triangle\chi = \boldsymbol{q} * \triangle(\tilde{n}\chi) & \text{in } \mathcal{D}'(R^3). \end{cases} \tag{26}$$

$$\begin{aligned} U_j(x) &= \int_{\Omega_2} \triangle_y \left(E_j^k(x-y) \cdot \tilde{n}_k(y) \right) d\Omega_y, \\ &= \int_{\Omega_2} \operatorname{div}_y \nabla_y \left(E_j^k(x-y) \cdot \tilde{n}_k(y) \right) d\Omega_y, \\ &= \int_\Gamma \frac{\partial}{\partial n_y} E_j^k(x-y) \cdot \tilde{n}_k(y) d\sigma_y, \\ &= \int_\Gamma \frac{\partial}{\partial n_y} E(x-y) \cdot \tilde{n}_j(y) d\sigma_y + \int_\Gamma \frac{\partial}{\partial n_y} F_j^k(x-y) \cdot \tilde{n}_k(y) d\sigma_y. \end{aligned} \tag{27}$$

The second term of the right hand side dose not contribute to make the jump because

$$F_j^k = E * \frac{\partial^2}{\partial x_k \partial x_j} \delta = \frac{\partial^2}{\partial x_k \partial x_j} E. \tag{28}$$

The first term is a double layer potential. Then U satisfies

$$[\boldsymbol{U}] = \boldsymbol{U}|_{\Gamma_+} - \boldsymbol{U}|_{\Gamma_-} = \boldsymbol{n}, \tag{29}$$

$$\boldsymbol{n} \cdot [\boldsymbol{U}] = \boldsymbol{n} \cdot \boldsymbol{n} = 1. \tag{30}$$

Similarly, P satisfies

$$P(x) = \int_{\Omega_2} \triangle \left\{ q^k(x, y) \cdot \tilde{n}_k(y) \right\} d\Omega_y, \tag{31}$$

$$= \int_\Gamma \frac{\partial}{\partial n_y} q^k(x, y) \cdot n_k(y) d\sigma_y, \tag{32}$$

$$= \frac{\partial}{\partial x_k} \int_\Gamma n_k(y) \frac{\partial}{\partial n_y} E(x-y) d\sigma_y. \tag{33}$$

Then there holds

$$[P] = P|_{\Gamma_+} - P|_{\Gamma_-} = 0. \tag{34}$$

Let us note that U and P are real analytic in $R^3 \backslash \Gamma$.

Step 2.

Let $u = U + \tilde{u}$ and $p = P + \tilde{p}$. Then $(\tilde{u}, \tilde{p})$ satisfies

$$(Pr)_{c2} \begin{cases} \triangle \tilde{u} - \nabla \tilde{p} = 0 \text{ in } \Omega, \\ \text{div} \tilde{u} = 0 \qquad \text{in } \Omega, \\ \tilde{u} = -U \qquad \in \left\{ H^{\frac{1}{2}}(\partial \Omega) \right\}^2. \end{cases} \tag{35}$$

where $\displaystyle\int_{\Omega} \text{div} U \, d\Omega = \int_{\Gamma} U \cdot n \, d\sigma = 0$. Note that there exists a unique solution $\{\tilde{u}, \tilde{p}\}$ for $(Pr)_{c2}$ satisfying

1. $\tilde{u} \in \left\{ H^1(\Omega) \right\}^2$,
2. $\tilde{p} \in L^2(\Omega) \backslash R$.

Then $n \cdot [u] = n[U] + n[\tilde{u}] = n[U] = 1$ in $\left\{ H^{\frac{1}{2}}(\Gamma) \right\}^2$.

Theorem 2

The solution of $(Pr)_c$ satisfies that $n \cdot [u] = 1$ in $\left\{ H^{\frac{1}{2}}(\Gamma) \right\}^2$.

The case of the jump condition for the type of $[\sigma_n] = a$

Let us consider the problem $(Pr)_a$ to generate $[\sigma_n] = a(=1)$.

$$(Pr)_a \begin{cases} \triangle u - \nabla p = \nabla \chi \text{ in } \Omega, \\ \text{div} u = 0 \qquad \text{in } \Omega, \\ u = 0 \qquad \text{on } \partial \Omega. \end{cases} \tag{36}$$

[1st step] Potential flow is defined to satisfy,

$$(Pr)_{a1} \begin{cases} \triangle U - \nabla P = \nabla \chi \quad \text{in } R^3, \\ \text{div} \, U = 0 \qquad \text{in } R^3. \end{cases} \tag{37}$$

It is obvious that

$$\begin{cases} U(x) = 0 \qquad \text{in } R^3, \\ P(x) = -\chi \quad \text{in } R^3. \end{cases} \tag{38}$$

is the solution of $(Pr)_{a1}$.

[2nd step]

$u = U$ and $p = P$ satisfies $(Pr)_a$. Then we have

Theorem 3

$$\left[\frac{\partial u_n}{\partial n} - p\right] = 1 \;\; \text{in} \;\; \left\{H^{-\frac{1}{2}}(\Gamma)\right\}^2. \tag{39}$$

The case of the jump condition for the type of $[\sigma_T] = b(= 1)$

Let us consider the problem $(Pr)_b$ to generate $[\sigma_T] = b$.

$$(Pr)_b \begin{cases} \triangle u - \nabla p = \Sigma \chi \;\; \text{in} \;\; \Omega, \\ \text{div} u = 0 \qquad\quad \text{in} \;\; \Omega, \\ u = 0 \qquad\qquad \text{on} \;\; \partial\Omega. \end{cases} \tag{40}$$

[1st step] Potential flow is defined to satisfy

$$(Pr)_{b1} \begin{cases} \triangle U - \nabla P = \Sigma \chi \quad \text{in} \;\; R^3, \\ \text{div} \, U = 0 \qquad\qquad \text{in} \;\; R^3. \end{cases} \tag{41}$$

Then we have

$$\begin{aligned} U_j(x) &= \int_{R^3} E_j^k(x-y) \cdot T_k(y) \cdot \delta(y - \Gamma) \, dy \\ &= \int_{\Gamma} E_j^k(x-y) \cdot T_k(y) \, d\sigma_y \end{aligned} \tag{42}$$

and

$$U_{\tilde{T}}(x) = \int_{\Gamma} E_j^k(x-y) \cdot T_j(x) \cdot T_k(y) \, d\sigma_y. \tag{43}$$

from which follows

$$(\tilde{n} \cdot \nabla) \, U_{\tilde{T}} = - \int_{\Gamma} \frac{\partial}{\partial n_y} E(x-y) \cdot T_j(x) \cdot T_j(y) \, d\sigma_y + \cdots . \tag{44}$$

By use of non contribution of F_j^k to make the jump, we have

$$\left[\frac{\partial}{\partial n} U_T\right] = -1. \tag{45}$$

[2nd step]

Repeating similar arguments as before, we have

Theorem 4

$$\left[\frac{\partial u_T}{\partial n}\right] = -1 \;\; \text{in} \;\; \left\{H^{-\frac{1}{2}}(\Gamma)\right\}^2. \tag{46}$$

The Stokes Equation with a Variable Viscosity

In this section, we shall deal with Stokes equation with a variable viscosity in place of the one with a constant viscosity discussed in 3.5.1 as follows;

$$
\begin{cases}
\dfrac{\partial}{\partial x_j}\left(\nu(x)\dfrac{\partial u_i}{\partial x_j}\right) - \dfrac{\partial p}{\partial x_i} = f_i \text{ in } \Omega \text{ for } i = 1, 2, \\[3mm]
\dfrac{\partial u_j}{\partial x_j} = 0 \qquad\qquad\quad \text{in } \Omega,
\end{cases}
\tag{47}
$$

where $\nu(x) = \nu_1(x)(1 - \chi) + \nu_2(x)\chi$ and $\nu_j(x) \in C(\Omega_j)$ $(j = 1, 2)$.

According to the replacement of the Laplacian with the divergence form, the jumped boundary conditions defined on Γ should be modified;

$$
[\sigma_n] = \nu_1 \frac{\partial u_{1n}}{\partial n} - p_1 - \left(\nu_2 \frac{\partial u_{2n}}{\partial n} - p_2\right),
\tag{48}
$$

$$
[\sigma_T] = \nu_1 \frac{\partial u_{1T}}{\partial n} - \nu_2 \frac{\partial u_{2T}}{\partial n},
\tag{49}
$$

$$
[u_n] = \nu_1 u_{1n} - \nu_2 u_{2n},
\tag{50}
$$

$$
[u_T] = \nu_1 u_{1T} - \nu_2 u_{2T}.
\tag{51}
$$

However, if the relation $u_{1n} - u_{2n} = c$ on Γ is required in place of $[u_n] = c$, then $n \cdot \triangle\chi$ should be replaced with $n \cdot \frac{\partial}{\partial x_j}\left(H\frac{\partial}{\partial x_j}\chi\right)$. In fact,

$$
\nu_1 u_{1n} - \nu_2 u_{2n} = \nu_1(u_{1n} - u_{2n}) + (\nu_1 - \nu_2)u_{2n} \text{ on } \Gamma.
\tag{52}
$$

Then H should be defined on Γ in the following way;

$$
H = c\nu_1 + (\nu_1 - \nu_2)u_{2n} \text{ on } \Gamma.
\tag{53}
$$

The additional terms except $n \cdot \triangle\chi$ in (S_2) brings about the same jumps as in the statement of Theorem 1.

Remark 2 *Distribution theoretic approach to multi-phase flow for Stokes equation with a variable viscosity is proved by treating the transmission problem defined on Γ under the weak formulation of the equation, that will be shown in the succeeding paper.*

Remark 3 In the case of time dependent Stokes equations, we can show the same result as obtained in the steady case.

3.6 Unified Flow Equations for Total Flow System

Coupling unified flow equations for all phases with interfacial interactions due to Discontinuous interface generating method.

- **Incompressibility condition** for multi-phase flow system;

$$\frac{\partial u_j}{\partial x_j} + \frac{2(\rho_w - \rho_f)}{\rho_w + \rho_f} k_f \chi_b \frac{\partial \chi^{(fl)}}{\partial n^{(fl)}} = 0, \quad \text{in } \Omega, \; t > 0. \tag{54}$$

An existence of the second term in the left hand side is due to the reactivity condition.

- **Conservation of momentum** for multi-phase flow system;

$$\rho \left(\frac{\partial u_i}{\partial t} + u_j \frac{\partial u_i}{\partial x_j} \right) + c(\chi_{ws} + \chi_{fs}) u_i = -\frac{\partial p}{\partial x_i}$$

$$- \chi^{(b)} \frac{\partial}{\partial x_j} \left(H \frac{\partial}{\partial x_j} \chi_{fs} \right) \cdot n_i^{(fs)} + \frac{\partial}{\partial x_j} \bigg\{ (\mu_{air} \chi_{air}$$

$$+ \mu_{wt} \chi_{wt} + \left(\mu_f + \frac{\eta}{\sqrt{4 D_\Pi + \varepsilon_b^2}} \right) \chi_{fl} \right) \left(\frac{\partial u_i}{\partial x_j} + \frac{\partial u_j}{\partial x_i} \right) \bigg\}$$

$$+ \rho \chi_s g_T \left(\frac{u_{T_1}}{\sqrt{|u_T|^2 + \varepsilon_g^2}} (\Sigma_1 \chi_{fl})_i + \frac{u_{T_2}}{\sqrt{|u_T|^2 + \varepsilon_g^2}} (\Sigma_2 \chi_{fl})_i \right) + \rho g \, \delta_{i,3}$$

$$\text{in } \Omega, \; t > 0, \tag{55}$$

where $H = -\mu_{wt}(k_f + k_w) + \left\{ \mu_{wt} - \left(\mu_f + \frac{\eta}{\sqrt{4D_\Pi + \varepsilon_b^2}} \right) \right\} u_j$.

The second term means the dipole moment distribution along the reaction surface included in the bacteria zone, which plays an important role to satisfy the reactivity condition. The fourth term in the right hand side represents the adhesion and sliding phenomena occured on the interface between oil and sand.

- **Motion equation of free surface for oil** is;

$$\frac{\partial \chi^{(fl)}}{\partial t} + u_j^{(fl)} \frac{\partial \chi^{(fl)}}{\partial x_j} = k_f \cdot \left| \nabla \chi^{(fl)} \right| \chi^{(b)} + (\text{Double well potential}); \tag{56}$$

$$\text{in} \Omega, \; t > 0, \tag{57}$$

- **Outer boundary conditions** are prescribed as follows;

$$\begin{cases} u_n = u_0 \sin \omega t, \\ u_T = 0, \end{cases} \text{on } \Gamma_{IN}(t), \qquad \begin{cases} u_n = 0, \\ \dfrac{\partial u_T}{\partial n} = 0, \end{cases} \text{on } \Gamma_R, \tag{58}$$

$$\begin{cases} u_n = 0, \\ u_T = 0, \end{cases} \text{on } \Gamma_Q, \qquad \begin{cases} \sigma_n = 0, \\ u_T = 0, \end{cases} \text{on } \Gamma_A. \tag{59}$$

3.7 Anti-smearing Device for Numerical Free Surface

Introduction of Double well potential to interfacial motion equations in order to avoid smearing of free surfaces (H. Kawarada and H. Suito, 1997)

$$\frac{\partial \chi_{wf}}{\partial t} + \frac{\partial}{\partial x_j}(\chi_{wf} u_j) = -G'(\chi_{wf}), \tag{60}$$

$$\frac{\partial \chi_{fl}}{\partial t} + \frac{\partial}{\partial x_j}(\chi_{fl} u_j) = -G'(\chi_{fl}), \tag{61}$$

where $G(\chi)$ is a double-well potential defined as follows;

$$G(\chi) = \begin{cases} +\infty & \chi < 0, \\ \frac{h^2}{16}\chi^2(\chi-1)^2 & 0 < \chi < 1, \\ +\infty & 1 < \chi. \end{cases} \tag{62}$$

By this potential, the value of χ is confined to the vicinities of 0 and 1. It should be noted that this technique should be used carefully so that mass conservation is satisfied.

4 Discretization of Mathematical Model

4.1 Discretized Mathematical Model Based on Finite Difference Approximation

- Third order upwind scheme for convection term in order to stabilize the computation
- GP-BiCG method for solving accurately the Poisson equation for the pressure

4.2 Harmonic Averaging Mapping (HAM) to Density Distribution

Modification of density distribution by HAM in order to avoid violation of incompressibility condition for two-phase flow with large density discrepancy (H. Kawarada, T. Kumaki and H. Suito, 2000)

5 Visualization of Numerical Results

- Various visualization methods for understanding mechanisms of phenomena
- Animations for understanding unsteady phenomena

References

1. G. Duvaut and J. L. Lions, (1976): Inequalities in Mechanics and Physics, Springer-Verlag.
2. J. Bebernes and D. Eberly, (1989): Mathematical Problems from Combustion Theory, Springer-Verlag.
3. A. Sasamoto and H. Kawarada (1993): Numerical Simulation of the Flow in a Vessel with the Free Boundary and the Sliding Wall, Mathematical Sciences and Applications, Vol. 1, Nonlinear Mathematical Problems in Industry, pp.175–187.
4. H. Fujita (1994): A Mathematical Analysis of Motions of Viscous Incompressible Fluid under Leak or Slip Boundary Conditions, Proceedings of International Conference on Mathematical Fluid Dynamics and Modeling, RIMS Kōkyūroku, Vol. 888, pp.199–216.
5. H. Fujita, H. Kawarada and A. Sasamoto (1995): Analytical and Numerical Approaches to Stationary Flow Problems with Leak and Slip Boundary Conditions, Proceedings of 2nd Japan-China Seminar on Numerical Mathematics, Kinokuniya LN in Numer. Appl. Anal. Vol. 14, pp.17–31.
6. H. Fujita, H. Kawahara and H. Kawarada (1995): Distribution Theoretic Approach to Fictitious Domain Method for Neumann Problems, East-West J. Numer. Math., Vol. 3, No.2, pp.111-126.
7. H. Suito and K. Kuwahara (1996): Volume Visualization of Thermal Convection, Physics of Fluid, Vol.8, No.9, p.S3.
8. H. Kawarada and H. Suito (1997): Numerical Method for a Free Surface Flow on the bases of the Fictitious Domain Method, East-West J. Numer. Math., Vol. 5, No. 1, pp.57-66.
9. H. Kawarada, H. Fujita and H. Suito (1998): Wave motion Breaking upon the Shore, GAKUTO International Series, Mathematical Sciences and Applications, Vol. 11, pp.145-159.
10. E. Baba, Y. Doi, M. Tamashima and H. Ogawa (1998): Visualization of internal flow of sea bed by use of transparent glass beads, Proceedings of First International Workshop on Coastal Ecosystems and Hydrodynamics, pp.22-39.
11. E. Baba and C. Cheong (1999): Effect of Spilled Oil on Exchange of Sea Water over the Sand Beach and Tidal-flat, Proceedings of Workshop on Environmental Fluid Mechanics for Coastal Ecosystems, p. 11, Massachusetts Institute of Technology, USA.
12. H. Kawarada, E. Baba and H. Suito (2000): Effect of spilled oil on Coastal Ecosystems, ECCOMAS2000 CD-ROM Proceedings,
13. E. Baba, C.-J. Cheong and M. Okada (2000): Effect of spilled oil on seawater infiltration into a tidal flat under wave and tide actions, Environmental Sciences, Vol. 7, No. 3, pp.139-148.
14. H. Kawarada, E. Baba and H. Suito (2001): Effects of Wave Breaking Action on Flows in Tidal-flats, Notes on Numerical Fluid Mechanics, Vol. 78, Computational Fluid Dynamics for the 21st Century, M. Hafes, K. Morinishi and J. Periaux, Eds., pp.275-289, Springer.
15. H. Kawarada, T. Kumaki and H. Suito (2001): Two phase free surface flow with large density discrepancy, 1st MIT conference on Computational Fluid and Solid Mechanics.
16. C. Bernardi, V. Girault, F. Hechet, H. Kawarada and O. Pironneau (2001): A finite element problem issued from fictitious domain techniques, Publications du Laboratoire d'Analyse Numérique, R-00041, Université Pierre et Marie Curie.

17. H. Kawarada and H. Suito: Multi-phase Flow with Reaction, Lecture Notes in Computational Science and Engineering, Vol.19, Mathematical Modeling and Numerical Simulation in Continuum Mechanics, I.Babuska, P.G.Ciarlet and T.Miyoshi, Eds., Springer, (To appear).
18. H. Kawarada, E. Baba and H. Suito: Effect of Spilled Oil on Coastal Ecosystems, Proceedings of the Fifth China-Japan Joint Seminar on Numerical Mathematics, Science Press in Beijing, (To appear).

An Ant System Heuristic for the Two-Dimensional Finite Bin Packing Problem: preliminary results

Marco A. Boschetti[2] and <u>Vittorio Maniezzo</u>[1]

[1] Department of Mathematics, University of Bologna, Bologna, Italy,
`maniezzo@csr.unibo.it`
[2] Department of Computer Science, University of Bologna, Bologna, Italy.

1 Introduction

Ant Colony Systems are constructive metaheuristics ([6],[10]), which iteratively expand partial solutions eventually achieving complete, feasible and possibly good quality problem solutions. When constructing, an estimation of the quality of each possible extension of the incumbent partial solution is to be computed. The issue arises of how to determine the quality of each extension given a partial solution, where the quality of one same extension could be different for any different partial solution, thus for an exponentially increasing number of cases. This issue is not relevant in the case of the Travelling Salesman Problem (TSP), the problem for which ant systems were originally presented [11], but arises in many cases, such as the Quadratic Assignment Problem (QAP) [22] or the Two-Dimensional Finite Bin Packing Problem (2BP). In this paper we use the 2BP as a benchmark for studying this important issue.

The 2BP consists of determining the minimum number of large identical rectangles, *bins*, that are required for allocating without overlapping a given set of rectangular items. The items are allocated with their edges parallel or orthogonal to the bin edges and for each of them it is given the size and a parameter specifying whether it has a fixed orientation or it can be rotated by $90°$.

The 2BP is a generalization of the well known one-dimensional Bin Packing Problem (1BP), where n items of given weight w_i have to be packed into the minimum number of bins of capacity W. Therefore, the 2BP is strongly NP-hard as well as the 1BP (see Garey and Johnson [19]).

The 2BP finds many practical applications as the cutting of standardized steel plates or wood sheets into smaller pieces to produce panels or furniture, the packing on shelves in warehousing or on truck beds in transportation, the paging of articles and advertisings in newspapers, etc.

In practical applications there exist many versions of the 2BP. The problem considered in this paper contains as special cases the oriented and non-oriented versions of the 2BP that, according to the classification of Lodi, Martello and Vigo [21], are called 2BP|O|F and 2BP|R|F, respectively. Following this classification scheme Boschetti and Mingozzi [3] denote with 2BP|M|F the problem considered in this paper.

For the 2BP|O|F heuristic algorithms have been proposed by Chung, Garey and Johnson [5], Frenk and Galambos [18], Berkey and Wang [2], Lodi, Martello and Vigo [21], Faroe, Pisinger and Zachariasen [16] and Boschetti and Mingozzi [4]. Lower bounds for the 2BP|O|F are described in Martello and Vigo [23], Fekete and Schepers [17] and Boschetti and Mingozzi [3]. Exact methods are presented by Martello and Vigo [23] and Pisinger and Sigurd [24].

For the 2BP|R|F heuristic algorithms have been proposed by Bengtsson [1], El-Bouri, Popplewell, Balakrishnan and Alfa [15], Lodi, Martello and Vigo [21] and Boschetti and Mingozzi [4]. While lower bounds for 2BP|R|F presented in the literature are due to Dell'Amico, Martello and Vigo [7] and Boschetti and Mingozzi [4].

For the more general 2BP|M|F a heuristic algorithm and a lower bound have been proposed by Boschetti and Mingozzi [4].

Extensive survey on cutting and packing problems can be found in Dowsland and Dowsland [12], Dyckhoff and Finke [13] and Lodi, Martello and Monaci [20]. Moreover, an annotated bibliography is given in Dyckhoff, Scheithauer and Terno [14].

In this paper we consider an Ant-based heuristic for the 2BP|M|F. Ant systems have been proposed in widely varied forms since their presentation in 1991 [6], in the case of this study we preliminarily consider a very basic ant heuristic, much alike that presented in 1991. The issue addressed by this paper is how to assess the quality of an extension of a partial solution, when there is the possibility of having an exponential number of partial solution which differently affect the quality of the same extension. It is obviously necessary to partition the partial solutions space into a manageable number of subsets of solutions, here we propose an ordering based possibility.

We designed an algorithm which constructs solutions essentially according to the ant schema proposed in [6], then uses the local optimization proposed by Boschetti and Mingozzi [4] to permit to search only in the space of local optima.

The paper is structured as follows. Section 2 briefly introduces ant-based systems and outlines the variant used in this paper. Section 3 describes the problem addressed by this research and the local search heuristic by Boschetti and Mingozzi [4]. Section 4 presents the specific issues of our ant system for the 2BP|M|F, section 5 shows the preliminary computational results so far obtained and, finally, section 6 proposes our conclusions drawn from the current state of this research.

2 Ant Colony Optimization

Combinatorial optimization algorithms based on ideas derived from the observation of ant colony behaviors were initially proposed by Colorni, Dorigo and Maniezzo [6], [8], [9] with their Ant System. The main underlying idea was that of parallelizing search over several constructive computational threads, all based on a dynamic memory structure incorporating information on the effectiveness of previously obtained results and in which the behavior of each single agent is inspired by the behavior of real ants.

A combinatorial optimization problem is defined over a set $\mathbf{C} = \{c_1, \ldots, c_n\}$ of basic components. A subset S of components represents a solution of the problem and $\mathbf{F} \subseteq 2^{\mathbf{C}}$ is the subset of *feasible solutions*. Thus a solution S is feasible if and only if $S \in \mathbf{F}$. A *cost function* z is defined over the solution domain, i.e., $z : 2^{\mathbf{C}} \to \mathbb{R}$, and the objective is to find a minimum cost feasible solution S^*, i.e., to find $S^* \in \mathbf{F}$ such that $z(S^*) \leq z(S), \forall S \in \mathbf{F}$. Failing this, the algorithm anyway returns the best feasible solution found, $\bar{S}$.

The Ant System (AS) was the first algorithm of a class, eventually named Ant Colony Optimization (ACO) algorithms. The main common feature of all algorithms of the class is to modify a constructive heuristic so that the ordering of the components could be recalculated at each iteration taking into account not only the a priori expectation, η_j, of the usefulness of a particular component c_j, but also an a posteriori measure, τ_j, of the goodness of solutions constructed using that particular component. The general framework can be schematized as follows.

General Framework

Step 1. (*Initialization*)
Initialize a set A of partial solutions: $a_i = \emptyset, i = 1, \ldots, m$.

Step 2. (*Expansion of the partial solutions*)
For each partial solution $i = 1, \ldots, m$ choose a component c_j to append to solution a_i with probability given as a function of a_i, η_j, τ_j.

Step 3. (*Check if all solutions are completed*)
If the solutions in A are not complete, go to step 2.

Step 4. (*Evaluation of the solution*)
Evaluate $z(a_i), i = 1, \ldots, m$, and update τ_j, for every component c_j, accordingly.

Step 5. (*End condition*)
If the end conditions are not satisfied go to step 1.

Step 4 can be conveniently integrated with a local search starting from each a_i solution in A.

The importance of this original Ant System resides mainly in being the prototype of a number of ant algorithms which have found interesting and successful applications.

An *ant* is the "computational agent" managed at step 2 of the general framework, which iteratively constructs a solution for the problem to solve. Partial problem solutions are seen as *states*; each ant *moves* from a state ι to another one ψ, corresponding to a more complete partial solution. At each step σ, each ant k computes a set $A_k^\sigma(\iota)$ of feasible expansions to its current state, and moves to one of these in probability, according to a probability distribution defined in the following.

For each ant k, the probability $p_{\iota\psi}^k$ of moving from state ι to state ψ depends on the combination of two values:

1. the attractiveness η of the move, as computed by some heuristic indicating the *a priori* desirability of that move;

2. the trail level τ of the move, indicating how proficient it has been in the past to make that particular move: it represents therefore an *a posteriori* indication of the desirability of that move.

Trails are *updated* at each iteration, increasing the level of those that facilitate moves that were part of "good" solutions, while decreasing all other ones. The specific formula for defining the probability distribution at each move makes use of a set $tabu_k$ which indicates a problem-dependent set of infeasible moves for ant k. Different authors use different formulae, according to [22] probabilities are computed as follows: $p_{i\psi}^k$ is equal to 0 for all moves which are infeasible (i.e., they are in the tabu list), otherwise it is computed by means of the following formula:

$$p_{\iota\psi}^k = \frac{\alpha \cdot \tau_{\iota\psi} + (1 - \alpha) \cdot \eta_{\iota\psi}}{\Sigma_{(\iota\nu) \notin tabu_k}(\alpha \cdot \tau_{\iota\nu} + (1 - \alpha) \cdot \eta_{\iota\nu})} \tag{1}$$

where parameter α is a user-defined parameter ($0 \leq \alpha \leq 1$) which defines the relative importance of trail with respect to attractiveness. After each iteration t of the algorithm, that is when all ants have completed a solution, trails are updated as follows:

$$\tau_{\iota\psi}(t) = \rho\tau_{\iota\psi}(t-1) + \Delta\tau_{\iota\psi} \tag{2}$$

where ρ is a user-defined coefficient and $\Delta\tau_{i\psi}$ represents the sum of the contributions of all ants that used move (ι, ψ) to construct their solution. The ants' contributions are proportional to the quality of the achieved solutions, i.e., the better an ant solution, the higher will be the trail contribution added to the moves it used. The general structure of an ACO algorithm is as follows.

ACO Framework

Step 1. (*Initialization*)
Initialize $\tau_{\iota\psi}$, for every move (ι, ψ).

Step 2. (*Construction*)
For each ant k repeat the following operations until ant k has completed its solution:
a) compute $\eta_{\iota\psi}$, for every move (ι, ψ);
b) choose the state to move into, with probability given by (1);
c) append the chosen move to the k-th ant's set $tabu_k$.
Carry each solution to its local optimum.

Step 3. (*Trail update*)
For each ant move (ι, ψ) compute $\Delta\tau_{\iota\psi}$ and update the trail matrix by means of equation (2).

Step 4. (*Terminating condition*)
If the end conditions are not satisfied go to step 2.

3 Problem Description

An unlimited stock of rectangular bins of size (W, H) are given and n rectangular items of sizes (w_j, h_j), $j \in J = \{1, ..., n\}$, are required to be placed into the bins. We assume that with each item $j \in J$ is associated an input parameter ρ_j, which is equal to 1 if item j can be rotated of $90°$ and is equal to 0 if the item j cannot be rotated.

Figure 1 shows an example where the four items at the left are to be positioned in the gray master shown at the right.

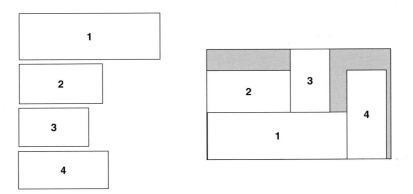

Fig. 1. Example of a 2BP instance.

The objective is to allocate without overlapping all items into the minimum number of bins. Notice that problem 2BP|M|F corresponds to 2BP|O|F when $\rho_j = 0$, $\forall j \in J$, while corresponds to 2BP|R|F when when $\rho_j = 1$, $\forall j \in J$.

We denote with $A = W \times H$ the area of the bin and with $a_j = w_j \times h_j$ the area of item $j \in J$.

We assume that each bin is located in the positive quadrant of the Cartesian coordinate system with its *origin*, the bottom left-hand corner, placed in position $(1, 1)$ and with its bottom and left-hand edges parallel to the x-axis and the y-axis, respectively. Moreover, we assume that the sizes of bins and of items are integers satisfying $w_j \leq W$ and $h_j \leq H$, for every item $j \in J$.

3.1 A local search for the 2BP|M|F

In this subsection we describe the local search of Boschetti and Mingozzi [4], henceforth called LS, for building a feasible solution for the problem 2BP|M|F. Given a list of items ordered using a given criterion, LS builds a feasible solution trying to allocate in turns the items, following the given order, considering a bin at a time. When no more items can be allocated in the current bin, LS *closes* such bin and *opens* a new one. The process stops when all items have been allocated. The heuristic proposed by Boschetti and Mingozzi repeat LS changing the order of the items until the number of bins used is equal to a known lower bound or a maximum number of iterations is reached. At each iteration the order of the items is changed applying some pricing rules.

A feasible solution is represented by specifying for each item $j \in J$ the bin index π_j where j is located, the coordinates (p_j, q_j) of its bottom left-hand corner, referred to as the *origin* of the item, and its rotation r_j, where $r_j = 0$, if the item is not rotated, and $r_j = 1$, if the item is rotated by $90°$.

The general framework of LS is the following.

Algorithm LS(J',AC)

Step 1. (*Erase the current solution*)
　　　　Let $i = 1$ be the index of the current empty bin.
　　　　Define $\pi'_j = n$, $\forall j \in J$.

Step 2. (*Filling up bin i*)
　　　　Consider the items of J' in sequence and for the current item $j \in J'$ find the best position (p'_j, q'_j) and rotation r'_j for placing j into the current bin i using the allocation criterion AC (see section 3.2).
　　　　If a feasible position exists, then place item j in solution by setting $\pi'_j = i$ and $J' = J' \setminus \{j\}$.

Step 3. (*Check if all items have been allocated*)
　　　　Let $z = i$ the cost of the emerging solution.
　　　　If $J' = \emptyset$, then Stop.
　　　　If $J' \neq \emptyset$ and $i < z^* - 1$ (where z^* is the best solution found so far), then open a new bin by setting $i = i + 1$ and go to step 2, otherwise Stop.

Procedure LS requires in input the ordered set of items J' and a parameter AC which specifies the allocation criterion to use in step 2. In the following section 3.2 we describe two criterion AC_1 and AC_2 for allocating an item into a bin.

3.2 Item Allocation

The general criteria used at step 2 of LS to allocate an item j into a bin is to choose a position where j does not overlap with the items already allocated into the bin and such that it is not possible to move it to the left and/or downward as its left-hand edge and its bottom edge are both adjacent to the edges of other items and/or to the bin edges.

In the following each feasible position for item j is represented by the triplet (p, q, r), where the pair (p, q) represents the coordinates of the origin of the item, while $r = 1$ if the item must be rotated by $90°$ and $r = 0$ if the item must not be rotated. We define two triplets $(p, q, 0)$ and $(p, q, 1)$ to indicate that item j having $\rho_j = 1$ can be located in position (p, q) either without rotation or rotating it by $90°$.

We denote with $F(j, B_i)$ the set of all feasible triplets, satisfying the criteria described above, for locating item j into bin i, where $B_i = \{j \in J : \pi'_j = i\}$ is the subset of items already located into the bin.

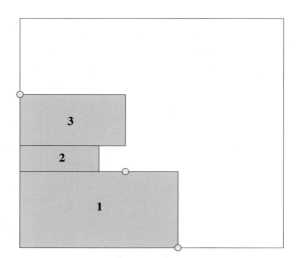

Fig. 2. Example of feasible positions for locating item j with $(w_j, h_j) = (5, 2)$ and $\rho_j = 1$: $F(j, B_i) = \{(1, 7, 0), (5, 4, 0), (5, 4, 1), (7, 1, 1)\}$.

In Fig. 2 it is shown the example reported in [4], where an item j of size $(w_j, h_j) = (6, 2)$ and $\rho_j = 1$ have to be placed into a bin i of size $(W, H) = (10, 9)$ containing items $B_i = \{1, 2, 3\}$ of sizes $(w_1, h_1) = (6, 3)$, $(w_2, h_2) = (3, 1)$ and $(w_3, h_3) = (4, 2)$ located in positions $(p'_1, q'_1, r'_1) = (1, 1, 0)$, $(p'_2, q'_2, r'_2) = (1, 4, 0)$ and $(p'_3, q'_3, r'_3) = (1, 5, 0)$. For the example reported in Fig. 2 we have $F(j, B_i) = \{(1, 7, 0), (5, 4, 0), (5, 4, 1), (7, 1, 1)\}$.

If only a feasible triplet exists, i.e. $|F(j, B_i)| = 1$, then item j is allocated in position $(p, q, r) \in F(j, B_i)$ by setting $p'_j = p$, $q'_j = q$ and $r'_j = r$. If $|F(j, B_i)| > 1$, then the best position is chosen according to the criteria that the bin layout which is more likely for item j is obtained by allocating item j in the position where the *not occupied* area under it and at its left hand side is minimum.

We denote with $f(p, q, r)$ the amount of area of the two rectangular regions under and at the left-hand side of the item j located in position (p, q, r) not occupied by the items in B_i. In Fig. 3 it is given an example of function $f(p, q, r)$ for the problem of Fig. 2.

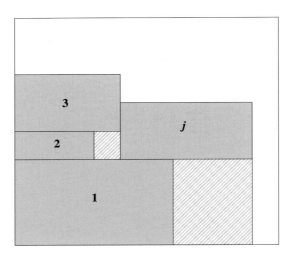

Fig. 3. Example: the hatched regions correspond to the not occupied area under and at the left-hand side of the item j in position $(5, 4, 0)$, i.e. $f(5, 4, 0) = 10$.

Let $f^* = min\{f(p, q, r) : (p, q, r) \in F(j, B_i)\}$ and $F'(j, B_i) = \{(p, q, r) \in F(j, B_i) : f(p, q, r) = f^*\}$. Procedure LS uses the following two rules AC_1 and AC_2 for allocating item j at step 2:

AC$_1$: Locate j in position (p'_j, q'_j, r'_j) by defining in sequence p'_j, q'_j and r'_j as follows: $p'_j = min\{p : (p, q, r) \in F'(j, B_i)\}$, $q'_j = min\{q : (p'_j, q, r) \in F'(j, B_i)\}$ and $r'_j = min\{r : (p'_j, q'_j, r) \in F'(j, B_i)\}$.

AC$_2$: Locate j in position (p'_j, q'_j, r'_j) by defining in sequence q'_j, p'_j and r'_j as follows: $q'_j = min\{q : (p, q, r) \in F'(j, B_i)\}$, $p'_j = min\{p : (p, q'_j, r) \in F'(j, B_i)\}$ and $r'_j = min\{r : (p'_j, q'_j, r) \in F'(j, B_i)\}$.

In the example reported in Fig.3 we have $f(1, 7, 0) = 7$, $f(5, 4, 0) = 13$, $f(5, 4, 1) = 13$ and $f(7, 1, 1) = 7$, therefore, the item j is located in position $(p'_j, q'_j, r'_j) = (1, 7, 0)$ by method AC_1, while it is located in position $(p'_j, q'_j, r'_j) = (7, 1, 1)$ by method AC_2.

4 An ant Metaheuristic Applied to the 2BP|M|F

As mentioned in section 2, in order to completely specify the ant heuristic it is necessary to define how to compute the visibility η and the trail update $\Delta\tau$.

Obviously the two are conceptually interlinked, and a major problem with respect to more straightforward TSP or related applications is that an item $j \in J$ is not *per se* more or less desirable than another item i, neither does the opportunity of the insertion of one or the other depend substantially from the item which was last inserted. Unfortunately, it is the whole of the partial solution to extend which determines the greater opportunity of one or the other choice.

This dependency from the whole of the partial solution is not a characteristic feature of the 2BP, but rather the opposite: the exception are the problems like the TSP or the sequential ordering, for which it is sufficient to know the last inserted elements (thus in the case of the TSP the last node of the path) in order to evaluate desirabilities.

In the case of the 2BP we addressed the problem as follows. The Boschetti-Mingozzi local search LS is based on an ordering of the items. We designed an ant algorithm which adaptively updates the ordering and leave to the LS algorithm the task to construct a feasible solution.

Specifically, we maintain implicit solutions in the form of permutations of the set of items, which will be directly translated into the corresponding explicit solutions as detailed in section 3.1. Unfortunately, we cannot lay trail as in the TSP, which would correspond to reinforcing pairs of items, as there is in general no advantage to consider items in pairs. However, since there is a loose correspondence between positions in the ordering and bin where the corresponding item will be allocated into, we decided to reinforce the item/position pairs. Trail τ_{ij} thus reflect the average quality of solutions where item j was positioned in i. Building blocks consisting of a subset of several items to be allocated together can emerge by reinforcing contiguous positions of the item subset.

Visibility values η_j are associated with each item $j \in J$, and are defined by means of the pricing rules described in section 4.1.

The complete ant algorithm is therefore the following.

Algorithm 2BPants

Step 1. (Initialization)
Initialize trail values to constant τ_{init} and visibilities η_j as in section 4.1.
Initialize the best solution S^* by allocating each item $j \in J$ into a different bin $\pi_j = j$ in position $(p_j, q_j) = (1, 1)$ and rotation $r_j = 0$. Set $z^* = n$ and $iter = 1$.

Step 2. (Set up ant solutions)
Initialize each ant solution $S_k = \emptyset$, $k \in K$.

Step 3. (Ant Solution Construction)
For each ant $k \in K$ repeat the following operations until ant k has completed its solution:
a) choose by means of formula (1) the next item to append to S_k;
b) append the chosen move to the k-th ant's set $tabu_k$.

Step 4. (Local Search)
For each solution S_k construct the corresponding 2BP solution $\hat{S}_k$, of cost $z(\hat{S}_k)$, by means of procedure LS.

Step 5. (Trail update)
Update trails by means of formula (2) using the positions of S_k evaluated according to the costs of $\hat{S}_k$.

Step 6. (Update the best solution)
Let k' be an ant for which $z(\hat{S}_{k'}) = \min\{z(\hat{S}_k) : k \in K\}$.
If $z(\hat{S}_{k'}) < z^*$ then $z^* = z(\hat{S}_{k'})$, and $S^* = \hat{S}_{k'}$.

Step 7. (Stop condition)
If z^* is equal to a known lower bound or $iter = MaxIter$ then Stop, otherwise go to Step 2.

The following algorithm *AntBP* has been used in the computational tests reported in section 5.

Algorithm AntBP

Step 1. Set $U_{BM} = \infty$.

Step 2. For each item allocation method AC_1 and AC_2 (see section 3.2), for each initial pricing rule IP_1, ..., IP_4(see section 4.1) perform step 3.

Step 3. Execute algorithm *2BPants* using the selected rules and update the upper bound $U_{BM} = max\{z^*, U_{BM}\}$.

4.1 Updating Visibilities

At each iteration of *AntBP* we use a different pricing function for defining the visibilities η_j, $j \in J$. These are the same as those introduced in [4] for defining item *prices*, which in fact play the same role as visibilities here, i.e., they are the basis for defining the item ordering which, in turns, will lead to a complete solution. The general idea is to give high prices to items which are more difficult to locate. We alternatively use four item elements for defining the initial visibilities η_j:

IP$_1$: The area, $\eta_j = a_j$;   **IP$_2$:** The width, $\eta_j = w_j$;
IP$_3$: The height, $\eta_j = h_j$;   **IP$_4$:** The perimeter, $\eta_j = 2w_j + 2h_j$.

Visibilities are never changed during an ants execution. At the end of each iteration of *2BPants* we update the trails τ_{ij}, $i, j \in J$, by means of formula (2). This corresponds to consider in turns each ant solution, read the ordering which originated that solution (item j in position i, for each $j \in J$) and increase or decrease the value τ_{ij} depending on the quality of the resulting ant solution.

Trails are initialized to a value τ_{init}, which in our application was defined to be a fraction of the average of all η values.

5 Computational Results

The algorithm presented in this paper was implemented in FORTRAN 77 and run on a Pentium 3 Intel at 933 MHz. Since no computational results for the 2BP|M|F have been presented in the literature, computational results are only given for the special case 2BP|O|F in order to compare the quality of the ants solutions with the best ones presented in the literature.

The benchmark instances are the same as those used in [4], that is ten classes of randomly generated test problems, where every class contains five groups of ten instances each. The first 6 classes of problems were originally proposed by Berkey and Wang [2], while the last four classes have been introduced by Martello and Vigo [23]. All test problems are available on the web page "http://www.or.deis.unibo.it/~research.html".

These instances were used to permit comparison with alternative problem-specific heuristics, even though they are not typical benchmarks for meta-heurstics. These last algorithms in fact find their reason for being in the need to devote substantial computational effort to explore the search space looking for elusive good quality solutions, therefore they are not on their own in finding solutions to a high number of instances, with minimal CPU time to devote to each of them. However, when comparing the results obtained by the ants with those obtained by the local search alone, without ants initialization, it appears that ants global search contributes to the quality of the solutions found, albeit inducing a much higher CPU cost.

Table 5 show our current preliminary results. The columns show:
- *class*: problem class;
- *cat*: problem category;
- *L*: master length;
- *W*: master width;
- *N*: number of items;

Table 1. Compared computational results of ants and LS

class	cat	L	W	N	GapLB	TimeLB	Ants	N.BinA	T.Ants	LS	N.BinLS	T.LS
I	1	10	10	20	1.000	0.000	10	6.60	0.044	9	6.70	0.006
	2	10	10	40	1.027	0.000	7	13.00	25.002	7	13.00	0.005
	3	10	10	60	1.016	0.011	7	19.80	50.658	7	19.80	0.011
	4	10	10	80	1.011	0.000	8	27.20	58.830	8	27.20	0.016
	5	10	10	100	1.016	0.017	6	31.50	205.516	6	31.50	0.028
					1.014	0.005	38		68.010	37		0.013
II	1	30	30	20	1.000	0.000	10	1.00	0.006	10	1.00	0.005
	2	30	30	40	1.100	0.005	9	2.00	96.619	9	2.00	0.022
	3	30	30	60	1.000	0.011	10	2.50	0.021	10	2.50	0.017
	4	30	30	80	1.033	0.012	9	3.20	191.261	9	3.20	0.055
	5	30	30	100	1.000	0.011	10	3.90	0.039	10	3.90	0.054
					1.026	0.007	48		57.589	48		0.030
III	1	40	40	20	1.000	0.000	10	4.70	0.137	7	5.00	0.011
	2	40	40	40	1.048	0.000	6	9.40	56.408	6	9.50	0.028
	3	40	40	60	1.047	0.016	4	13.80	172.580	4	13.80	0.050
	4	40	40	80	1.040	0.016	4	18.80	294.298	4	18.80	0.056
	5	40	40	100	1.050	0.031	3	22.50	542.449	3	22.50	0.079
					1.037	0.012	27		213.174	24		0.049
IV	1	100	100	20	1.000	0.004	10	1.00	0.004	10	1.00	0.011
	2	100	100	40	1.000	0.021	10	1.90	0.032	10	1.90	0.033
	3	100	100	60	1.100	0.054	8	2.50	422.827	8	2.50	0.099
	4	100	100	80	1.066	0.062	8	3.20	641.094	8	3.20	0.133
	5	100	100	100	1.033	0.092	9	3.80	718.030	9	3.80	0.170
					1.040	0.047	45		356.397	45		0.089
V	1	100	100	20	1.000	0.016	10	5.90	0.148	9	6.00	0.016
	2	100	100	40	1.025	0.022	8	11.50	41.764	6	11.70	0.038
	3	100	100	60	1.062	0.059	3	17.90	291.823	2	18.00	0.109
	4	100	100	80	1.029	0.120	5	24.20	357.157	5	24.20	0.164
	5	100	100	100	1.054	0.184	1	28.60	993.969	1	28.60	0.259
					1.034	0.080	27		336.972	23		0.117
VI	1	300	300	20	1.000	0.005	10	1.00	0.011	10	1.00	0.011
	2	300	300	40	1.200	0.051	8	1.70	381.310	8	1.70	0.089
	3	300	300	60	1.000	0.132	10	2.10	34.910	10	2.10	0.175
	4	300	300	80	1.000	0.248	10	3.00	0.293	10	3.00	0.279
	5	300	300	100	1.066	0.371	8	3.40	1757.315	8	3.40	0.527
					1.053	0.161	46		434.768	46		0.216
VII	1	100	100	20	1.110	0.010	5	5.20	23.793	4	5.30	0.012
	2	100	100	40	1.106	0.032	1	10.70	169.565	0	10.90	0.055
	3	100	100	60	1.112	0.062	0	15.60	404.272	0	15.80	0.127
	4	100	100	80	1.121	0.109	0	22.20	702.463	0	22.30	0.182
	5	100	100	100	1.122	0.176	0	26.80	1091.646	0	27.00	0.275
					1.114	0.078	6		478.348	4		0.130
IIX	1	100	100	20	1.080	0.010	6	5.30	18.910	5	5.40	0.0220
	2	100	100	40	1.113	0.024	1	10.70	163.800	1	11.00	0.061
	3	100	100	60	1.112	0.078	0	15.70	400.528	0	16.20	0.115
	4	100	100	80	1.112	0.121	0	21.90	700.273	0	22.10	0.198
	5	100	100	100	1.124	0.192	0	27.20	1086.338	0	27.30	0.290
					1.108	0.085	7		473.970	6		0.137
IX	1	100	100	20	1.000	0.010	10	14.30	0.010	9	14.40	0.011
	2	100	100	40	1.000	0.021	10	27.50	0.026	8	27.70	0.034
	3	100	100	60	1.000	0.039	10	43.50	0.046	9	43.60	0.055
	4	100	100	80	1.000	0.072	10	57.30	0.077	10	57.30	0.077
	5	100	100	100	1.000	0.103	10	69.30	0.124	9	69.40	0.142
					1.000	0.049	50		0.057	45		0.063
X	1	100	100	20	1.100	0.000	8	4.10	17.008	7	4.20	0.010
	2	100	100	40	1.061	0.027	6	7.30	104.832	6	7.30	0.033
	3	100	100	60	1.076	0.069	3	10.10	385.102	3	10.10	0.098
	4	100	100	80	1.064	0.114	2	13.00	720.517	2	13.10	0.161
	5	100	100	100	1.045	0.165	3	16.00	989.090	3	16.00	0.240
					1.069	0.075	22		443.310	21		0.108

- *GapLB*: gap of the lower bound;
- *TimeLB*: time to compute the lower bound;
- *Ants*: cost of the ants solution;
- *N.BinA*: number of bins used in the ants solution;
- *T.Ants*: time needed by the ants solution;
- *LS*: cost of the local search solution (1 iteration);
- *N.BinLS*: number of bins used in the LS solution (1 iteration);
- *T.LS*: time needed by the LS solution (1 iteration).

As mentioned, results are still preliminary in that the ants algorithm used is still very basic. We believe that using a more sophisticates framework, such as for example ANTS [22], will further improve the results obtained. From table 5 it appears in fact that ants help to improve the quality of the solutions proposed by the local search alone (both in terms of average solution quality and of number of optimal solutions found), but the results proposed here are still inferior to those obtained by the sophisticated tailored heuristic presented by Boschetti and Mingozzi [4].

6 Conclusions

In this paper we have described a basic ant system for solving the two-dimensional finite bin-packing problem where items can be rotated for allocation. The problem was chosen for its high industrial interest and for being representative of combinatorial problems for which the whole of a partial solution affects the evaluation of the quality of each further item as a new insertion.

The drawback to face is that for this problem very effective tailored heuristics exist, and standard benchmark consist of typical industrial size instances, which can usually be solved to optimality in short CPU time.

Face to this, the proposed ant system proved able to devise solutions of quality comparable to that of those reported in [4], which represent the state of the art for the 2BP.

The results reported here must be considered still preliminary in that the ant system used is a very basic one. Better results, albeit paying more computational efforts, can be obtained by means of more advanced ant algorithms, such as for example ANTS [22].

References

1. Bengtsson, B.: 1982, 'Packing rectangular pieces - a heuristic approach'. *The Computing Journal* **25**, 353–357.
2. Berkey, J. and P. Wang: 1987, 'Two dimensional finite bin packing algorithms'. *Journal of Operational Research Society* **38**, 423–429.

3. Boschetti, M. and A. Mingozzi: 2003a, 'The two-dimensional finite bin packing problem. Part I: New lower bounds for the oriented case'. *4OR* **1**(1), 27–42.

4. Boschetti, M. and A. Mingozzi: 2003b, 'The two-dimensional finite bin packing problem. Part II: New lower and upper bounds'. *4OR* **1**(2), 135–148.

5. Chung, F., M. Garey, and D. Johnson: 1982, 'On packing two-dimensional bins'. *SIAM Journal on Algebraic and Discrete Methods* **3**, 66–76.

6. Colorni, A., M. Dorigo, and V. Maniezzo: 1991, *Distributed Optimization by Ant Colonies*, Proceedings of ECAL'91, European Conference on Artificial Life. Elsevier Publishing.

7. Dell'Amico, M., S. Martello, and D. Vigo: 2001, 'A lower bound for the non-oriented two-dimensional bin packing problems'. *to appear in Discrete Applied Mathematics.*

8. Dorigo, M.: 1992, *Optimization, Learning and Natural Algorithms*, Ph.D. Thesis. Milano: Politecnico di Milano.

9. Dorigo, M., A. Colorni, and V. Maniezzo: 1991, 'Positive Feedback as a Search Strategy'. *Technical Report TR91-016, Politecnico di Milano.*

10. Dorigo, M. and L. M. Gambardella: 1997, 'Ant Colony System: a cooperative learning approach to the Traveling Salesman Problem'. *IEEE Transaction on Evolutionary Computation* **1**, 53–66.

11. Dorigo, M., V. Maniezzo, and A. Colorni: 1996, 'The Ant System: optimization by a colony of cooperating agents'. *IEEE Transactions on Systems, Man, and Cybernetics-Part B* **26(1)**, 29–41.

12. Dowsland, K. and W. Dowsland: 1992, 'Packing problems'. *European Journal of Oprational Research* **56**, 2–14.

13. Dyckhoff, H. and U. Finke: 1992, *Cutting and packing in production and distribution*. Physica Verlag, Heidelberg.

14. Dyckhoff, H., G. Scheithauer, and J. Terno: 1997, 'Cutting and packing (C & P)'. *in: M. Dell'Amico, F. Maffioli, S. Martello (Eds.), Annotated Bibliographies in Combinatorial Optimization, Wiley, Chichester* pp. 393–413.

15. El-Bouri, A., N. Popplewell, S. Balakrishnan, and A. Alfa: 1994, 'A search based heuristic for the two-dimensional bin-packing problem'. *INFOR* **32**, 265–274.

16. Faroe, O., D. Pisinger, and M. Zachariasen: 2003, 'Guided local search for the three-dimensional bin packing problem'. *INFORMS Journal on Computing* **15**, 267–283.

17. Fekete, S. P. and J. Schepers: 2000, 'On more-dimensional packing II: Bounds'. Technical Report 97.289, Universität zu Köln, Germany.

18. Frenk, J. and G. Galambos: 1987, 'Hybrid next-fit algorithm for the two-dimensional rectangle bin packing problem'. *Computing* **39**, 201–217.

19. Garey, M. and D. Johnson: 1979, *Computer and intractability, a guide to the theory of NP-completeness*. Freeman, New York.

20. Lodi, A., S. Martello, and M. Monaci: 2002, 'The two-dimensional packing problems: a survey'. *European Journal of Operational Research* **141**, 241–252.

21. Lodi, A., S. Martello, and D. Vigo: 1999, 'Heuristic and Metaheuristic Approaches for a class of two-dimensional bin packing problems'. *INFORMS Journal on Computing* **11**, 345–357.

22. Maniezzo, V.: 1998, *Exact and Approximate Nondeterministic Tree-search Procedures for the Quadratic Assignment Problem*, Vol. (to appear on: INFORMS Journal of Computing) of *Technical Report CSR 98-1*. University of Bologna:.

23. Martello, S. and D. Vigo: 1998, 'Exact solution of the two dimensional finite bin packing problem'. *Management Science* **44**, 388–399.

24. Pisinger, D. and M. Sigurd: 2003, 'Using decomposition techniques and constraint programming for solving the two-dimensional bin packing problem'. Technical Report 03/01, DIKU, University of Copenhagen, Danmark.

Distributed Multidisciplinary Design Optimisation in Aeronautics using Evolutionary Algorithms, Game Theory and Hierarchy

Eric J. Whitney[1], Luis F. Gonzalez[1], and Jacques Périaux[2]

[1] School of Aerospace, Mechanical and Mechatronic Engineering
    The University of Sydney, Sydney, NSW 2006, Australia
    {eric,gonzalez}@aeromech.usyd.edu.au
[2] Pole Scientifique - Dassault Aviation - UPMC
    Dassault Aviation, 78 Quai Marcel Dassault, 99214 Saint-Cloud, France
    jacques.periaux@dassault-aviation.fr

This paper is dedicated to the memory of Jacques-Louis Lions for his many contributions to Applied Mathematics and Applied Sciences.

1 Introduction

No problem in engineering can be solved exactly. As engineers we are forced to introduce simplifying assumptions into the nature of problems that are posed to us, so that the problems can then be solved by the methods available. The primary solution tool used by engineers today is still intuition; that is to say that when posed with a certain problem (be it in design, analysis, optimisation, control, manufacture or some other field), a certain understanding of the problem is brought to bear and a potential solution is arrived at rather rapidly. When intuition fails, or when it is insufficient in determining the solution in finer detail, we are then faced with the dilemma of how to proceed.

Today, many ingenious automated processes can be utilised that are able to significantly refine a design from its starting point. Together with this optimisation, modern automated processes require fewer simplifying assumptions when determining the nature of the problem, and this is evident in the increasing sophistication of available analysis tools.

In this paper a portion of a relatively new field of automated design and optimisation processes called evolutionary algorithms (EAs) is addressed. Broadly speaking, an EA is essentially an automated designer or optimiser that operates in a manner which mimics a natural evolutionary process. These algorithms have been arrived at after considering the way in which nature works to solve the difficult problem which Charles Darwin originally described as natural selection, or 'the survival of the fittest'. The reason why such algorithms are considered today is that they have been able to automatically solve design and optimisation problems which *could not be hitherto solved by any previously available mathematical technique*. Evolutionary algorithms exhibit many unique features that make them extremely useful in the solution of a wide range of different engineering problems.

The major drawback to the widespread use of evolutionary algorithms today is that they remain computationally expensive. In other words, the solution to a given problem often requires the evaluation of a large number of candidate designs. In most cases, for similar convergence to a solution, this evaluation cost is at least two orders of magnitude larger than that of a conventional optimiser. However there are methods for reducing this penalty, and it will be seen that the newer evolutionary methods available can now make this cost at least bearable. Conversely, the benefits of using evolution techniques for design or optimisation are that they are robust towards multimodal fitness landscapes (problems with more than one local minima) and noise (whether derived via experiment or simulation error). Further, they generally only require payoff information to run, that is, they only require a relative fitness to be assigned to each candidate solution and *not supplemental information such as first or higher order derivatives* of the fitness function.

A question arises as to when evolutionary algorithms should be used. Evolutionary computation should not be used in design or optimisation when the problem can be posed in a closed form, when it can have its merit evaluated directly without error, and when only one local optima (corresponding to a unique global optima) is known to exist. This implies that the usage of EAs in design or optimisation should be considered a possibility when the following conditions exist:

- The problem can not be posed in a closed form or even in an entirely mathematical form.
- The evaluation of each candidate solution is subject to noise or error. These may be experimental noise, error involved in a computational approximation if such a computational scheme is used, or some other form of uncertainty which may be known to exist.
- The problem is known to have more than one (or possible many) local optima; possible types of candidate solution where the gradient of the fitness function is zero. One or more of these local optima will correspond to one or more global optima.
- The problem may change over time. This is also known as a 'non-stationary environment'.

Overview

After an introduction to evolutionary algorithms (section 1.1) and aerodynamic optimisation (section 1.2), we will introduce some new concepts relating to parallel asynchronous function evaluation (section 2.1), hierarchical optimisation (section 2.2) and multiobjective optimisation using game theory (sections 2.3 and 2.4). All these additions extend and enhance the applicability of evolutionary methods. Following this, we present five test cases in aeronautics:

- A two-objective inverse aerofoil design (section 3.1).
- A two-objective aerofoil direct design for cruise and loiter conditions (section 3.2).
- A two-objective aerofoil direct design for cruise and takeoff conditions with a flap (section 3.3).
- A transport aircraft conceptual design with one objective (section 3.4).
- An Unmanned Combat Air Vehicle (UCAV) conceptual design with two objectives, computed with Pareto and Nash approaches (section 3.5).

Finally we present a conclusion, including perspectives on the work.

1.1 Evolutionary Algorithms

Evolutionary algorithms are design and optimisation algorithms that mimic the natural process of 'survival of the fittest'. Broadly speaking they operate simply through the iterated mapping of one population of solutions to another population of solutions. This is contrasted with conventional deterministic search techniques which proceed from one given sub-optimal solution to another, until an optimum solution is reached. Evolutionary algorithms fall into the category of *stochastic* (randomised) optimisation methods whereby repeated application of the method to the same initial starting conditions may yield different results. Evolutionary algorithms work by exploiting population statistics to some extent, so that when newer individual solutions or *offspring* are generated from parents, some will have inferior characteristics and some will have superior characteristics. The general working principles of the iterated mapping then reduces to generating an offspring population, removing a certain number of inferior individuals, and obtaining the subsequent population.

1.2 Aerodynamic Optimisation

Regarding aerodynamics, the necessity of optimisation is clear, given that even very small improvements in performance can yield large reductions in operating costs, lowering of undesirable environmental pollution (such as nitrous oxides, carbon dioxide and noise), shorten travel times, improve payload capacity and improve performance. Because aircraft are being operated over longer and longer lifetimes, any improvements achieved during the design and development stage are rewarded with large gains.

Aerodynamic optimisation is an intensive area of research, however the tools most commonly used in industry today are still intuition and conventional optimisers. An expert in aerodynamics can often decide upon a very efficient configuration or geometry using only analysis tools, with no optimisation process whatsoever. Aircraft have been designed in this way until quite recently. The conventional optimisation techniques used today fall into a number of categories:

- Trade studies, which simply consider a large number of proposed configurations, and of these the best is selected. In this case, normally Computational Fluid Dynamics (CFD) is used to 'screen' a number of good configurations that are then wind tunnel tested.
- Conventional optimisers, which work on well understood mathematical principles derived from numerical analysis. In this field there are derivative estimation methods by perturbation, parameter sensitivity analyses, solver-optimiser coupled schemes, inverse methods (where a good pressure or Mach distribution is known in advance), multi-point conventional methods and variable fidelity methods.
- The adjoint method, which is derived from control theory [Jam95]. This is probably the most popular area of research at the present time – it allows a deterministic-type optimisation using very few iterations (or equivalent iterations) of the flow analysis code. This can often require less than ten complete flowfield analyses for good convergence. This technique has produced very low drag solutions to constrained problems when provided with a good starting point.

For many aerodynamic optimisation problems, the wide bounds required and novel constraints would be prohibitive to conventional optimisers. In many cases it has been shown that the evolutionary approach often replicates the result found by a conventional optimiser, albeit in a much longer time. However, when unusual configurations or multiobjective tradeoffs are explored, the answer is supplied forthwith – whereas a conventional optimiser would be unable to return any answer at all.

2 Modern Techniques and Extensions

In this section, some extensions to standard evolutionary algorithms are discussed. These extensions allow the algorithm to:

- Utilise a cheap, readily available parallel processing capability running variable-time iterative solvers on desktop computers, through *asynchronous solution.*
- Allow for the exploitation of variable-fidelity or multi-physics solvers, through a *hierarchical population topology.*
- Be applied to more varied types of engineering problems in one or many objectives, through *Pareto tournament selection.*

2.1 Asynchronous Solution

The design of an optimisation algorithm for asynchronous solution of candidate problems can be broadly defined as the non-dependence of an optimisation procedure on the speed of a given solver. The need for asynchronous fitness function evaluation arises from the fact that many methods of solution used in engineering today may take variable times to complete their operation. The classic example of this is the modern CFD solver. In a typical industrial code used for external aerodynamic analysis of aeroplanes, the time for the residual of the solution to converge to a specified level (either machine zero or an arbitrarily selected higher value) can vary over a significant range. If as an example, we are given an unstructured mesh based Navier-Stokes solver, a number of factors have a significant effect on the execution time:

- The number of cells involved, and the simplicity of their connectivity.
- The skewness and aspect ratio of the cells, especially considering that in our case we are normally forced to use automatic mesh generation.
- The presence of unforeseen large gradients in the flow such as confluent boundary layers, merged shocks and flow separation.
- The degree of mesh anisotropy in the area of large flow gradients, which will probably have to be automatically adapted.

With all of these factors considered, it should seem obvious that the optimisation algorithm used to drive the design process should be designed with this possibility in mind. The previous generation of evolutionary algorithms have generally used a generation based approach, where a newly created offspring population is sent to a parallel computer to be solved as a unit. A problem with generational models is that they create an unnecessary *bottleneck* when used on parallel computers. If the population size is approximately equal to the number of processors, and most candidate offspring sent for solution can be successfully evaluated, then some processors will complete their task quickly with the remainder taking more time. With a generational approach, those processors that have already completed their solutions will remain idle until all processors have completed their work.

A final and most important need for asynchronous evaluation is that it provides an ideal method of using existing desktop computers for problem solving. The need for asynchronicity arises due to the fact that many of these machines will have different operating speeds, and may be added or removed from the parallel task when they are needed for other work; meaning that no correct number or combination of resources can be known in advance.

Implementation and Advantages

The approach used in this paper, is to ignore any concept of generation based solution. This approach is similar to work done by Wakunda and Zell [WZ00], however the method for choosing good offspring (the selection operator) is quite different from normal evolutionary algorithms (refer to section 2.3). Whilst a parent population exists, offspring are not sent as a complete 'block' to the parallel slaves for solution. Instead one candidate is generated at a time, and is sent to any idle processor, where it is evaluated at its own speed. When candidates have been evaluated, they are returned to the optimiser and either accepted by insertion into the main population or rejected. This requires a new method for selecting superior offspring, because they cannot now be compared *one against the other*, which is the standard selection technique used in generational algorithms. In fact, a single offspring must compete against a previously established benchmark and if successful must replace (according to some rule) an individual pre-existing in the population.

We implement this benchmarking via a separate evaluation buffer, which provides a statistical 'background check' on the comparative fitness of the solution. The length of the buffer should represent a reasonable statistical sample size, but need not be too large; approximately twice the population size is more than ample. When an individual has had a fitness assigned, it is then compared to past individuals (both accepted and rejected) to determine whether or not it should be inserted into the main population. If it is to be accepted, then some acceptance rule is invoked and it replaces a member of the main population.

A Short Simulation

To demonstrate the effectiveness of the asynchronous evaluation method, we give a short simulation. We solve the simple sphere function $f = \sum_{i=1}^{N} x_i^2$ on a single computer. In simulation, we offer two methods of evaluation; for the *asynchronous* case we assign a small fictitious delay to each function evaluation, and for the *synchronous* case we assign the same delay to all individuals in advance but we wait until the slowest evaluation has completed – in exactly the same manner that this would occur in practice over a cluster of computers. The result is presented for both asynchronous and synchronous runs in figure 1. The results are scaled vertically so that when no asynchronicity exists ($\frac{t_{slowest}}{t_{fastest}} = 1$), the workload factor is also unity; in any case it can be seen that both algorithms perform identically in this situation just as expected. Configurations up to $\frac{t_{slowest}}{t_{fastest}} = 5$ are explored. It is obvious from figure 1 that the synchronous approach fares poorly when a higher variability exists in execution time, when compared to the asynchronous approach.

2.2 Hierarchical Population Topology

A hierarchical population topology, when integrated into an evolution algorithm, means that a number of separate populations are established in a hierarchical layout to solve the given problem, rather than a single 'cure-all' type single population layout. This method was first proposed by Sefrioui [SP00, SSP00], and is shown in figure 2.

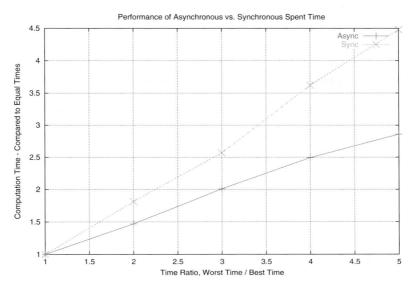

Fig. 1. Execution time variation between asynchronous and synchronous evaluation.

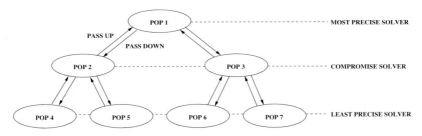

Fig. 2. Hierarchical population topology.

The purpose of utilising a hierarchical topology is to exploit the possibility of using incomplete or rapid function evaluation information during optimisation progression. In other words, perfect (or the most compete available) evaluation of the fitness function is not required all the time in order to solve a given problem. The advantage of the hierarchical method is predominantly speed; to achieve a given quality level for the solution a much shorter wall-clock time is required.

Implementation and Advantages

The advantage of the hierarchical layout is that it is quite suitable for engineering type problems, where the fitness function can be evaluated with varying levels of precision. The layers are arranged so that the uppermost layer uses the most precise solver available for fitness function optimisation.

The next layer (or layers) are assigned intermediate quality solvers, and the lower-most layer uses the least accurate solver available. The task of the lower layers is predominantly *exploration*, where large jumps in design variables are expected and the population is not too particular about small improvements in fitness. The middle layer is a compromise position between the top and bottom layers. The top layer is tasked with *exploitation*, using very small steps in the search space and fine-tuning the final result. As an example we can take the aerodynamic optimisation of a transonic aircraft wing. A first attempt at dividing the workload into three layers might proceed as follows:

1. Use a full compressible Navier-Stokes solver with turbulence modeling on the top layer;
2. Use an Euler solver with coupled boundary layer on a second layer;
3. Use a full potential solver on the lowest layer.

The full potential solver would provid initial guesses at good solutions to the Euler layer, where they can be rechecked and further evolved. Good solutions from the Euler layer would progress up to the top level where they are again checked and further evolved. In this manner no time is wasted using the Navier-Stokes solver to explore large parts of the search space, which would be almost impossible in any reasonable time.

2.3 Game Theory – Pareto Tournament Selection

The purpose of a multiobjective optimiser is to provide answers to problems which can not or should not be posed in single objective form. To this end, a number of approaches have been developed for use in evolutionary algorithms, and the two that we consider are:

- *Pareto fronts:* Originally proposed by Vilfredo Pareto, a solution to a given multiobjective problem is the Pareto optimal set found using a cooperative game, and this spans the complete range of compromise designs between each objective.
- *Nash equilibria:* Originally proposed by John F. Nash [Nas50, Nas51], a Nash equilibrium point is the position in a competitive (or non-cooperative) game whereby no player can improve his position at the expense of the other, with each player optimising one fitness function over a subset of the objective variables.

The merit of our approach is in applicability; deterministic optimisers have difficult computing concave Pareto fronts, and methods utilising Pareto front gradients are still incapable of computing mixed-type and discontinuous fronts [Deb98]. A Pareto optimal set does however give the practicing engineer great insight into the variations in a design that exist between optima for each design objective. Formally, the Pareto optimal set can be defined as the set of solutions that are *non-dominated* with respect to all other points in the search space, or that they dominate every other solution in the search space except fellow members of the Pareto optimal set. For two solutions **x** and **y** (in minimisation form):

$$rel(\mathbf{x}, \mathbf{y}) = \mathbf{x} \text{ dominates } \mathbf{y} \; if : \; f_i(\mathbf{x}) \leq f_i(\mathbf{y}) : \; \forall i \in 1 \ldots M$$
$$rel(\mathbf{x}, \mathbf{y}) = \mathbf{x} \text{ nondominated w.r.t. } \mathbf{y} \; if : \; f_i(\mathbf{x}) \leq / \geq f_i(\mathbf{y}) : \; \forall i \in 1 \ldots M$$
$$rel(\mathbf{x}, \mathbf{y}) = \mathbf{y} \text{ dominates } \mathbf{x} \; if : \; f_i(\mathbf{x}) \geq f_i(\mathbf{y}) : \; \forall i \in 1 \ldots M$$

$$(1)$$

For a problem in M objectives. This is called the 'relationship' operator. In practice we compute an approximation to the continuous set, by assembling $ParetoSet = \left[\mathbf{x}_1^*, \mathbf{x}_2^*, \ldots, \mathbf{x}_\mu^*\right]$. Figure 3 shows (for a two objective case) the regions where a particular point either dominates other solutions, is dominated by other solutions or has non-dominance with other solutions.

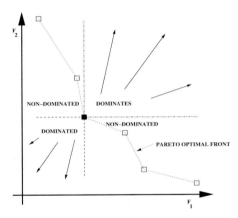

Fig. 3. Pareto relationship between one point and other regions of the fitness space (minimisation problem).

To implement an optimisation algorithm that is equally applicable to both single and multiobjective problems, a suitable selection operator capable of handling either situation must be developed. We propose the Pareto tournament selection operator, which is an extension of the standard tournament operator popular in many approaches [Gol98, Mic92]. Most evolutionary algorithms configured for multiobjective optimisation currently used the nondominated sorting approach; This is a straightforward way to adapt an algorithm that is designed as a single objective optimiser into a multiobjective optimiser, and is used by many researchers.

The problem with sorting approaches is that the method is not a fully integrated one. Briefly, a sorting method works by computing the set of nondominated solutions amongst a large statistical sampling (either a large population or previous data), and assigning these solutions rank 1. Then ignoring these points, the process is repeated until a 'second' Pareto front is found, and this is assigned rank 2. This process continues until all points are ranked, and then the value of the rank is assigned to the individual as a now single objective fitness. An example of this ranking is shown in figure 4.

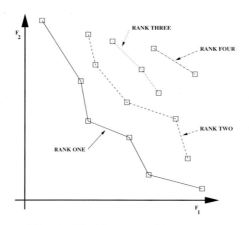

Fig. 4. The Pareto ranking process.

A problem arises when considering whether it is fair to assign individuals in the second rank half the fitness of the first, and whether the third rank deserves a third of the fitness of the first. This poses a dilemma about the level of equality present amongst the solutions, as often solutions with excellent information may lie adjacent to, but not in, rank 1. To solve this 'artificial parameter' problem, it is possible to introduce scaling, sharing and niching schemes, however all of these introduce problem-specific parameters. It is of course always desirable to compose an algorithm that does not introduce such unnecessary parameters.

The current operator is a novel approach in that it requires no additional 'tuning' parameters, works seamlessly with the asynchronous selection buffer (section 2.1), and is very easy to encode. Simply, to determine whether a new individual $\mathbf{x}'$ is to be accepted into the population μ, we compare it with the selection buffer B by assembling a small subset of the buffer called the tournament $Q = [\mathbf{q}_1, \mathbf{q}_2, \dots, \mathbf{q}_Q]$. We assemble Q by selecting individuals from the buffer, exclusively at random, until it is full. We then simply ensure that the new individual is not dominated by any in the tournament:

$$rel(\mathbf{x}', \mathbf{q}_j) \neq (\mathbf{q}_j \text{ dominates } \mathbf{x}') : \; \forall j \in Q \qquad (2)$$

If this is the case, then it is immediately accepted, and is inserted according to given replacement rules. The only parameter that needs to be determined in advance is the tournament size, a parameter which would exist in a single objective optimisation anyway. Selection of this parameter requires a small amount of problem specific knowledge, and should vary between $Q = \frac{1}{2}B$ (strong selective pressure) and $Q = \frac{1}{6}B$ (weak selective pressure). Optimisation success is not overly sensitive to this value, provided the user errs on the side of weak selective pressure (smaller tournaments) in the absence of better information. The egalitarian approach to the tournament (by selecting individuals at random) ensures good diversity amongst the selected individuals, and no niching or forced separation of individuals has been found necessary. It can also be seen that in the event that the fitness vectors have only one element (a single objective optimisation), this operator simplifies to the standard tournament selection operator.

2.4 Game Theory – Nash Solutions

A well known concept from game theory for multiobjective or multidisciplinary optimisation is the Nash equilibrium [Nas50]. The Nash equilibrium is determined by n players competing symmetrically for n criteria, where each player optimises a unique set of optimisation variables, and all other variables are determined by the other players. For example, for player i the vector of problem variables is $X = ((x_1, \bar{x_2}, \bar{x_3}, ..., \bar{x_i}, x_{i+1}, ...\bar{x_n})$ where all parameters $\bar{x_j}$, $j \neq i$ are 'locked' by the other players. Player i is interested only in the objective, namely $f_i = f_i(X)$ where $F(X) = (f_1(X), f_2(X), ..., f_n(X))$ is the entire multi-objective problem.

A Nash solution to a given problem can only find a single point solution, rather than the entire Pareto front. While this point may form part of the Pareto front, in most cases it will be suboptimal in the Pareto sense. However, the advantage of Nash computation is that the answer is arrived at rather quickly compared to the computation of the complete Pareto front. In addition, it is useful to consider the manner in which the variables are split between the players. Generally, the variables for each player can be selected due to their expected influence on the objective for that player, and thus it is clear from the outset that the solution can match the expectations of the design engineer.

Using evolutionary algorithms, we implement this using one population (or population hierarchy) for each player, whereby information is exchanged between the EAs after a migration period has occurred, and this is shown in figure 5.

There are two migrations present when using the hierarchical EA-Nash scheme, first there is a circulation of solutions up and down (as per usual hierarchical solution), whereby the best solutions progress from the bottom layer to the top layer where they are refined. After a certain predefined larger number of function evaluations, there is a 'Nash' migration where information (the 'locked' variables) between the players is exchanged between corresponding nodes on each hierarchical tree.

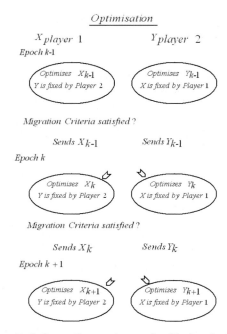

Fig. 5. Information exchange for Nash solutions.

3 Test Cases

3.1 Case I: Two Dimensional Subsonic Inverse Aerofoil Reconstruction

In this test case, we present an inverse aerofoil reconstruction where we compute a geometry which best fits two given pressure distributions. The pressure distributions were previously generated by a Navier-Stokes solver, but in this case we recompute the problem using a panel method with a coupled viscous boundary layer. The fact that the original and reconstruction solvers are different implies that we will not be able to find the solution exactly, and also that we will in fact finish with a Pareto front of solutions rather than a single correct and unique point.

Problem Definition

We are given target pressure distributions for the two flow conditions at discrete surface points i: P_i^{*1} and P_i^{*2}. To compute the two fitness functions, we use a least-squares approach applied separately for each objective:

$$
\begin{aligned}
f_1 &= \frac{1}{N_1} \sum_{i=1}^{N_1} \left(P_i^{*1} - P_i^1 \right)^2 \\
f_2 &= \frac{1}{N_2} \sum_{i=1}^{N_2} \left(P_i^{*2} - P_i^2 \right)^2
\end{aligned}
\tag{3}
$$

Where N_1 and N_2 are the number of prescribed pressure points for each case. Two two flow conditions are given in table 1.

Table 1. Inverse aerofoil design flow conditions..

	Objective One	Objective Two
M_∞	0.18	0.54
Re	14.0×10^6	14.0×10^6

Flow Solver

The flow solver used is *XFOIL* written by M. Drela [Dre01]. This code is a panel method potential solver with a sophisticated two-equation boundary layer model that handles laminar, turbulent and thinly-separated regions (including finite separation bubbles). Free or fixed boundary layer transitions can be specified, and in this case we allow for free transitions assuming a relatively smooth construction method will be used in practice. Because the code can only correctly model subsonic flows, after the solution of every candidate geometry the reported surface c_p values are inspected to ensure that no portion of the surface is supersonic. In other words, we ensure that $\forall i : \ c_{p_i} < c_p^*$ where c_p^* is the sonic value, given by:

$$c_p^* = \frac{2}{\gamma M_\infty^2} \left\{ \left[\frac{1 + \frac{1}{2}(\gamma - 1)M_\infty^2}{1 + \frac{1}{2}(\gamma - 1)} \right]^{\frac{\gamma}{\gamma - 1}} - 1 \right\} \tag{4}$$

Where $\gamma = 1.4$ for standard air. If any candidate is found to have a surface c_p exceeding this value, it is rejected immediately. In addition, any individual that does not have a converged boundary layer solution (produced by the internal iterative solver in *XFOIL*) is also rejected immediately. Since the solver is command-line based, it was implemented 'as is' by automatically generating a *UNIX* script an calling the solver directly. No internal modification of the source code was necessary.

Individual Representation

The aerofoil geometry is represented by two Bézier curves, one for the mean line and one for the thickness distribution. The mean line-thickness distribution is a standard method for representing aerofoils [AD80], as it closely couples the representation with the results; the mean line has a powerful effect on cruise lift coefficient and pitching moment, while the thickness distribution has a powerful effect on the cruise drag. Put simply, the aerofoil is obtained by perpendicular offset of the thickness distribution about the mean line. For a given mean line point (x_m, y_m) and matching thickness distribution height y_t, an upper and lower surface point can be obtained:

$$x_{u,l} = x_m \mp y_t \sin(\theta) \tag{5}$$

$$y_{u,l} = y_m \pm y_t \cos(\theta) \tag{6}$$

Where θ is the angle of the mean line at (x_m, y_m). This is shown in figure 6. We select the $x-$positions of the Bézier control points in advance; the $y-$positions remain as the unknowns. The only restrictions are that the first and last points are fixed to $(0, 0)$ and $(1, 0)$ to provide leading and trailing edges respectively, and that the first control point on the thickness distribution must be directly above the leading edge (i.e. $(0, y_{c,1})$) to provide a rounded geometry[3]. We bound the vertical heights to range $y_c \in [0.01, 0.12]$ giving a very wide range of possible geometries (theoretically spanning aerofoils from 2% to 24% thick). The advantage of using single high-order Bézier curves for the representation rather than piecewise splines or others is their geometric stability. A Bézier curve must by definition always be contained within the bounding envelope of control points. Furthermore, if the bounding envelope is not re-entrant, then the curve will also have this property. Also, Bézier curves do not 'kink' like a piecewise spline, and the defining equations are not stiff. Therefore, a small change in control point location will always result in a small change in surface representation. This provides a favourable interface between the optimiser and the flow solver.

Implementation

We implement this case using a single population of size 20. We use six evenly spaced control points on the thickness distribution and six on the mean line. Remembering that the angle of attack for each objective is unknown, the problem therefore consists of fourteen unknowns.

Results

The solver was allowed to converge for a fixed period of time. If we track the convergence of the objective one optimal point as shown in figure 7, we can see that the optimisation ran for over 60000 function evaluations, but converged after approximately 35000. Figure 8 shows the Pareto front, with most points lying bunched in a compromise position, with just a few outside points representing the complete span. We select one of the aerofoils from the middle of the Pareto front for analysis, and this is shown in figure 9. The surface pressures of this aerofoil for both flow conditions given by *XFOIL*

---

[3] Bézier curves are by definition always tangent to the extreme edges of their defining envelopes.

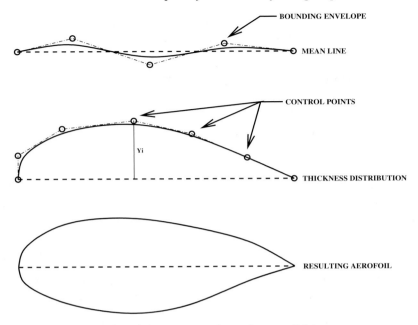

Fig. 6. Aerofoil representation using two Bézier curves.

are shown in figure 10. It is important to note that this conventional aerofoil geometry was obtained with no initial guess or prior knowledge of camber, thickness, style or any other geometric property. It can be seen that the pressures match very well except at two particular points; the leading edge lower surface for objective one, and the leading edge upper surface for case two. The cause of this is the different way in which the boundary layer is captured by the Navier-Stokes solver as compared to the panel method, which results in the reconstructed aerofoil operating at a slightly incorrect lift as well as the difference in leading edge spike resolution.

3.2 Case II: Two Dimensional Subsonic Aerofoil With Transit and Loiter Objectives

Problem Definition

Unmanned Aerial Vehicle (UAV) systems are ever increasingly becoming important topics for aerospace research and industrial institutions. There are difficulties in these new concepts because of the compromising nature of the missions to be performed, like high- or medium-altitude surveillance, combat environments (UCAV) and many others. Particular care must especially be taken in aerodynamic optimisation, due to the often very long endurance and high speed dash requirements; even small improvements in drag can represent large weight savings over an entire mission.

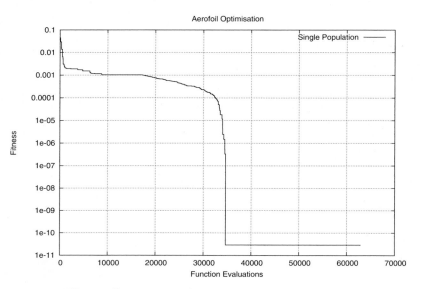

Fig. 7. Convergence of objective one optimal point.

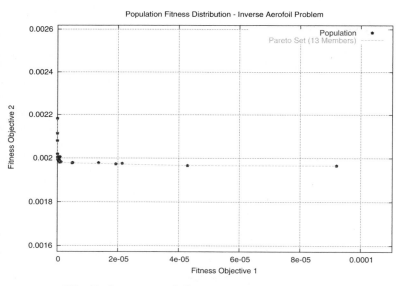

Fig. 8. Inverse aerofoil reconstruction Pareto front.

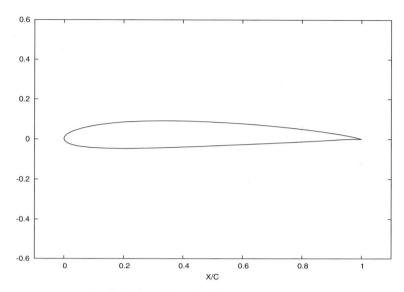

Fig. 9. Selected aerofoil from reconstruction.

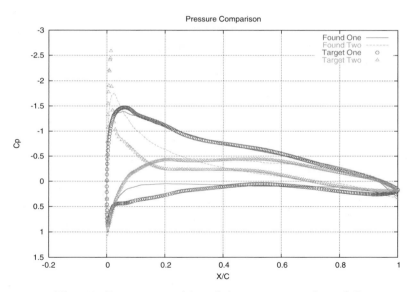

Fig. 10. Pressure matching of the reconstructed aerofoil.

This case represents an aerofoil design study for a hypothetical low-cost UAV. The objective is to design an aerofoil that provides low drag at two separated design points; both high speed transit and loiter. The high speed transit condition is representative of the flow conditions that would ordinarily be encountered by a UAV when moving rapidly from takeoff to a reconnaissance loiter area, or from one reconnaissance area to another. The loiter condition is representative of the flow conditions that would be encountered once at the given reconnaissance area. A compromise situation exists, because the high speed transit requires a low operating lift coefficient at high Reynolds and Mach numbers, whereas the low speed loiter requires moderate a lift coefficient at low Reynolds and Mach numbers. The operating conditions are given in table 2.

Table 2. UAV Transit and loiter flow conditions.

	Transit	Loiter
c_l	0.05	0.78
M_∞	0.60	0.15
Re	14.0×10^6	3.5×10^6

The form of the problem is stated simply:

$$\min(f_1): \quad f_1 = c_{d_{transit}} \tag{7}$$
$$\min(f_2): \quad f_2 = c_{d_{loiter}} \tag{8}$$

In addition, the thickness of the aerofoil must exceed 12% ($\frac{t}{c} \geq 0.12$) and the pitching moment must not be more severe than -0.065 ($c_m \geq -0.065$). Both constraints are applied by equally penalising both fitness values via a linear penalty method. In addition, aerofoils generated outside the thickness bounds of 10% to 15% are rejected immediately, before analysis.

Implementation

This case is solved via the hierarchical method, using a three level tree with varying panel densities. The three levels are established as follows:

- A population size of 20 and 119 panels used on the aerofoil surface.
- A population size of 20 and 99 panels used on the aerofoil surface.
- A population size of 10 and 79 panels used on the aerofoil surface.

Four this case we use four control points on the mean-line and six on the thickness distribution. This problem was solved on a single laptop computer running at 1.0 GHz. The case was halted when no further progress could be seen; 5300 head node function evaluations were completed, and the run took approximately four hours.

Results

The resulting Pareto set is shown in figure 11. It is clear that three discontinuous regions have been found for the Pareto set, a fact alone that would make application of a conventional optimiser to this problem difficult.

The ensemble of aerofoils comprising the Pareto front are shown in figure 12. It can be seen that classical aerodynamic shapes have been evolved, even considering that the optimisation was started completely from random and the evolution algorithm had no problem specific knowledge of appropriate solution types. We select three aerofoils for consideration from the Pareto front of 20 members (numbers 2, 10 and 20) to illustrate the two objective extremes and a compromise geometry. Figure 13 shows an objective one optimal aerofoil in the transit flow regime, and it can be seen that it has evolved a conventional low-drag pressure distribution and standard sleek form. Figure 14 and 15 show the compromise aerofoil, having a very pronounced S-shaped camber distribution. The pressure distribution is again seen to be relatively conventional, with a marked favourable gradient on the lower surface in both flow regimes. Figure 16 shows the objective two optimal aerofoil in the loiter regime, and finally it can be seen that the pressure distribution is of the classical 'rooftop' type on the upper surface while having an almost constant favourable pressure gradient on the lower surface.

Concluding this case, it is observed that all aerofoils easily satisfy the design constraints. Without any problem specific knowledge, the evolution algorithm has discovered forms (figures 13 and 16) that would have been designed by an expert in aerodynamics, as well as an unusual but effective compromise form (figures 14 and 15).

3.3 Case III: Two Dimensional Subsonic Aerofoil With Transit and Takeoff Objectives

Problem Definition

For the second case we again consider the design of a single element aerofoil for the same low-cost UAV using the same flow solver, but considering both the transit cruise and takeoff points together. Because the UAV is a low-cost, simplified project, only a plain flap is considered. This facilitates the design and optimisation process, because there is no deployed flap the same single element aerofoil code can be applied directly. The first design point is the same as proposed in subsection 3.2. The design points are given in table 3.

The problem is again:

$$\min(f_1): \ f_1 = c_{d_{transit}} \tag{9}$$
$$\min(f_2): \ f_2 = c_{d_{takeoff}} \tag{10}$$

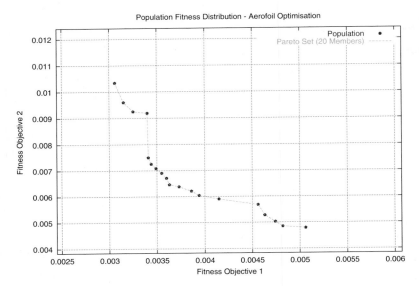

Fig. 11. Pareto set of UAV Transit-Loiter design.

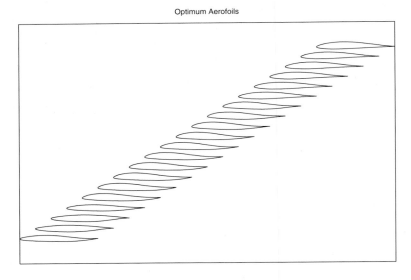

Fig. 12. Ensemble of Pareto set aerofoils – UAV Transit-Loiter design.

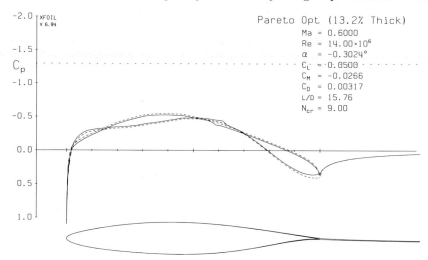

Fig. 13. Objective one optimal aerofoil – Transit flow conditions.

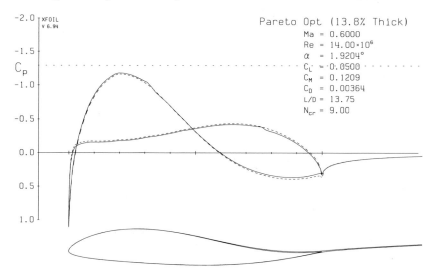

Fig. 14. Compromise aerofoil – Transit flow conditions.

Table 3. UAV Transit and takeoff flow conditions.

	Transit	Takeoff
c_l	0.05	1.40
M_∞	0.60	0.11
Re	14.0×10^6	2.46×10^6
$Flap$	None	30% Chord, Deflected $\delta_f = 10°$

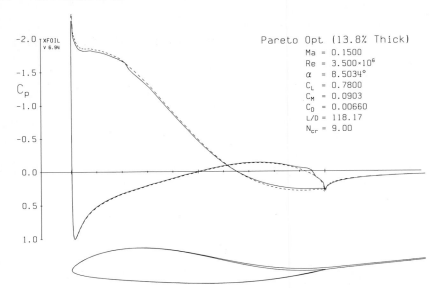

Fig. 15. Compromise aerofoil – Loiter flow conditions.

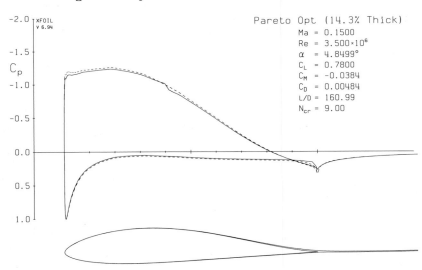

Fig. 16. Objective two optimal aerofoil – Loiter flow conditions.

The thickness of the aerofoils must exceed 12% ($\frac{t}{c} \geq 0.12$), however the pitching moment requirement ($c_m \geq -0.065$) is only enforced for the flaps-up (objective one) case.

Individual Representation

The two Bézier curve approach is used, similar to subsection 3.1, however for more variability in the surface geometry that may be required due to the specialised nature of the flap, the number of control points in increased. In this case we use six free control points on the mean line and eight on the thickness distribution, giving a problem in fourteen unknowns.

Implementation

The optimiser is configured as a single population with a population size of 20, intermediate recombination is used between two parents and 139 panels used by the solver. The higher number of panels is indicative of the resolution required to correctly model the flap-joint region, where there is a rapid direction change (a kind of pseudo-discontinuity) on the body surface. This problem was solved on a single laptop computer running at 1.0 GHz. The case was halted after 7700 function evaluations (only one node exists), and took approximately eight hours to run.

Results

The resulting Pareto set is shown in figure 17. The aerofoils comprising the Pareto front are shown in figure 18. In this case, some unusual aerodynamic forms are found, all possessing rather pronounced S-shaped mean lines and thin trailing edges. In a similar manner, we select three aerofoils representing objective one optimal (number 1, figure 19), objective two optimal (number 17, figure 22) and compromise (number 13, figures 20 and 21) geometries.

While some quite low drag shapes have resulted, they are quite different to the ones found in the first problem. In fact, the evolution algorithm has discovered a small inherent modeling limitation in the solver. It has been exploited because the Pareto method is a cooperative game, and all geometries must by definition be checked against both objectives (have both fitnesses fully computed). Because the boundary layer model in the solver can often have difficulty converging when there are severe (separation inducing) adverse pressure gradients, the aerofoils associated with the first test case would ordinarily have been rejected (subsection 3.2) if they had a deflected flap. The pressure spike with the flap down is plainly evident in figures 21 and 22 at the 70% chord position. Instead, the evolution process has produced shapes that almost return to normal aerofoil geometries when the flap is deflected $\delta_f = 10°$, in effect *bypassing* a true flaps-down analysis.

In concluding this case, it is seen that low drag aerofoils have been produced that satisfy the design constraints, but that would otherwise not be considered optimal. They result purely because of difficulty in converging the boundary layer, and this problem can be considered as specific to high-curvature surface perturbations (common to plain flaps) and the software resolving only thinly separated boundary layers. The correct way to approach this case in the future would be to use a full Navier-Stokes approach with turbulence modeling; bearing in mind that this is considerably more computationally expensive.

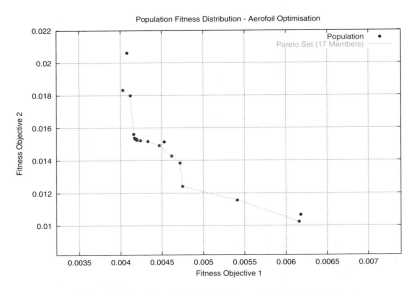

Fig. 17. Pareto set of UAV Transit-Takeoff design.

3.4 Case IV: Subsonic Transport Design

The objective in this case is to find an optimum set of design variables for a subsonic medium size transport aircraft as given in [Mcc03]. The aircraft has two wing mounted engines, and the number of passengers and crew is fixed to 200 and 8 respectively. The design requirements are specified in table 4. The mission profile is represented in figure 23.

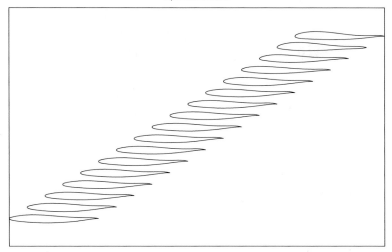

Fig. 18. Ensemble of Pareto set aerofoils – UAV Transit-Takeoff design.

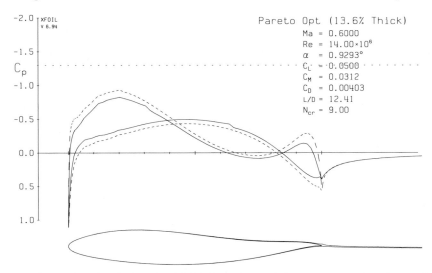

Fig. 19. Objective one optimal aerofoil – Transit flow conditions.

Table 4. Design Requirements.

Description	Value
Range [R, Nm]	2500
Ultimate Load Factor [n_{ult}]	4.2
Maximum Takeoff Field Length [s_{to}, ft]	6000

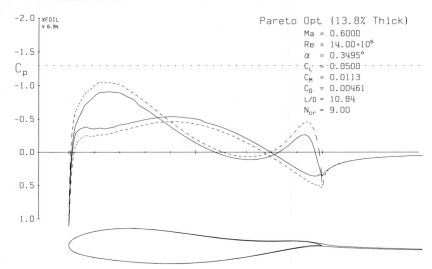

Fig. 20. Compromise aerofoil – Transit flow conditions.

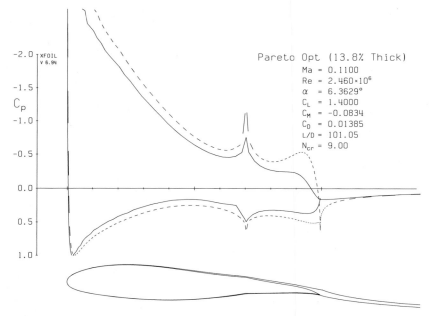

Fig. 21. Compromise aerofoil – Takeoff flow conditions.

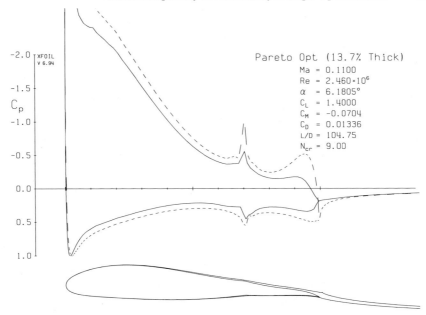

Fig. 22. Objective two optimal aerofoil – Takeoff flow conditions.

Design Variables

The design variables for optimisation and their upper and lower bounds are represented in table 5.

Table 5. Design variables for a subsonic medium size transport aircraft.

Description	Lower Bound	Upper Bound
Aspect Ratio $[AR_w]$	7.0	13.1
Engine Thrust $[T,\ \text{lbf}]$	30500	50000
Wing Area $[S_w,\ \text{sq ft}]$	1927	2872
Sweep $[\Lambda_w,\ \text{deg}]$	25	40
Thickness $[\frac{t}{c}]$	0.091	0.235

Fitness Functions and Design Constraints

This case is a single objective minimisation problem, the fitness function devised for this problem is toward minimum fuel weight required to complete the mission.

$$f = min(W_f)$$

Constraints in this case are minimum takeoff distance, moment coefficient for stability and control and range required. Violation of these constraints are treated with an adaptive penalty criteria.

Solver

The solver used to evaluate the aircraft configuration is FLOPS (FLight OPtimisation System). FLOPS [Mcc03] is a workstation based code with capabilities for conceptual and preliminary design and evaluation of advanced design concepts. The sizing and synpaper analysis in FLOPS are multidisciplinary in nature. It has a numerous modules for noise, detailed takeoff, performance, structures, control, aerodynamics and other capabilities, it is used in some universities for MDO development as well as aerospace firms and government. The aerodynamics module uses a modified version of the the EDET (Empirical Drag Estimation Technique), and modifications to include smoothing of the drag polars. Different hierarchical levels could be adapted for drag build up using higher fidelity models that evaluate full Navier-Stokes flow and wake. FLOPS has capabilities for optimisation but in this case it was used only for analysis and adapted to the EAs optimiser. Details on the solver can be found in [Mcc03].

Implementation

The solution to this problem has been implemented using a single population and parallel asynchronous evaluation, varying number of population sizes and EAs specific parameter were investigated. A small population size of 10 produced good results.

Results

The algorithm was allowed to run for 1500 functions evaluations. Table 6 shows the design variables and results for the best configuration found as compared to a conventional conjugate gradient based (Polak-Ribiere) algorithm and a conventional Broyden-Fletcher-Goldfarb-Shano (BFGS) algorithm [Mcc03]. Through this approach the EA gives a 3.5% and a 2.4% improvement in gross weight respectively.

3.5 Case V: Air Superiority Unmanned Combat Air Vehicle (UCAV) With Gross Weight and Cruise Efficiency Objectives

Problem definition

The goal in the following is to address the issue of multidisciplinary aircraft conceptual design optimisation. Aerodynamic performance, cost minimisation

Table 6. Design variables and results for best configuration found after optimisation.

Description	EA Best	BFGS	CG
Aspect Ratio [AR_w]	13.1	13.0	12.8
Engine Thrust [T, lbf]	34770	38852	39021
Wing Area [S_w, sq ft]	1929	2142	2218
Sweep [Λ_w, deg]	27.0	28.4	27.32
Thickness [$\frac{t}{c}$]	0.091	0.112	0.096
Fuel Weight [W_f, lbs]	*34337*	37342	36092
Gross Weight [W_g, lbs]	*216702*	222154	224618

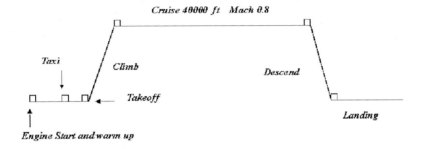

Fig. 23. Mission profile for a subsonic medium size transport aircraft.

and range might be improved if a multi-criteria multi-point optimisation can be developed that considers numerous separate design points. In this case the objectives are maximisation of the cruise efficiency $(M_\infty \times \frac{L}{D})_{cruise}$ and minimisation of gross weight. Two different approaches were run for comparison, one using the concept of Pareto optimality and the other using Nash equilibrium. The mission profile is represented in figure 24.

Table 7. Design Requirements for UCAV

Variable	Requirement
Range [R, Nm]	1000
Cruise Mach Number [M_{cruise}]	1.6
Cruise Altitude [h_{cruise}, ft]	40000
Ultimate Load Factor [n_{ult}]	12
Takeoff Field Length [s_{to}, ft]	7000

Design Variables

The design variables after optimisation and the upper and lower bounds are represented in table 8.

Table 8. Design Variables for optimisation of the UCAV.

Description	Lower Bound	Upper Bound	Best
Aspect Ratio [AR_w]	3.1	5.3	5.14
Engine Thrust [T, lbf]	32000	37000	32923
Wing Area [S_w, sq ft]	600	1400	625
Sweep [Λ_w, deg]	22	47	30.7
Thickness [$\frac{t}{c}$]	0.02	0.09	0.023
Taper Ratio [λ_w, deg]	0.15	0.55	0.19

Fitness Functions

This case is a multi-objective problem where the fitness functions to be optimised are gross weight and cruise efficiency:

$$f_1 = min(W_g)$$

$$f_2 = min(1/(M_\infty \times \tfrac{L}{D})_{cruise})$$

Design constraints

The performance constraints are: 6 G sustained at Mach 0.6 at 10000 feet, 5 G sustained at Mach 0.9 at 30000 feet, an acceleration from Mach 0.9 to 1.5 at 20000 feet in 30 seconds, maintaining a turn rate of 15 deg/sec at Mach 0.9 at 20000 feet and an excess energy = 50 ft/sec at Mach 0.9, 30000 feet and 4 G's. Violation of these constraints is again treated with an adaptive penalty criteria.

Implementation

In the Nash approach we split the variables between two players; Player One optimises for $(M_\infty \times \tfrac{L}{D})_{cruise}$ maximisation using variables, $(X_1) = (AR_w, \tfrac{t}{c}, \Lambda_w)$, while Player Two optimises W_g using variables $(X_2) = (S_w, T, \lambda_w)$. We exchange locked variables between the EAs after an epoch equal to five times the population size in evaluations. The Pareto optimality approach involves one level and population size 15.

Results

The algorithm was allowed to run for 600 functions evaluations but converged after 300. Table 8 shows the parameters for the best configuration found. Figure 25 shows an example of convergence history of one of the Players. The final population (including the Pareto optimal front) and comparison with the Nash equlibrium result are shown in figure 26, as we can see the point obtained by the Nash equilibrium is a suboptimal in the Pareto sense.

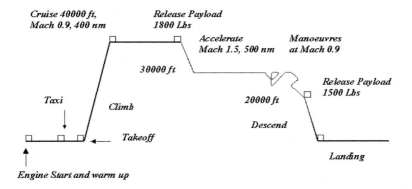

Fig. 24. Mission profile for the UCAV.

4 Conclusions

This paper has presented an single evolutionary algorithm for the solution of a wide range of problems in aerodynamics. The fundamental reason for the use of an evolutionary algorithm is because a single conventional (deterministic) optimiser is incapable of solving the wide variety of problems stated herein. Deterministic optimisers are not easily parallelised, are difficult and often impossible to apply to multiobjective problems and do not handle noise or multimodal problems well. Therefore, the need for a fast, robust and broad evolutionary technique has become apparent. All of the extensions are introduced to improve the efficiency, robustness and applicability of the methods to problems in aeronautics.

The algorithm has been shown to find both conventional and novel aerodynamic solutions to the stated problems, and in each case this was without any problem specific knowledge, using a *vanilla-flavoured* solver without modification. None of the solutions presented herein take more than half-a-day to compute on a modest parallel cluster.

In the future, it will be necessary to couple real solvers from different disciplines to compute true multidisciplinary solutions. Especially important

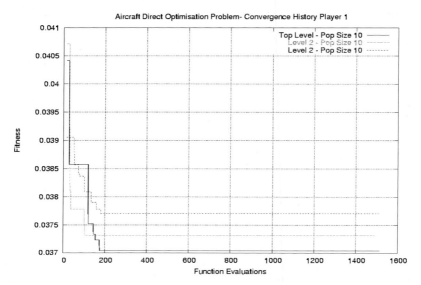

Fig. 25. Optimisation progress of player one.

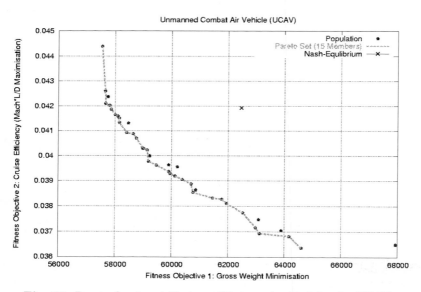

Fig. 26. Pareto front and Nash equilibrium obtained for the UCAV.

are the coupling of aerodynamic solvers to structural methods for weight reduction, as well as electromagnetic methods for radar cross-section (RCS) minimisation. The only hinderance to this at the present is the effective integration of several solvers with a single model – essentially a computer science issue. The future is bright.

Acknowledgements

We would like to offer many thanks to Zdenek Johann of Dassault Aviation for providing the inverse test case, Arnie McCullers at NASA LaRC for providing the FLOPS code, Mark Drela for providing the *XFOIL* software and Mourad Sefrioui of Dassault Aviation and K. Srinivas of The University of Sydney for enlightening discussions on these topics.

References

[AD80] Abbott, I. H., Von Doenhoff, A. E.: Theory of Wing Sections. Dover (1980)

[Deb98] Deb, K.: Multi-Objective Genetic Algorithms: Problem Difficulties and Construction of Test Problems. In: Technical Report CI-49/98, Dept. of Computer Science/LS11, University of Dortmund, Germany (1998)

[Dre01] Drela, M.: *XFOIL* 6.95 User Guide. Technical Report, MIT Computational Aerospace Sciences Laboratory (2001)

[Gol98] Goldberg, D.: Genetic Algorithms in Search, Optimization and Machine Learning. Addison-Wesley (1989)

[Jam95] Jameson, A.: Optimum Aerodynamic Design Using Control Theory. In: Computational Fluid Dynamics Review (1995)

[Mcc03] McCullers, A.: FLOPS User's Guide – Release 6.02. NASA Langley Research Center (2003)

[Mic92] Michalewicz, Z.: Genetic Algorithms + Data Structures = Evolution Programs. Artificial Intelligence. Springer-Verlag (1992)

[Nas50] Nash, J. F.: Equilibrium Points in N-Person Games. In: Proceedings of the National Academy of Science, number 36, pages 46–49 (1950)

[Nas51] Nash, J. F.: Noncooperative Games. In: Annals of Mathematics, number 54, page 289 (1951)

[SP00] Sefrioui, M., Périaux, J.: A Hierarchical Genetic Algorithm Using Multiple Models for Optimization. In: Proceedings of the Sixth International Conference "Parallel Problem Solving from Nature" (*PPSN-VI*), pages 879–888. Springer (2000)

[SSP00] Sefrioui, M., Srinivas, K., Périax, J.: Aerodynamic Shape Optimization Using a Hierarchical Genetic Algorithm. In: European Congress on Computational Methods in Applied Sciences and Engineering (*ECCOMAS*), Barcelona, Spain, September (2000)

[WZ00] Wakunda, J., Zell, A.: Median Selection for Parallel Steady-State Evolution Strategies. In: Proceedings of the Sixth International Conference "Parallel Problem Solving from Nature" (*PPSN-VI*), pages 405–414. Springer (2000)

Multi Objective Robust Design Optimization of Airfoils in Transonic Field

L. Padovan[1], V. Pediroda[1], and Carlo Poloni

Dipartimento di Energetica, Università Degli Studi di Trieste, Italy

1 Introduction

The present research propose optimization methods that are robust in the sense that they produce solutions insensitive to changes in the input parameters: these optimization methods are known as *Robust Design Techniques*. The need for Robust Design method appears in many contests: during the preliminary design process, the exact value of some input parameters is not known, consequently the aim is try to look for a solution as less dependent on the unknown input parameters as possible. The concept of robust optimization is demonstrated by using a 2-D airfoil shape optimization problem. It has been observed [2] that minimizing drag at a single design point has unintended consequences at nearby off-design points. Hicks and Vanderplaats demonstrate that a direct optimization approach that minimizes drag at one mach number (e.g., M=0,75) actually increases drag at nearby Mach number (e.g., M=0,70). To avoid this consequence the airfoil drag minimization problem could be faced by means of an inverse optimization approach but the final result could be almost the same of the single point design [7]. The conclusion is that the latter techniques produce solutions that perform well in correspondence of the design point but have poor off-design characteristics.

The present research shows a optimization method that look for solutions which are insensitive to fluctuations of the operative conditions. Starting from the statistical definition of stability (related to the definition of variance), the method finds, at the same time, good solutions for performance and stability. The goal of robust optimization is to find the airfoil shape that minimizes the mean of the drag coefficient over a range of free-stream Mach numbers and angle of attack and keeps the performance fluctuations as low as possible. The robust optimization model of minimizing the mean and the variance can be used for this purpose:

$$\min_{\Delta D, \Delta \alpha} (E(c_d), \sigma^2(c_d)) \tag{1}$$

subject to:

$$E(c_l) \geq E(c_l)^* \atop \sigma^2(c_l) \leq \sigma^2(c_l)^* \Big\} \tag{2}$$

for $\alpha_{min} \leq \alpha \leq \alpha_{max}$ & $M_{min} \leq M \leq M_{max}$.

$E(c_l)^*$ and $\sigma^2(c_l)^*$ are respectively the mean and the variance of the lift required, D is a set of m airfoil geometric design variables, (c_l) and (c_d) are respectively the lift and drag coefficient (of the current airfoil shape). The mean and variance of (c_d) are defined as follows:

$$E(c_d) = \int_\alpha \int_M c_d(D, \alpha, M)p(M)p(\alpha)dMd\alpha \tag{3}$$

$$\sigma^2(c_d) = \int_\alpha \int_M (c_d(D, \alpha, M) - E(c_d))^2 p(M)p(\alpha)dMd\alpha \tag{4}$$

Where $p(M)$ and $p(\alpha)$ are respectively the probability density function of M and α. In this research we choose a uniform distribution (e.g., $p(M) = 1/(M_{max} - M_{min})$): it means that any fluctuation of the free-stream Mach number has the same probability to happen. Furthermore in the optimization process we consider a discrete evaluation of the mean and variance, so that the robust optimization model is the same proposed in 1 and 2, but the mean and variance of (c_d) are:

$$\overline{c_d} = \frac{\sum\limits_{i=1}^{n} c_d(D, \alpha, M)_i}{n} \tag{5}$$

$$\sigma^2(c_d) = \frac{\sum\limits_{i=1}^{n} (c_d(D, \alpha, M)_i - E(c_d))^2}{(n-1)} \tag{6}$$

Setting the robust optimization method as mentioned above in eqs. (5) and (6) a very high time machine could be required in order to get a good estimation of the objective functions (mean and variance); the methods must be able to find solutions by using a moderate number of high-fidelity disciplinary analysis. This last requirement acknowledges the fact that disciplinary analysis (e.g., computational fluid dynamics (CFD)) can be computationally expensive: for this purpose an interpolation modelling technique known as Kriging is used. Sacks, et al. [11] proposed the Kriging method, developed in the fields of spatial statistics and geostatistics, in order to approximate the results of deterministic computer analysis. With the Kriging method the interpolation of the sampled data is carried out using a maximum likelihood estimation (MLE) procedure [13], which allows for the capturing of multiple local extrema. Just as any approximation model, the accuracy of a Kriging model depends greatly on the number of sample data points used and their locations in multidimensional space. In order to fully exploit the advantages of Kriging models using as small number of samples as possible without significantly sacrificing the accuracy of the approximation, an adaptive approach of the location of the samples is proposed. These methods are deeply related to the techniques for improving and quantifying the accuracy of the approximations used in various Multidisciplinary Design Optimization (MDO) procedures, while reducing the amount of design space information required to develop the approximations.

In order to develop the robust design techniques described so far, a real multi-objective approach is performed. A Multi Objective Genetic Algorithm (MOGA, [10]) is used during the optimization phase. In Ref. [9] and [6] an attractive option for solving the 2-D airfoil problem with robust design is presented, which uses the CFD analysis of Ref. [1]. The former researches demonstrate the potential importance of robust airfoil optimization, but the Kriging method proposed [3] does not exploit a Gaussian stochastic process in order to get the maximum likelihood estimation.

Overview of Kriging Method

Originally developed and used in mining engineering and geostatistics, the Kriging method is an approach for curve fitting and response surface approximation. In the 1980s, statisticians developed Design and Analysis of Computer Experiments (D.A.C.E.) for deterministic computer-generated data based on the Kriging method [11]. The Kriging method used in this study is based on the D.A.C.E. approach.

Suppose we have evaluated a deterministic function of k variables at n points. Denote sampled point i by $\boldsymbol{x}^{(i)} = (x_1^{(i)}, \ldots, x_k^{(i)})$ and the associated function value by $y^{(i)} = y(\boldsymbol{x}^{(i)})$, for $i = 1, \ldots, n$, the Kriging (D.A.C.E.) technique is based on the follow stochastic process model:

$$d(\boldsymbol{x}^{(i)}, \boldsymbol{x}^{(j)}) = \sum_{h=1}^{k} \theta_h |x_h^{(i)} - x_h^{(j)}|^{p_h} \qquad (\theta_h \geq 0, p_h \in [1,2]) \qquad (7)$$

$$Corr[\epsilon(\boldsymbol{x}^{(i)}), (\boldsymbol{x}^{(j)})] = exp[-d(\boldsymbol{x}^{(i)}), \boldsymbol{x}^{(j)})] \qquad (8)$$

$$y(\boldsymbol{x}^{(i)}) = \mu + \epsilon(\boldsymbol{x}^{(i)}) \qquad (i = 1, \ldots, n) \qquad (9)$$

where eq. (7) is the weighted distance formula between the sample points $\boldsymbol{x}^{(i)}$ and $\boldsymbol{x}^{(j)}$, eq. (8) is the correlation between the errors at $\boldsymbol{x}^{(i)}$ and $\boldsymbol{x}^{(j)}$. Eq.

(9) is the model we use in the stochastic process approach: μ is the mean of the stochastic process, $\epsilon(\boldsymbol{x}^{(i)})$ is Normal$(0,\sigma^2)$; the latter term is the realization of a stationary Gaussian random function that creates a localized deviation from the global model [4]. The parameter θ_h in the distance formula (7) can be interpreted as measuring the importance or 'activity' of the variable x_h. The exponent p_h is related to the smoothness of the function in coordinate direction h, with $p_h = 2$ corresponding to smooth functions and values near 1 corresponding to less smoothness [8]. The stochastic process model in Eqs. (7)-(9) is essentially a generalized least squares (GLS) [12] model with a simple set of regressors (just a constant term) and a special correlation matrix that has unknown parameters and depends upon distances between the sampled point.

The Kriging approximation presented by Schonlau [12] used the best linear unbiased predictor (BLUP) of y at the point at which we are predicting, $\boldsymbol{x}^*$. Let $\boldsymbol{r}$ denote the $n - vector$ of correlations between the error term at $\boldsymbol{x}^*$ and the error at the previously sampled points. That is, element i of $\boldsymbol{r}$ is $r_i(\boldsymbol{x}^*) \equiv Corr[\epsilon(\boldsymbol{x}^*), (\boldsymbol{x}^{(i)})]$, computed using the formula for the correlation function in Eqs. (7) and (8). The estimated model of Eq. (9) can be expressed by the BLUP of $y(\boldsymbol{x}^*)$:

$$y(\boldsymbol{x}^*) = \hat{\mu} + \boldsymbol{r}^T \boldsymbol{R}^{-1}(\boldsymbol{y} - \mathbf{1}\hat{\mu}) \tag{10}$$

where $\boldsymbol{y} = (y^{(1)}, \ldots, y^{(n)})^T$ denote the n-vector of observed function values, $\boldsymbol{R}$ denote the $n \times n$ matrix whose (i,j) entry is $Corr[\epsilon(\boldsymbol{x}^{(i)}), (\boldsymbol{x}^{(j)})]$, and $\mathbf{1}$ denote an n-vector of ones. The value for $\hat{\mu}$ is estimated using the generalized least squares method as:

$$\hat{\mu} = (\mathbf{1}^T \boldsymbol{R}^{-1} \mathbf{1})^{-1} \mathbf{1}^T \boldsymbol{R}^{-1} \boldsymbol{y} \tag{11}$$

The estimates of $\hat{\theta}_h$ and $\hat{p}_h$, end hence an estimate of the correlation matrix $\boldsymbol{R}$, are obtained by maximizing the following function:

$$-\frac{n \ln(\hat{\sigma}^2) + \ln|\boldsymbol{R}|}{2} \tag{12}$$

where

$$\hat{\sigma}^2 = \frac{(\boldsymbol{y} - \mathbf{1}\hat{\mu})^T \boldsymbol{R}^{-1}(\boldsymbol{y} - \mathbf{1}\hat{\mu})}{n} \tag{13}$$

The mean squared error (MSE) of $\hat{y}(\boldsymbol{x})$ can be derived as:

$$s^2(\boldsymbol{x}^*) = \sigma^2[1 - \boldsymbol{r}^T \boldsymbol{R}^{-1} \boldsymbol{r}] + \frac{(1 - \mathbf{1}^T \boldsymbol{R}^{-1} \boldsymbol{r})^2}{\mathbf{1}^T \boldsymbol{R}^{-1} \mathbf{1}} \tag{14}$$

eq. (14) provides an estimate of the variance of the stochastic process component of the Kriging approximation.

Earlier studies imply that including the parameter p_i as part of maximum likelihood estimation did little to improve to the Krigin approximation, thus in the current study $p_i = 2$ was used for all design variables. Regarding the MLE we used MOGA [10]: in fact, the likelihood function can be very discontinuous. Initially we tried with Simplex but many times this algorithm stops the searching prematurely. Even if MOGA needs a lot of computations, given a set of samples the time required for the MLE is negligible.

Adaptive Kriging

Before we build an approximation we require a systematic means of selecting the set of inputs (called Design Of Experiments, or DOE) at which to perform a computational analysis. One popular choice for generating experimental design for computational experiments is the Latin Hypercube [5]. Instead to use this latter technique we propose an adaptive arrangement of the available set of samples (data base) exploiting the value of the MSE. The value of MSE depends on the correlation of the landscape as well as on the local density of points.

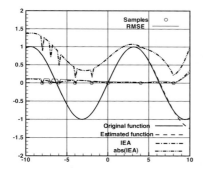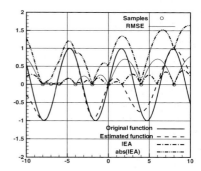

Fig. 1. Compare of the extrapolation of sin(x/2) (left) and sin(x) (right) by means Kriging

In Fig. (1) has been illustrated an example of extrapolation by means Kriging. It is possible to note that the higher is the number of maxima and minima of the function and the lower is the accuracy of Kriging (exploiting the same data base of samples). We have reported the behavior of RMSE too (Root Mean Squared Error): the RMSE tell us the accuracy of the prediction and it is very low corresponding to the coordinates of the samples. It is possible to understand that the approximation is more precise in regions with high point density. In the same figure we represent further functions:

$$IEA = y(\boldsymbol{x}) \cdot RMSE \qquad (IEA \ = \ Index \ of \ Absolute \ Error) \qquad (15)$$

$$abs(IEA) = |(y(\boldsymbol{x}) - Y_{min})/(Y_{max} - Y_{min})|$$
$$+RMSE_{y(\boldsymbol{x})}/(RMSE_{max} - RMSE_{min}) \qquad (16)$$

Eqs. (15) and (16) represents the index we use to set the adaptive arrangement of the samples. In fact we try to exploit the value of RMSE to understand where the extrapolation is not accurate, taking care at the same time of the extrapolated value associated. For example, applying Kriging given a data base of samples, according to the values of IEA or abs(IEA) it is possible to know where the extrapolation is not accurate. Eq. 16 has the same meaning of Eq. 15 but it is normalized. The Y_{max} and Y_{min} values are respectively the highest and lowest values of the extrapolated function while the same meaning have $RMSE_{max}$ and $RMSE_{min}$ regarding RMSE. In the same way it is possible to define:

$$IER = RMSE/y(\boldsymbol{x}) \qquad (IEA \ = \ Index \ of \ Relative \ Error) \qquad (17)$$

$$abs(IER) = |(y(\boldsymbol{x}) - Y_{max})/(Y_{max} - Y_{min})|$$
$$+RMSE_{y(\boldsymbol{x})}/(RMSE_{max} - RMSE_{min}) \qquad (18)$$

The Eqs. 15-18 can be used separately or together coupled (15 and 17 or 16 and 18). For example eq. 15 assumes high values corresponding to high values of the function and RMSE, instead eq. 16 assumes high values in correspondence of low values of the function and high values of RMSE. In this way we can localize the accuracy of the extrapolation.

An example on how the adaptive method works is shown in Fig. (2): the studied function is the Test 1 function [10]. We show how Kriging perform using a Latin Hypercube and adaptive distribution of the samples. It is possible to note that both distributions allow a good estimation of the function studied but in this case higher accuracy is reached trough the adaptive Kriging. We have used abs(IEA) and abs(IER) alternatively: starting from the samples that belongs to the corner of the dominion studied, we add a new samples in the current data base iteratively. The use of the indexes in Eqs. (15-18) could be exploit in order to get an higher local accuracy: in fact if we have used the IEA (or abs(IEA)) the extrapolation would have been more accurate corresponding to the neighborhood of the maxima value of the function. In the same way we could exploit IER or abs(IER).

Results

The 2-D airfoil optimization problem stated in Eqs. 1 and 2 is used to test the robust optimization approach. The achievable configurations have been

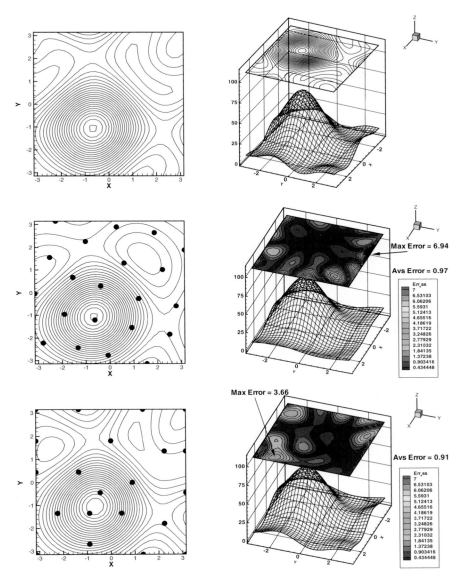

Fig. 2. Test 1 function predicted by means different data bases by comparison. Correct function (1st row), predicted function by means Latin Hypercube distribution of the samples (2nd row), predicted function by means adaptive distribution of the samples (3rd row).

reached by modifying an initial baseline configuration that correspond to the supercritical airfoil RAE2822 designed by the Royal Aircraft Establishment. The present research is tested by solving a fully turbulent Navier-Stokes CFD analysis code and in particular the MUFLO code and AIRFOIL [1] as mesh generator have been used. The Reynolds number is set to $Re = 1,5 \cdot 10^6$. Regarding the parameterization the baseline configuration has been modified exploiting two B-splines, one for the intrados and one for the extrados (Figure 3). Eight-teen bounded geometric design variables, D_j, shown as circles, are the B-spline control points used to create a wide variety of 2-D airfoil shapes; a geometric constraint is set. Since the RAE2822 has an maximum thickness value slightly higher than the 12% of the chord length, in the optimization process the thickness is fixed to be higher than 11.95%.

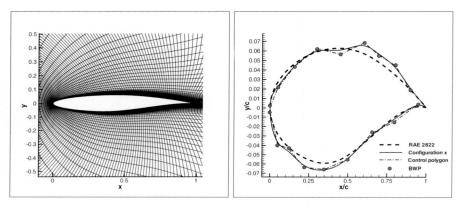

Fig. 3. One of the possible meshes around the RAE2822 achievable with airfoil (left side) and parameterization (right side)

The aim of the Robust Design Optimization shown in this paper is to find an airfoil shape that behaves better than the RAE2822 within a range of Mach number and angle of attack corresponding to $M = 0.73 \pm 0.03$ and $\alpha = 2° \pm 0.5°$. Altogether, according to the principles of Robust Design described previously, we take care of four objective functions during the optimization process: $\overline{c_l}$ and $STDc_l$ are constrained, while $\overline{c_d}$ and $STDc_d$ will be minimized. MOGA [10] (Multi Objective Genetic Algorithm) is used to solve the Multi Objective Robust Design Optimization of airfoils in transonic conditions and modeFRONTIER is the software used to implement MOGA. The general strategy to achieve the objective functions given a configuration (airfoil shape) by means of the Kriging method consists of the following steps:

(1) set an starting data base of 4 or 5 training data (evaluating CFD analysis);

(2) extrapolate the c_l and c_d functions by means of Kriging and evaluate the objective functions ($\overline{c_l}$, $\overline{c_d}$, $STDc_l$, $STDc_d$); get the location (*star-location*) of the extrapolated value where the index (Eqs. 15-18) assumes the highest value (step 1 is concluded: now iteratively);

(3) evaluate an CFD analysis corresponding to the *star-location* and update the data base;

(4) achieve the studied functions through the data base updated and compare the objective functions of two consecutive steps; get the new *star-location* (step n);

(5) the process stops if the differences between the objective values achieved in two consecutive steps is lower than an defined error;

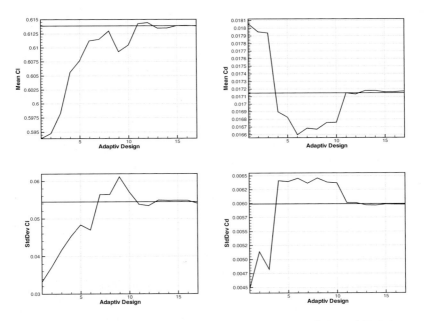

Fig. 4. Convergency of the objective functions of the RAE2822

Paying attention to the former scheme it is possible to understand that thanks to the adaptive method, in automatic way, we can define the minimum number of the samples in order to achieve the best extrapolated function. The more complicate the function is and the more samples we need in order to extrapolate a correct function. Facing any kind of problem we don't know how complex the function is, but through the adaptive method it is possible to recognize when it is worth to apply many high fidelity analysis. Fig. (4) shows how many samples kriging needs to reach a good accuracy of the objective functions regarding the RAE2822: by means not so high number of samples a good accuracy is reached. Fig. (5) shows the histories of the objective functions during the optimization process: it is possible to note the improvements of the RAE2822s' performances underlined through an horizontal line. Fig. (6) (left) shows the Pareto front: there's no high spread of the solutions, in fact seems that the solutions collapses in to one point. Probably the complexity of the problem needs a longer seeking in order to get a spreader Pareto front; the target of the next research will be to find a spreader Pareto front within a reasonable number of high fidelity analysis. Fig. (7) (left) shows the c_d functions of the RAE2822 and solution 1 within the dominion of operating conditions studied: it is possible to see the meaning of the stability of one solution at a glance. Fig. (7) (right) shows the pressure distributions on RAE2822 and one good solution.

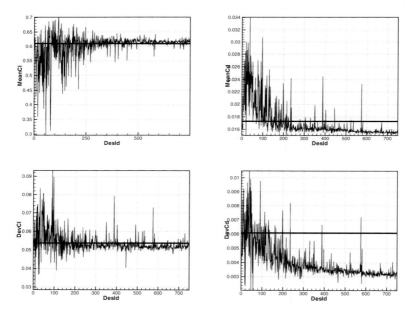

Fig. 5. Histories during the optimization process

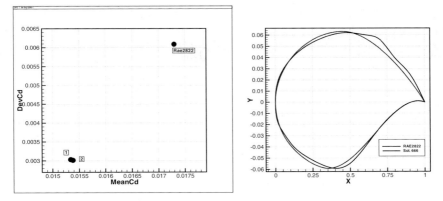

Fig. 6. Pareto front (left); RAE2822 and best configuration geometries by comparison (right)

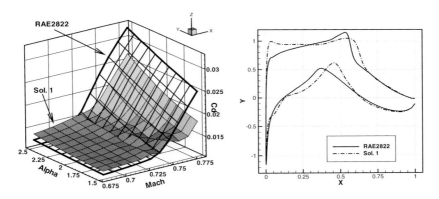

Fig. 7. Performance of RAE2822 and solution 1 by comparison (left); pressure distribution on RAE2822 and solution 1 for highest Mach number design point and $\alpha=2,5$ (right)

Conclusions

This paper uses a 2-D airfoil optimization problem to illustrate how it is possible to set an optimization under uncertainty: such kind of problems are recognized as Robust Design Optimization. In fact the aim is to find solutions that are insensitive to fluctuations of the input parameters, and consequently Robust solutions. According to the principles of Robust Design, the stability of the magnitude is associated to each magnitude that define the performances of the subject studied (in this case the airfoil). Consequently the Robust Design problem is always characterized by more than one objective function, and GA is recommended in order to reach all the possible solutions. In this paper MOGA has been used and 4 objective function have been considered.

To reduce the high number of expensive disciplinary analysis necessary to implement Robust Design, an extrapolator method known as Kriging has been used. Through Kriging, thanks to an adaptive method, it has been possible to define the minimum number of direct simulations to achieve good values of the objective functions. This method is very helpful above all when the uncertain parameters have a strong nonlinear effect on the objective functions. Consequently for each configuration, in automatic way, it is possible to understand when it is worth or no to compute several computations.

It is concluded that Robust optimization is an important tool for multidisciplinary design. It is needed when some of the design variables, like operating conditions, are fluctuating in randomly way or under uncertainty of the design specifications and above all when the uncertain parameters have a strong nonlinear effect on the objective functions. The authors believe that this strategy could be extended to a wide range of problems characterized by non linear behavior of the objective functions subject to fluctuations of the input parameters.

References

1. Haase, W.;Wagner, B.;Jamenson, A., *Development of a Navier-Stokes method based on a finite volume technique for unstady Euler equations*, in Proceeding of the 5th GAMM conference on Numerical Methods in Fluid Mechanics, Friedrich Viewer und Sohn, Braunschweigh Wiesbaden, 1983.
2. Hichs, R. M.;Vanderplaats, G. N.; 1977 *Application of Numerical Optimization to the Design of Supercritical Airfoil Without Drag-Creep*. SAE Paper 770440, Business Aircraft Meeting, Wichita, Kansas, March 29-April 1, 1977.
3. Isaaks and Srivastava *An Introduction to Applied Geostatistics*, Oxford University Press, 1989.
4. J. R. Koehler and A. B. Owen *Computer Experiments*, Handbook of Statistics, Vol. 13, pp. 261-308, Elsevier Science, New York, eds. S. Ghosh and C. R. Rao.
5. McKay, M.D.; Beckman, R. J. and Conover, W. J., *A comparison of three methods for selecting values of input variables in the analysis of output from a computer code*, Technometrics 21, pp. 239-245, 1979.
6. Padovan L., Pediroda V., Poloni C., *Multi Objective Robust Design Optimization of airfoils in transonic conditions*, International Congress on Evolutionary Methods for Design, Optimization and Control with Applications to Industrial Problems, EUROGEN 2003, G. Bugeda, J.A.- Désidéri, J. Periaux, M. Schoenauer and G. Winter (Eds), ©CIMNE, Barcelona, September 2003.
7. Padula, S. L.; Wu Li 2002 *Options for robust design optimization under uncertaintly*. AIAA Conference 2002.
8. Parzen, E.(1963) *A new approach to the synthesis of optimal smoothing and prediction systems*, in R. Bellman, (ed.), Mathematical Optimization Techniques, pp. 75-108, University of California Press, Berkeley.
9. Poloni C., Pediroda V., Padovan L., *Multi Objective Robust Design Optimization of airfoils in transonic field (M.O.R.D.O.)*, 9th MPS Symposium (IPSJ Symposium Series Vol. No.2), Kyoto, January 2003.

10. Poloni C., Pediroda V., 1997, *GA coupled with computationally expensive simulations: tool to improve efficiency*, in Genetic Algoritms and Evolution Strategies in Engeneering and Computer Science, edited by Wiley Sons, 1997.

11. J. Sacks, W. J. Welch, T. J. Michell, and H. P. Wynn, *Design and Analysis of Computer Experiments* Statistical Science, Vol 4 No.4, pp. 409-453,1989.

12. Schonlau. M. *Computer Experiments and Global Optimization*, Ph D. Dissertation, University of Waterloo, 1997.

13. Timothy W. Simpson, Timothy M. Mauery, John J. Korte and Farrokh Mistree, *Comparison of Response Surface and Kriging Models in the Multidisciplinary Design of an Aerospike Nozzle*, 7th AIAA/USAF/NASA/ISSMO Symposium on Multidisciplinary Analysis and Optimization, St. Louis, Missouri, AIAA 98-4755, September 1998.

Printing: Krips bv, Meppel
Binding: Litges & Dopf, Heppenheim